IRON POWDER METALLURGY

PERSPECTIVES IN POWDER METALLURGY
Fundamentals, Methods, and Applications

PERSPECTIVES IN POWDER METALLURGY
Fundamentals, Methods, and Applications

Editors:

Henry H. Hausner
Adjunct Professor
Polytechnic Institute of Brooklyn
Consulting Engineer

Kempton H. Roll
Executive Director
Metal Powder
Industries Federation

Peter K. Johnson
Assistant Director
Metal Powder
Industries Federation

Volume 3

IRON POWDER METALLURGY

With an Introductory Chapter by Kempton H. Roll and Peter K. Johnson
and a
Foreword by Henry H. Hausner

Springer-Science+Business Media, B.V. 1968

The editors gratefully acknowledge permission to reprint the following articles:

Relations Between Pore Structure and Densification in the Compacting of Iron Powders, by G. Bockstiegel, *International Journal of Powder Metallurgy* 2(4):13-28 (1966) (Copyright 1966, American Powder Metallurgy Institute, New York).

Some Properties of Engineering Iron Powders, by C. J. Leadbeater, L. Northcott, and F. Hargreaves, *Symposium on Powder Metallurgy, Special Report No. 38, Iron and Steel Institute,* pp. 15-36 (1947).

The Pressing and Sintering Properties of Iron Powders, by G. Zapf, *Powder Metallurgy*, No. 7, pp. 218-248 (1961) (Published by the Iron and Steel Institute and the Institute of Metals, London).

High Velocity Compaction of Iron Powder, by E. M. Stein, J. R. van Orsdel, and P. V. Schneider, *Metal Progress* 85(4):83-87 (April 1964) (Copyright 1964, American Society for Metals, Ohio).

Study on the Continuous Rolling Fabrication of Iron Strip from Metal Powders (Part I), by T. Kimura, H. Hirabayashi, and M. Tokuyoshi, *Review, Electrical Communication Laboratory* 12(3/4):215-232 (March-April 1964) (Copyright 1964, Electrical Communication Laboratory, Nippon Telegraph and Telephone Corporation, Tokyo).

The Effect of Die Wall Lubrication and Admixed Lubricant on the Compaction of Sponge-Iron Powder, by P. M. Leopold and R. C. Nelson, *International Journal of Powder Metallurgy* 1(3):37-44; 1(4):37-40 (1965) (Copyright 1965, American Powder Metallurgy Institute, New York).

Hot Pressing of Iron Powders, by O. H. Henry and J. J. Cordiano, *AIME Tech. Pub. No. 1919, Metals Technology*, pp. 1-10 (October 1945) (Copyright 1945, American Institute of Mining and Metallurgical Engineers, New York).

Some Factors Influencing the Properties of Sintered Iron, by S. Lennart Forss, *Planseeber. Pulvermet.* 11:136-145 (1963) (Copyright 1963, Metallwerk Plansee, Reutte, Austria).

Study of the Mechanical Properties of Porous Iron in Tension and Torsion, by G. S. Pisarenko, V. T. Troshchenko, and A. Ya. Krasovskii, *Soviet Powder Metallurgy and Metal Ceramics*, No. 6(30), pp. 467-472 (June 1965) (Copyright 1966, Plenum Publishing Corporation, New York).

The Effect of Minor Additions of Sulphur on Pore Closure and Case-Hardenability of Sintered Iron, by G. Bockstiegel, *Powder Metallurgy*, No. 10, pp. 171-189 (1962) (Published by the Iron and Steel Institute and the Institute of Metals, London).

The Effect of HCl-H$_2$ Sintering Atmospheres on the Properties of Compacted Iron Powder, by R. D. McIntyre, *ASM Transactions Quarterly* 57:351-354 (1964) (Copyright 1964, American Society for Metals, Ohio).

Controlled Oxidation Prior to Sintering of Iron Compacts, by H. T. Harrison and C. G. Johnson, *Progress in Powder Metallurgy* 17:29-34 (1961) (Copyright 1961, Metal Powder Industry Federation, New York).

Densification and Grain Growth in the Later Stages of Sintering of Alpha-Iron, by H. F. Fischmeister, *Symposium sur la Métallurgie des Poudres*, Editions Métaux, Paris, pp. 115-128 (1964) (Published by Editions Métaux, Paris).

The Effect of Powder Particle Size on the Grain Size of Sintered Iron, by H. H. Hausner and R. King, *Planseeber. Pulvermet.* 8:28-36 (1960) (Copyright 1960, Metallwerk Plansee, Reutte, Austria).

Structure Formation in Iron Graphite Compositions, by V. A. Dymchenko and Yu. F. Morozov, *Soviet Powder Metallurgy and Metal Ceramics*, No. 6(30), pp. 449-453 (June 1965) (Copyright 1966, Plenum Publishing Corporation, New York).

The System Iron–Carbon in Powder Metallurgy, by P. Ulf Gummeson, *Tech. Bulletin of the Hoeganaes Corporation*, pp. 1-10 (March 1966) (Published by Hoeganaes Corporation, Riverton, N. J.).

Effect of Different Furnaces and Sintering Practices on Strength of Sintered Iron Powder, by E. Geijer, *Materials Research and Standards* 2:124-126 (February 1962) (Copyright 1962, American Society for Testing and Materials, Philadelphia).

The Fatigue Strength of Sintered Iron Compacts, by J. M. Wheatley and G. C. Smith, *Powder Metallurgy*, No. 12, pp. 141-158 (1963) (Published by the Iron and Steel Institute and the Institute of Metals, London).

Temperature Dependence of Mechanical Properties in Porous Iron, by N. A. Filatova, *Soviet Powder Metallurgy and Metal Ceramics*, No. 1, pp. 22-24 (January-February 1962) (Copyright 1963, Plenum Publishing Corporation, New York).

ISBN 978-1-4899-6225-6   ISBN 978-1-4899-6467-0 (eBook)
DOI 10.1007/978-1-4899-6467-0
Library of Congress Catalog Card Number 67-17375

Foreword

During the last few years powder metallurgy has grown substantially in many new directions, including nuclear and space applications. Hardly any other branch, however, has grown faster than that of ferrous powder metallurgy. New types of iron powders with special and improved properties have been developed and have conquered the market. There are many more P/M parts and parts of larger dimensions produced from iron alloy powders than ever before. Only a few years ago an American automobile contained 60 to 70 P/M parts, whereas most of today's automobiles contain 80 to 100 P/M parts. Business machines have been developed which include more than 400 P/M parts. During the last few years not only the quantities but also the quality of ferrous powder metallurgy parts has steadily increased.

All the development in production has been paralleled by the rapidly growing literature on iron powder metallurgy. A selected Bibliography on Iron Powder Metallurgy published in 1965* listed more than 440 annotated references from the five-year period 1960-1964. Most of the literature on iron powder metallurgy is published in a variety of technical and scientific journals in many countries. In view of this it is somewhat surprising that the powder metallurgy books published during the last ten years contain rather little information on iron powder metallurgy, although there is definitely a need and demand for more extensive coverage of this important branch of the subject. It is the purpose of this book to present a rather detailed review of iron powder metallurgy.

* By Hoeganaes Corp., Riverton, N. J.

This book contains a collection of papers dealing with various aspects of iron powder metallurgy. It is divided into 20 chapters, which represent mostly reprints of previously published papers by authors from the United States, the United Kingdom, the Soviet Union, Japan, Sweden, and Germany. Some of these chapters contain already classical information on the subject, such as the chapter by J. C. Leadbeater et al., or the chapter by O. H. Henry and J. J. Cordiano. The information given in this book concerns powder production, compacting methods, and the process of sintering. It mostly concerns pure iron, but some of the iron alloys, dense and porous iron parts and their characteristics, as well as some design problems are also covered.

The bibliography in the appendix covers the methods for the production of iron powders and includes 142 selected annotated references. Ninety-two of these references concern the production of iron powders by various reduction processes.

Since the chapters were written by different authors, some duplication of information was unavoidable. Needless to say, a volume of less than 400 pages does not cover the entire field of iron powder metallurgy. Nevertheless it is believed that the information given in this book represents a valuable review of iron powder metallurgy which might be of interest not only to practical powder metallurgists but also to physical metallurgists, mechanical engineers, designers, and others. It is sincerely hoped that some of the information given in this volume may stimulate additional development work and new applications for iron P/M parts.

Henry H. Hausner

Contents

Iron Powder Metallurgy Spearheads Industry Growth

Kempton H. Roll

Executive Director
Metal Powder Industries
Federation

and

Peter K. Johnson

Assistant Director
Metal Powder Industries
Federation

Iron powder represents the largest tonnage of raw materials used in powder metallurgy fabricating. Not simply chips or scraps of wrought materials, iron powders are highly engineered metallurgical end products that must meet strict chemical and physical specifications.

They are made by many different processes, each resulting in materials having different chemical and physical properties. In fact, the end products' properties are the function of its method of manufacture. Powder metallurgists, when describing a particular powder on recommending its end applications, look for such things as composition, microstructure, oxide content, acid insolubles,

particle size, particle-size distribution (sieve analysis), particle
shape and surface area, flow rate, and apparent density. But, because of this broad spectrum of characteristics, metal powder can
be varied within wide limits to meet specific requirements.

Shipments in 1966 of 100,000 tons, according to the Metal
Powder Industries Federation, established iron powder metallurgy
as a "tonnage" industry, transforming it from one whose output was
formerly measured in pounds. This emergence has been spear-
headed by P/M parts, iron powder's major market. It is estimated by MPIF that by 1970, iron powder shipments should surpass
250,000 tons with at least 60% of this slated for structural applications.

Powder metallurgy's current expansion and healthy future is
supported by (1) the steady increase in present P/M parts markets —
75% of the parts now produced go into automobiles and major appliances; (2) many new markets (farm equipment, aerospace, ord-
nance, power tools, small appliances, hardware, etc.); (3) a strong
trend toward larger P/M parts (35-lb parts are in production); (4)
better powders, including premixes, high compressibility powders,
and new alloy systems, all made to exacting quality specifications.
These iron powder systems are prevalent today: iron-carbon;
iron-copper; iron-copper-carbon; iron infiltrated with copper;
iron-nickel-carbon; iron-nickel-copper-carbon; (5) better engineered compacting equipment made specifically for P/M — not only
larger, but more complicated shapes are now economically practical; and (6) the development of new powder consolidating techniques,
such as high energy rate forming, isostatic pressing, and direct
rolling of mill shapes and strip.

Because of the increasing markets for P/M parts as well as
other iron powder applications, and the optimistic expectations for
future demand, all major iron powder producers either have already or are in the process of expanding their production capacities
and technical service facilities. Most producers are undergoing expansion programs that will more than double the industry's current
output capability.

The technologies for converting iron powder into useful products have also been advancing. Besides enjoying the benefits of
compacting systems that can insure better control in complex,
multilevel, larger parts, the P/M parts makers are leading a trend
to more precise quality assurance concepts, which in turn are

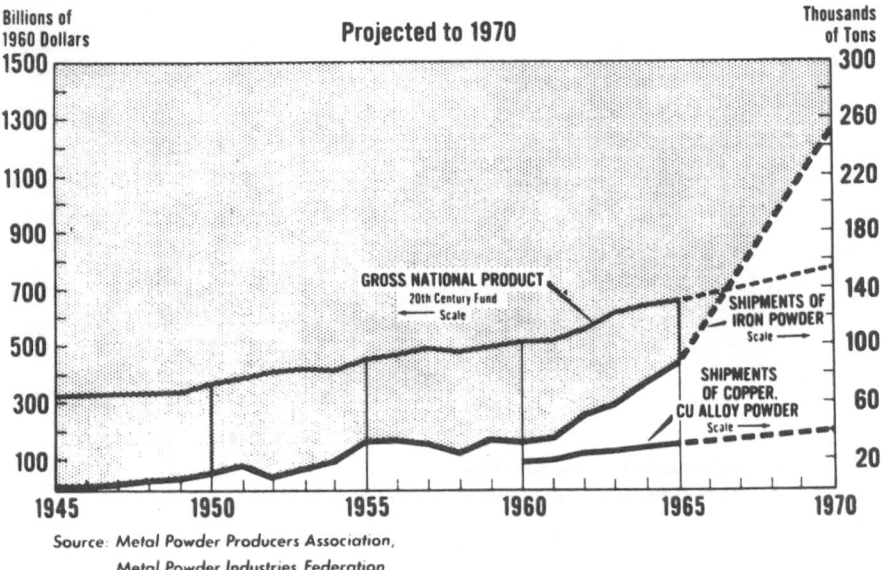

Fig. 1. Growth of metal powder production.

fostering tighter tolerance control, higher physical properties,
closer property uniformity, and better inspection techniques for
finished products. All of these are leading to the establishment
and application of improved materials standards and specifications
through the Powder Metallurgy Parts Association. Thus the P/M
parts industry's products are now numbered among the most reliable components manufactured today.

The production of iron P/M parts, a high volume, precision
process for mechanical and structural components, competes effectively with other metal-forming methods, such as die casting,
sand casting, screw machining, investment casting, stamping, and
forging. While offering economy in most instances, powder metallurgy also provides many unusual advantages that are not so apparent: (1) elimination of machining operations, (2) combining the
functions of several components into one single unit, (3) mass reproducibility, (4) good surface finish, (5) intricate design configu-
rations, (6) self-lubrication if desired, and (7) "tailored" properties—properties custom suited to the needs of the application, or,
properties that combine the features of materials, otherwise im-
miscible, i.e., iron "alloyed" with high precentages of copper or
lead, or metals with nonmetals.

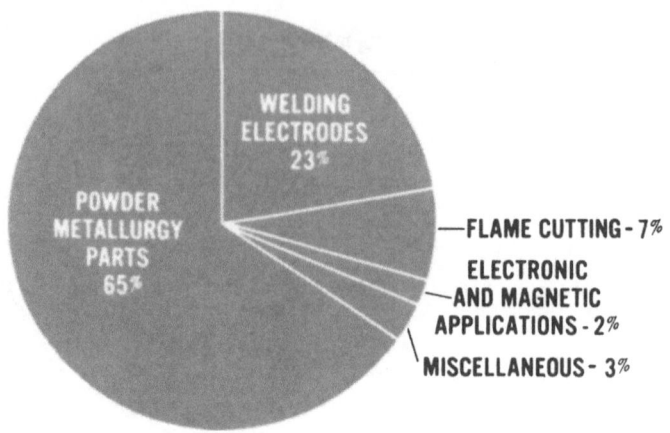

Source: Metal Powder Producers Association,
Metal Powder Industries Federation

Fig. 2. Iron powder shipments — 1965.

The physical properties of sintered iron structures now range from low density, porous bearings having tensile strengths as low as 14,000 psi to high density structural parts with tensile strengths exceeding 180,000 psi. Premixed iron–carbon powders with as–sintered tensile strengths ranging from 35,000 to 68,000 psi, depending on density, presently find the widest application for structural P/M parts. Heat treating, following sintering, brings

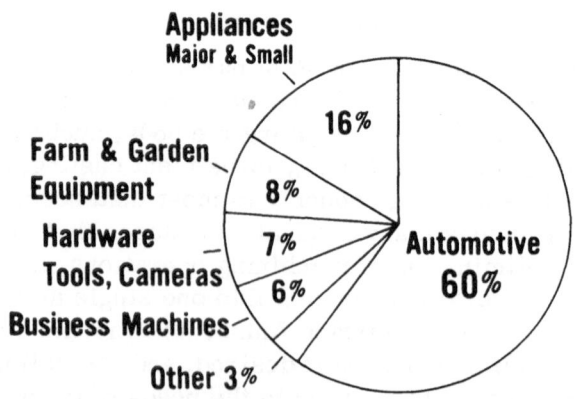

Source: Metal Powder Industries Federation

Fig. 3. P/M parts consumption — 1965 (% of tonnage distribution).

the tensiles up markedly—157,000 psi for iron containing 7-Ni, 2-Cu, 1-C, and with additional processing even this can be increased further, to 185,000 psi.

Iron powder has taken the lead in the commercial application of P/M because it is inexpensive compared to the nonferrous metals; it possesses properties, particularly strength, that other metals or nonmetals do not; it has one of the most favorable strength-to-weight-to-cost ratios; it alloys easily, especially with carbon, and thus possesses all the unique attributes of the Fe – C system including heat-treatability (most so-called iron powder P/M parts actually are steel by virtue of the carbon added as graphite prior to sintering); and it is readily available in most any quantity. The powder metallurgist literally creates his own materials in the manufacture of a part—starting with the basic powders, blending them and, in the sintering process, evolving a metallurgical structure to suit his needs.

Of course, iron powder is not limited to the manufacture of P/M parts. Almost one-quarter of all the iron powder produced is used in the manufacture of welding electrodes. These are wire rods coated with a mixture of iron powder, fluxes and alloying ingredients and used in contact arc welding. They offer a greater percentage of available metal and faster welding. The rod tip can be laid directly on the metal to be welded, and melting takes place from resistance without the necessity of maintaining a precise arc.

Flame cutting and scarfing represent another important use for iron powder, accounting for about 7,000 tons per year. Here the powder is introduced into the flame of an oxyhydrogen or acetylene torch, where it "burns" exothermically, releasing additional heat and raising the torch temperature so that it can cut through the most refractory of materials—concrete, for example. The molten iron oxide formed also acts as a fluxing agent to combine with refractory oxides, such as those found in cutting stainless steels, flushing them away to expose fresh metal surfaces. Flame scarfing serves the same function except that the iron powder-fed flame acts more like a "broom," sweeping away the oxide layers and surface inclusions on steel billets prior to further working.

The electronics industry also lays claim to some important uses for iron powder: 2,000 tons of highly refined, high purity powder in 1966. Much of this is made by the decomposition of iron pentacarbonyl to form what is commonly termed "iron carbonyl."

Powder made by other methods, including hydrogen reduction of
selected iron oxides, also find their way into iron powder cores
that are used in the tuning of radios and television circuits. Cores
are made by coating the powder particles with an insulating plastic
layer which also serves as a binder and then compacting the mixture to form the shape desired, ejecting, and baking to fuse the
binder.

     There are several applications that take advantage of the
magnetic properties of iron. For flaw detection in steels, iron
powder is coated on the magnetized steel where it reveals the lo-
cation of invisible cracks by tracing the irregularities in the magnetic lines of force. Another utilizes iron powder's ability to cling
to a magnet as a novel but commercially practical method of clean-
ing seeds, i.e., separating weed or broken seeds from the good
ones. The mixture of powder and seeds are tumbled together and
then passed over magnets where the good are separated from the
bad. Weed seeds normally are "hairy" and hold particles of iron.
Broken seeds are sticky and thus cause particles of iron to stick
to them. The good seeds are smooth and dry, no iron adheres and
they pass by the magnets ready for packaging.

     Remember back in chemistry class when you copperplated
a nail by dipping it in a solution of copper sulfate? The iron displaced the copper in solution, precipitating it on the iron nail in
the form of a bright coating. This technique is widely used as a
commercial method for recovering copper from mine waters containing copper salts, only now they use iron powder because of its
greater surface area and thus faster rate of reaction.

     Electrochemical displacement is also the basis for mirror
silvering. Iron powder is used to deposit a layer of copper backing
on the silvered mirror surface.. Its position in the electromotive
force series coupled with its high surface area and purity make
iron powder the most practical and efficient method for this purpose.

     It is the same factor—high specific surface area — that accounts
for another more spectacular use of iron powder: pyrotechnics.
The bright stars and streamers that float down from exploding
fireworks lighting up the heavens on summer holidays owe their
brilliance and starlike effect to iron powder as it burns in the air;

just as does the lowly "sparkler," hand-held and dazzling to the young eye.

While most everyone is exposed to iron powder through such uses as those cited above, few people realize that they are being exposed in an even more direct manner. Nearly everyone literally is a consumer of iron powder by virtue of iron-enriched foodstuffs. Most of the flours and cereals used in the baking of iron-fortified breads, breakfast foods, etc., contain iron in the form of special, high purity, pharmaceutical grade iron powder. Americans each year "eat" about 80 tons of iron powder!

Thus from the iron P/M parts in our autos and appliances to the pharmaceutical iron powder in our foods we are all becoming more and more involved with and dependent on iron powder and the science that makes this product possible, powder metallurgy.

Commercial Methods for Powder Production

A.B. Backensto

Glidden Co.
Hammond, Indiana

Just as each human being is marked individually to some extent by the family that produced him and the environment he developed in, each metal powder is marked by the process used in its production. This variation is fortunate in some respects because it gives a wide range of properties from which to choose a powder. The same variation from process to process has an unfortunate aspect because it makes the development of standard grades of metal powders rather difficult.

A great variety of mechanical, physical, and chemical methods are employed for the production of metal powders. Seven of the commonly used commercial methods are atomization, decomposition, electrolysis, gaseous reduction of solutions, grinding reduction of oxides with carbon, and reduction of oxides with gases.

Atomization

Atomization, as applied to the production of metal powders is a process which breaks up a stream of molten metal into individual droplets which solidify. A typical atomizing process is shown in Fig. 1. A jet of high pressure liquid (usually water) or gas, such

8

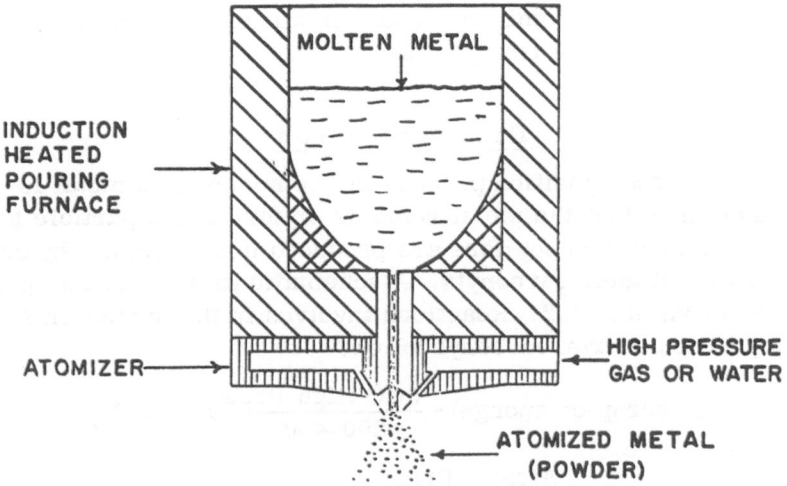

Fig. 1. Schematic drawing of typical atomizing process.

as air or inert gas, is directed at the molten stream either head on
or at right angles to break it up into fine particles. Cooling and
solidification are very rapid and the powder particles occur in the
shape of the metal droplets. Another technique is to let the stream
fall on a spinning disc which throws the metal out in a cascade of
droplets. Atomization is the main process for producing alloy

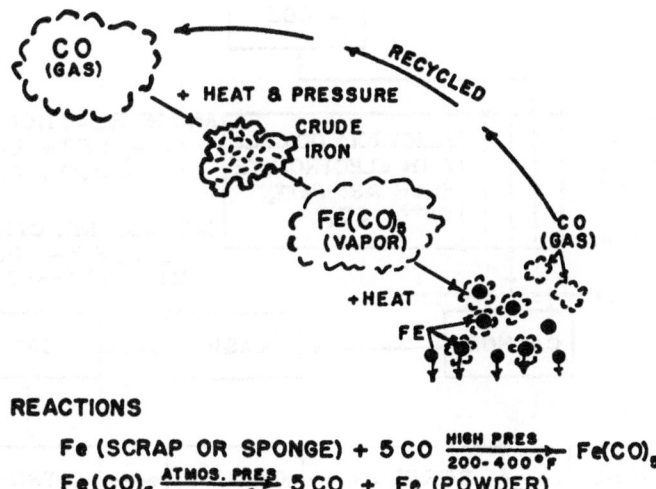

REACTIONS

$$Fe \text{ (SCRAP OR SPONGE)} + 5\,CO \xrightarrow[\text{200-400°F}]{\text{HIGH PRES}} Fe(CO)_5$$

$$Fe(CO)_5 \xrightarrow[\text{300-750°F}]{\text{ATMOS. PRES}} 5\,CO + Fe \text{ (POWDER)}$$

Fig. 2. Schematic drawing of decomposition using carbonyl process.

powders as well as aluminum, lead, tin, and zinc. It is also used in the production of pure iron powder.

Decomposition

     In the decomposition process the compound of a metal is broken down so that the metal remains or forms in a particle form. Some nickel and iron powders are produced commercially by decomposition of their carbonyls. A schematic drawing of this process is shown in Fig. 2. Reactions involved in the formation and decomposition of iron carbonyl follow:

$$\text{Fe (scrap or sponge)} + \frac{5CO\ \text{High Pres}}{200\text{--}400°F} \longrightarrow Fe(CO)_5$$

$$Fe(CO)_5 \xrightarrow[\ 300\text{--}750°F\ ]{\text{Atoms. Pres}} 5CO + Fe\ \text{(powder)}$$

If iron carbonyl is mixed with an inert gas and the decomposition takes place away from the walls of the decomposition vessel, spherical particles are produced which have an onionskin-layer structure. The powders are of very high purity.

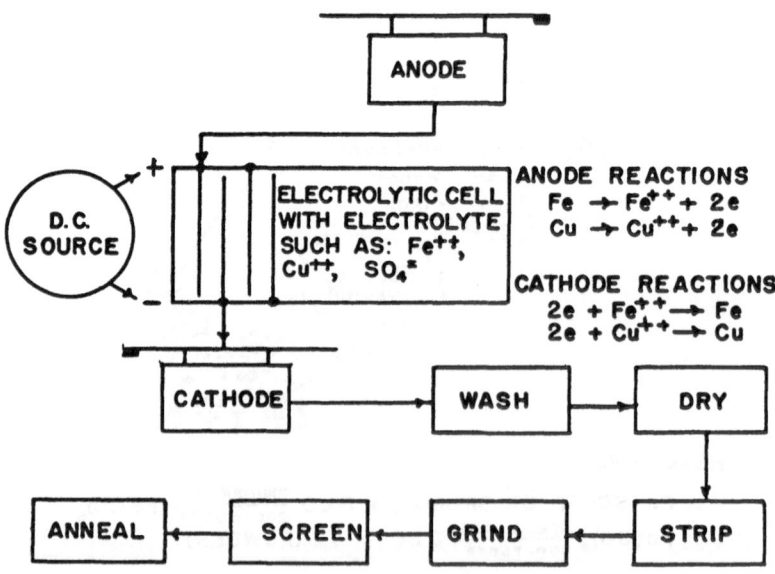

Fig. 3. Schematic of electrolytic process for making powders.

Electrolysis

Iron and copper powders are produced commercially by electrolytic as well as by other methods discussed. The process is relatively simple and reasonably easy to control. It usually involves the use of soluble anodes of the metal involved and stainless steel cathodes from which the cathode deposit is stripped as a brittle sheet for grinding into powder. The deposit may also be in powder form. In present commercial practice, iron is deposited as a brittle sheet which is ground to powder, whereas copper is deposited in powder form. The powder is then annealed in a reducing atmosphere to soften the particles and to lower the oxide content of the powder.

These electrolytic processes are really electrorefining in nature, and there is a considerable purification as anode material is transferred to cathode. Figure 3 shows a schematic outline of electrolysis for iron powder. Reactions involved for iron and copper are:

At Anode	At Cathode
$Fe \rightarrow Fe^{++} + 2e$	$2e + Fe^{++} \rightarrow Fe$
$Cu \rightarrow Cu^{++} + 2e$	$2e + Cu^{++} \rightarrow Cu$

Electrolytic powders are noted for their purity and good compressibility which is especially important in production of high density parts.

Gaseous Reduction of Solutions

A relatively new process involves the reduction of certain metals from solutions. This process is practiced primarily for the production of powdered nickel, cobalt, and copper. The metal-bearing solution is obtained from leaching of ores or scrap. The metal is reduced from the solution in an autoclave at elevated temperatures and pressures. The reaction is as follows:

$$M^{++} + H_2 \rightarrow M + 2H^+$$

where M^{++} represents Cu^{++}, Ni^{++}, or Co^{++}. (Nickel strip for coinage is made from powders produced by this process.)

Grinding

Grinding can be used for powder manufacture whenever a material is brittle enough so that it can be broken into finer particles.

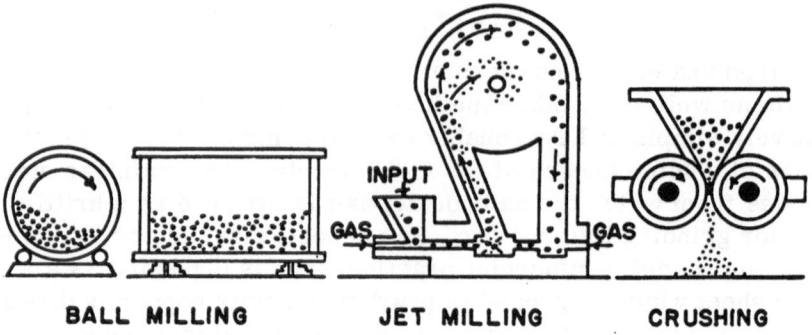

BALL MILLING   JET MILLING   CRUSHING

Fig. 4. Schematic representations of grinding processes.

If it is not brittle but, instead, soft or ductile, the resulting product
will be balls or flakes which are normally undesirable. Frequently,
a preliminary embrittling operation is used so that normally ductile materials can be ground. Grinding is done in ball, jet, or hammer mills. Some grinding processes are shown schematically in
Fig. 4. Examples of brittle materials which are ground to powder
are electrolytic iron cathodes (discussed previously), high sulfur
nickel shot, beryllium, antimony, bismuth, and manganese. Except for electrolytic iron powder, these powders do not find extensive use in pressing and sintering operations.

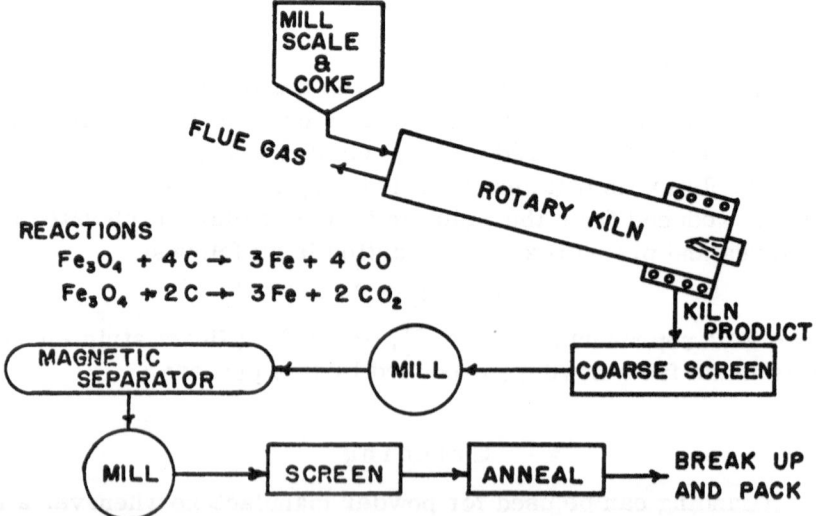

Fig. 5. Schematic outline of reduction with carbon in rotary kiln.

Reduction of Oxides with Carbon

Removal of oxygen from metal oxides with carbon is probably better known as a producer of liquid metal as practiced in blast furnace and cupola operations for pig iron. If the oxygen is removed without melting, a powder of spongy texture is produced. The outstanding example of this method is the reduction of iron ore in a tunnel kiln with carbon in the form of coke or coal. Reactions involved are:

$$Fe_3O_4 + 4C \rightarrow 3Fe + 4CO$$
$$Fe_3O_4 + 2C \rightarrow 3Fe + 2CO_2$$

This is one of the oldest processes and currently produces more iron powder than any other process.

A rotary kiln process as demonstrated in Fig. 5 also has been used for many years in the production of iron powder by carbon reduction of mill scale.

Iron powder is the only powder of interest to us made by this process today.

Reduction of Oxides with Gases

A simple and common commercial method is applied for powders whose oxides can be reduced by the reducing gases, H_2 and CO.

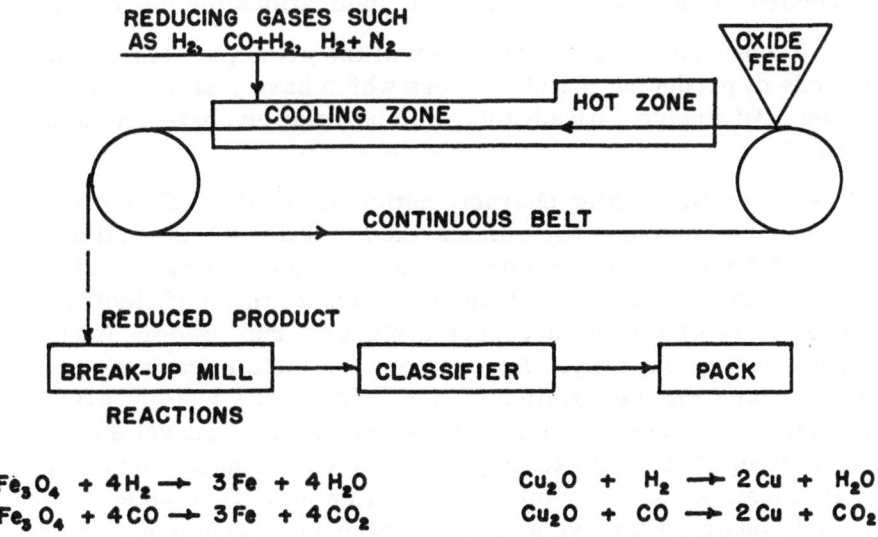

REACTIONS

$$Fe_3O_4 + 4H_2 \rightarrow 3Fe + 4H_2O \qquad Cu_2O + H_2 \rightarrow 2Cu + H_2O$$
$$Fe_3O_4 + 4CO \rightarrow 3Fe + 4CO_2 \qquad Cu_2O + CO \rightarrow 2Cu + CO_2$$

Fig. 6. Schematic outline of gaseous reduction process.

Most of these reductions are carried out in a continuous belt
furnace as shown in Fig. 6. One commercial process involves the
use of a fluid bed instead of a belt furnace. A marked disadvantage
of the reduction method is the fact that no impurities other than oxygen are removed. Therefore, the ore or oxide must be carefully
chosen for the desired purity.

Reactions involved in the reduction of Fe_3O_4 or Cu_2O with hydrogen (H_2) and carbon monoxide (CO) follows:

$$Fe_3O_4 + 4H_2 \rightarrow 3Fe + 4H_2O$$
$$Fe_3O_4 + 4CO \rightarrow 3Fe + 4CO_2$$
$$Cu_2O + H_2 \rightarrow 2Cu + H_2O$$
$$Cu_2O + CO \rightarrow 2Cu + CO_2$$

The gaseous reduction process is noted as one of the major processes for producing copper powder and, if hydrogen is used, for
production of iron powder with low growth when mixed with copper
powder and sintered. Tables 1 to 3 which conclude this chapter
contain rather concentrated information on powder reduction methods and give some properties of the powders as well as applications
for these powders.

Tables 1–3 which conclude this chapter contain rather concentrated information on powder reduction methods and give some
properties of the powders as well as applications for these powders.

Table 1 shows the comparison of the principal commercial
methods of producing metal powders which have just been discussed. Advantages, disadvantages, and relative costs are also
covered.

Some distinguishing characteristics of metal powders made by
the described methods are summarized in Table 2. Special attention is called to such characteristics as the fineness and spherical
shape of carbonyl powders, high purity and softness of electrolytic
powders, low growth-with-copper content of iron powders made by
gaseous reduction, range of properties available in carbon-reduced
powders, and the generally high apparent density and low green
strength of atomized powders. These are rather general statements
and, like most generalizations, there are exceptions. For example,
an atomization nozzle or melt composition may be adjusted under
special conditions to give an irregular-shaped particle with high
green strength.

TABLE 1. Comparison of Principal Commercial Methods of Producing Metal Powders

Method	Raw materials	Powders produced	Advantages	Disadvantages	Relative cost
Atomization	Scrap or virgin melting stock or metal or alloy powder desired	Stainless steel, brass, bronze, other alloy powders, Al, Sn, Pb, Fe, Zn	Best method for alloy powders. Applicable to any metal or alloy melting below 3000°F	Wide range of particle sizes, not all salable. Particles too spherical for some applications	Low to medium
Gaseous reduction of oxides	Oxides of metals such as Cu_2O, NiO, Fe_3O_4	Fe, Cu, Ni, Co, W, Mo	Easy to control particle size of powder. Good compacting powder	Requires high grade oxides. Restricted to reducible oxides	Low
Gaseous reduction of solutions	Ore for leaching or other metal salt solution	Ni, Co, Cu	Ore can be used. Purification during leaching. Fine particles	Applicable to few metals such as Ni, Co, Cu	Medium
Reduction with carbon	Ore or mill scale	Fe	Low cost. Control of particle size, controlled variation in properties possible	Requires high grade ore or mill scale. Applicable mainly to iron	Low
Electrolytic	Generally soluble anodes of iron and copper	Fe, Cu, Ni, Ag	High purity of product. Easy to control	Limited to few metals, cost	Medium
Carbonyl decomposition	Selected scrap, sponge, mattes	Fe, Ni, Co	Produces fine pure powders	Limited to few powders, high cost	High
Grinding	Brittle materials such as Be, high sulfur nickel, high carbon iron, Sb, Bi, Fe, Mn cathodes	Fe, Be, Mn, Ni, Sb, Bi	Controlled size of powder	Limited to brittle or embrittled materials. Quality of powder limits use. Slow	Medium

TABLE 2. Some Distinguishing Characteristics of Metal Powders Made by Various Commercial Methods

Method of production	Typical purity (1) (est.)	Particle characteristics Shape	Particle characteristics Meshes available	Compressibility (softness)	Apparent density	Green strength	Growth-with-copper of iron (2)
Atomization	High 99.5+	Irregular to smooth, rounded dense particles	Coarse shot to 325 mesh	Low to high	Generally high	Generally low	High
Gaseous reduction of oxides	Medium 98.5 to 99.+	Irregular, spongy	Usually 100 mesh and finer	Medium	Low to medium	High to medium	Low
Gaseous reduction of solutions	High 99.2 to 99.8	Irregular, spongy	Usually 100 mesh and finer	Medium	Low to medium	High	Iron not produced by this method
Reduction with carbon	Medium 98.5 to 99.+	Irregular, spongy	Most meshes from 8 down	Medium	Medium	Medium to high	Medium
Electrolytic	High+ 99.5+	Irregular, flaky to dense	All mesh sizes	High	Medium to high	Medium	High
Carbonyl decomposition	High 99.5+	Spherical	Usually in low micron ranges	Medium	Medium to high	Low	?
Grinding	Medium 99.+	Flaky and dense	All mesh sizes	Medium	Medium to low	Low	High

Notes: (1) Purity varies with metal powder involved.
(2) Growth-with-copper of iron during sintering is increase in radial dimension of compacted iron-plus-copper powders.

TABLE 3. Relation of Metal Powder Production Method
to Typical Applications of the Powders

Production method	Typical powders	Typical applications
Atomization	Stainless steel	Filters, mechanical parts, atomic reactor fuel elements
	Brass	Mechanical parts, flaking stock, infiltration of iron
	Fe	Mechanical parts (medium to high density), welding rods, cutting and scarfing, general
	Al	Flaking stock for pigment, solid fuels, mechanical parts
Gaseous reduction of oxides	Fe	Mechanical parts, welding rods, friction materials, general
	Cu	Bearings, motor brushes, contacts, iron-copper parts, friction materials, brazing, catalysts
Gaseous reduction of solutions (Hydrometallurgy)	Ni	Iron-nickel sinterings, fuel cells, catalysts, Ni strip for coinage
	Cu	Friction materials bearings, iron-copper parts, catalysts
Reduction with carbon	Fe	Mechanical parts, welding rods, cutting and scarfing, chemical, general
Electrolytic	Fe	Mechanical parts (high density), food enrichment, electronic core powders
	Cu	Bearings, motor brushes, iron-copper parts, friction materials, contacts, flaking stock
Carbonyl decomposition	Fe	Electronic core powders, additive to other metal powders for sintering
	Ni	Storage batteries, additive to other metal powders for sintering
Grinding	Mg	Welding rod coatings, pyrotechnics
	Ni	Filters, welding rods, sintered nickel parts
	Fe	Waterproofing concrete, iron from electrolytic cathodes (see Electrolytic above)

The relation of powder production method to typical applications of powders is shown in Table 3. This table is self-explanatory, but attention is called particularly to the numerous applications of metal powders made possible by the various methods of powder production.

Seven commercial methods for producing metal powders have been briefly discussed. They are:

Atomization
Decomposition
Electrolysis
Gaseous reduction of solutions
Grinding
Reduction of oxides with carbon
Reduction of oxides with gases

The powders produced by each of these processes have distinguishing characteristics. As a result, there are available for commercial application metal powders with a wide range of properties from which the powder user can select those which best suit his needs.

For more specific data on the production, and pressing, and sintering properties of various commercial metal powders, the reader is referred to the literature of the various powder producers and to the bibliography in the back of this book.

References

1. "Properties of Ferrous Powder Metallurgy Parts," Precision Metal Molding, May 1964, p. 83.
2. "Properties of Non-Ferrous Powder Metallurgy Parts," Precision Metal Molding, April 1964, p. 56.
3. Everhart, John L., "Designing for Metal Powder Structural Parts," Materials in Design Engineering, April 1959, p. 113.
4. Siegrist, Fred L., "Making Structural Parts by Powder Metallurgy," Metal Progress, June 1965, p. 89.
5. Alves, Alexander L., "Double Pressed, Double Sintered Parts," Precision Metal Molding, June 1965, p. 45.
6. Jones, W. D., Fundamental Principles of Powder Metallurgy. Edward Arnold, London (1960), pp. 1-241.
7. Dieter, George, "Powder Fabrication," International Science and Technology, December 1962, p. 58.

8. Cockburn, K. O., et al., "The Production and Characteristics of Chemically Precipitated Nickel Powder," Proc. Metal Powder Assoc., 1957, p. 10.

9. Ryan, V. H., and Tschirner, H. J., "Production and Characteristics of Chemically Precipitated Copper Powder," Proc., Metal Powder Assoc., 1957, p. 25.

10. "Symposium on Developments in the Production and Quality of Metal Powders," Powder Met., 1/2: 13-78 (1958).

11. Powder Met., Technical Assistance Mission Report No. 141, O.E.E.C., 1955, pp. 145-164, 235-237.

Iron Powder:
Production—Characteristics— Fabrication

Arnold R. Poster

Head, Powder Metallurgy
Conoral Aniline and Film Corporation
Easton, Pennsylvania

Applications, science, and technology concerning the use of iron powders in powder metallurgy are presented in the following chapters of this book. It is the purpose of this chapter to deal with the various methods which are used to manufacture iron powders, the properties of the material resulting from the various methods, and the influence of these properties upon the compaction and sintering processes.

There are a number of methods by which iron powders are produced. Each method gives a powder having different characteristics with regard to the properties of the individual particle and the mass of the particles. The applications to which these powders are put depend upon three main criteria. First is economics. Cost of raw material, as it affects the cost of the application, is basic. There are only a few applications where the cost of the powder does not affect the cost of the finished product. The second criterion is the properties of the finished item for which the powder is a raw material. In most instances the finished properties are greatly dependent upon the properties of the starting iron powder. In only a very few instances is there no or little correlation between these factors. Third is the compatibility of an iron powder with respect to established manufacturing facilities. For example, most established sintering furnaces are designed to operate at 2100°F maximum temperature. An iron powder requiring 2300°F for sintering would not be applicable in most instances regardless of price or properties.

Iron powders are manufactured by various methods in order to fulfill these basic requirements. In some instances applications do not have strict requirements and materials made by a number of manufacturing processes can be used for the same application. In other instances very special powder manufacturing procedures are used to achieve materials having special properties.

The Production of Iron Powder

The various methods used to manufacture iron powder are as follows:

1. Reduction
   (a) Carbon reduction of ore concentrates
   (b) Hydrogen reduction of oxidized mill scale
2. Electrolysis
3. Atomization
4. Carbonyl process
5. Pulverization
6. Shotting

The major share of iron powder is produced by either direct reduction of iron ore concentrates, hydrogen reduction of mill scale, or some other suitable form of iron oxide, or by atomization. Electrolysis and carbonyl processes are used to produce significant amounts of iron powders for special purposes. Pulverization and shotting techniques were used extensively at one time but are used today on a limited basis.

1. Reduction Methods

The reduction of a metal compound depends upon the ability of the reducing agent to react more readily with the nonmetallic radical of the compound. The driving force is the decrease in

standard free energy of the system. The reducing reagent may be
a pure solid, liquid, or gas, or a solution. The commercially important processes for the manufacture of iron powder utilize carbon monoxide and hydrogen as reduction reagents and iron oxides
as the metal compound.

The product resulting from the carbon monoxide reduction of
iron ore concentrates is commonly known as sponge iron. However
the product of an air atomization plus reduction process is also
called sponge iron. In addition, the product of the hydrogen reduction of mill scale or purified iron oxide, although not commonly
called sponge iron, contains spongy-looking particles and in some
instances is also thought of as being a sponge iron. The term
"sponge" actually refers to the low density sinter cake which forms
from the treatment of low grade ore concentrates with a reducing
gas. Confusion results from the fact that this is also a good word
with which to describe the appearance of the iron particle. It,
therefore, would seem correct to refer to iron powder resulting
from any reduction process as being sponge iron. In order to avoid
conflict, however, in this discussion the word "sponge" will not be
used.

(A) Carbon Reduction of Iron Ore Concentrates

Iron powder is produced by the direct reduction of iron ore
concentrates using carbon monoxide. In Europe the process is
known as the Sieurin process while in this country it is more commonly known as the Höganäs process. It is extensively covered in
the literature.

In this process powdered magnetite concentrates are reduced
by carbon in tunnel kilns. The concentrates are placed in layers
with coke in containers called saggers. Limestone is added to the
mixture to react with the sulfur contained in the coke. The saggers
are stacked on flat bed rail cars which are pulled through the kiln.
The kiln is approximately 565 ft long, divided as follows: 165 ft
for heating to reduction temperature, 180 ft for reduction, and
195 ft for cooling. Heating is by natural gas, the reduction takes
place at approximately 1200°C and requires several days. The following chemical reactions occur:

$$CaCO_3 \text{ (limestone)} \rightarrow CaO + CO_2$$
$$CO_2 + C \rightarrow 2CO$$
$$Fe_3O_4 + 4CO \rightarrow 3Fe + 4CO_2$$

The products of this reaction are carbon dioxide which passes from the furnace, and a sinter cake consisting mainly of iron but with some impurities from the ore and an appreciable amount of oxide.

The cake is then ground and the powder is magnetically separated. The resulting powder is further reduced in a continuous belt-type hydrogen-atmosphere furnace. This last treatment serves to anneal as well as further purify the powder.

(B) Reduction of Mill Scale

Although a number of plants are set up to produce iron powder by reduction of mill scale, the only one to receive extensive treatment in the literature is that using the Pyron process. The raw material is carefully selected mill scale, which may be rolling and drawing scale or oxidized scrap (known as synthetic mill scale). This is blended, dried, ground in a ball mill to approximately -100 mesh, and magnetically separated. To achieve a uniform oxide this ground powder is fed into a rotary gas-fired heating unit operating at 870-980°C with an air atmosphere. The powder is moved directly from the rotary furnace onto a lime-washed, mild steel belt for transport through a 50-ft-long reduction furnace. The powder height is approximately one inch, the furnace temperature is approximately 980°C and the reducing atmosphere is hydrogen.

The reduced powder forms a sinter cake which is disintegrated at the exit end of the furnace by a rotating wire brush which reciprocates across the belt; or the sinter cake may be broken up by crushing and milling. Normal reduction time is approximately 5 hours.

Mill scale may also be reduced with carbon monoxide by mixing it with some form of carbon, such as coke produced from coal or petroleum. This is also generally followed by a hydrogen reduction.

2. Electrolytic Process

Probably more has been written about the electrolytic process of producing metal powders than all other powder manufacturing processes. This process is capable of producing a wide range of iron powders having globular, nodular, irregular, or dendritic particles, although the usual particle shape is dendritic.

The starting material may be ferrous waste such as scrap mill scale, low grade ores, steel, cast iron, compacted sponge iron, or Armco iron. The electrolytic cell consists of one of the above materials as an anode, a cathode which is made of highly polished stainless steel and upon which the material is deposited from solution, and an electrolyte consisting of an aqueous solution of an iron sulfate or iron chloride.

Current passing through the cell causes iron ions to be deposited at the cathode. Depletion of metallic ions in the immediate area of the cathode causes a migration of ions by diffusion. The bath is continually supplied with metal ions from the anode material. In general, the production of powder on a cathode is favored by:

  a) High current densities,
  b) Weak metal concentration,
  c) Addition of acids,
  d) Low temperature,
  e) Avoidance of agitation.

Iron deposits on the cathode sheet as a spongy or solid mass and is removed from time to time by flexing the sheet. The deposit must be washed to remove all electrolyte, milled to the desired screen size, and the resulting powder heated in hydrogen or cracked ammonia to anneal out the work hardening due to milling and to remove any hydrogen entrapped by the powder at the cathode.

Fused salts may be used instead of aqueous solutions for certain applications. Another variation involves the use of liquid metal cathodes. Generally mercury is used, although experiments have been carried out with other molten metals. The liquid metal cathode process produces a very fine, dendritic elongated particle which is most desirable for use in certain magnet applications.

3. Atomization

Atomization is used as part of a process in which a subsequent reduction step actually plays a dominant role in determining the characteristics of the powder.

A commercially important process employing atomization for iron powder is the Mannesmann Process. In this process low carbon steel scrap with a maximum silicon content of 0.015%, plus approximately 20% pig iron low in silicon, sulfur and phosphorus, is melted in a cupola or electric furnace. The molten iron is held to 3.2-3.4% carbon, 0.1% maximum silicon, manganese, and phosphorus and approximately 0.03% sulfur, before tapping. Atomization is done by air and the resulting powder is collected in a water tank. The cooling rate of the powder in the water tank is adjusted so that the outside of each particle undergoes oxidation and decarburization. A controlled amount of carbon is left in the core of each particle.

The atomized powder is continuously dried by steam heated dryers so that the water content of the powder is less than 1%. The powder is then screened with the oversize material going to vibrating mills.

For further processing it is desirable to obtain a powder at this stage, when it has a certain carbon-to-oxygen ratio. Because of the difficulty in controlling the oxidation of the powder in the water bath each lot of powder is kept separate and tested for both carbon and oxygen. Powders from various charges are then mixed so that a 1:1.5 ratio of carbon to oxygen is present.

This material is then loaded into 14×14 in. trays approximately 3/4 in. deep and heated in 45-ft-long walking-beam-type furnaces having three heating zones of approximately 900-1100-1250°F.

As a result of reactions between Fe_3O_4, Fe_3C, and CO, the carbon and oxygen in the powder form carbon monoxide which blows the iron particles into the form of hollow spherical shapes. An external reducing gas is not used. Instead, the reaction produces an atmosphere inside the furnace consisting of approximately 75% CO and 25% CO_2.

During reduction a sinter cake is formed which is crushed, and further prepared as a powder by sieving and blending.

4.   Carbonyl Process

Iron carbonyl is a chemical compound produced by reacting low grade iron powder with carbon monoxide. The powder which results from the thermal decomposition of iron carbonyl is generally referred to as "carbonyl iron powder." This term does not indicate the chemistry of the material but rather its method of production.

The following reactions are involved:

$$1. \quad Fe_{(s)} + 5CO_{(g)} \rightarrow Fe(CO)_{5(g)}$$
$$2. \quad Fe(CO)_{5(g)} \rightarrow Fe_{(s)} + 5CO_{(g)}$$

The process is done in two stages. In the first stage preheated carbon monoxide is passed over low grade iron powder in a converter to form iron pentacarbonyl (Reaction 1). This takes place at 150-200 atmospheres and temperatures of 200-300°F. This compound, a gas at reaction temperature, passes out of the converter, is liquefied in a heat exchanger, and pumped to a storage tank. From the storage tank the liquid is then sprayed into a heated chamber, called a decomposer, where "cracking" of the carbonyl takes place at approximately 460°F, resulting in carbon monoxide and iron powder (Reaction 2).

Formation of the iron particle is by nucleation and growth in free space. The reaction $2CO \rightarrow C + CO_2$ also takes place during decomposition of $Fe(CO)_5$. The result of this is the deposition of free carbon upon the growing iron particle. As a result of many heating and cooling cycles in the reactor, layers of iron and carbon are deposited. Each time a new layer of iron is formed it is deposited upon the preceding layer of carbon and, therefore, adopts a different orientation than the underlying iron layer. This process causes an "onion skin" structure which is characteristic of as-decomposed material. Approximately 0.8% nitrogen is also introduced into the powder because a nitrogen counter-gas flow is used in the decomposer. Carbonyl iron powder may be used in this condition or it may be further treated in hydrogen at 800-1000°F. During this treatment decarburization takes place, the "onion skin" structure is destroyed, nitrogen is eliminated and the powder is stress-relieved.

5. Other Powder Production Methods

A number of variations of some of the preceding processes are of commercial importance. Notable among these are the variations in the atomization process. In one of these variations the liquid metal is water-atomized and an oxide layer forms on the surface of each particle. The oxygen in this layer is used to decarburize the powder in a subsequent heating operation. In another variation the atomization is done by air. Purified iron oxide, generally mill scale, mixed with the atomized powder as a source of oxygen and a subsequent heating operation is also carried out for decarburization and stress-relieving purposes.

A number of other iron powder manufacturing methods are also worthy of mention. In the carburization method of General Motors Corporation, steel scrap which is usually too soft for pulverizing, is first embrittled by a carburization treatment. The carburized material is ground to a powder which is then decarburized. It is generally not suitable for use by itself; however, it is reported that, when mixed with sponge iron in certain proportions, compacts which are especially suitable for impregnation work can be formed.

During World War II a considerable amount of iron powder was made by the DPG process. This process employs a rotating horizontal disc against which a stream of molten metal falls after first passing through a water cone. This method produces very compact particles. Shotting is another method which is used to a limited extent. In the shotting process low carbon steel liquid is blown against a rotating steel wheel in a shotting tank. The resulting powder contains porous disc-shaped particles which are crushed and pulverized in impact mills and annealed in a reducing atmosphere. Flake iron powder can be made by ball milling a soft iron powder made by any of the processes described, or flaking a thin sheet of wire.

Characteristics of a Powder

Properties of the Particle

Particle shape: To a large extent the process by which a metal powder is produced is mirrored in the properties of the particle. This is also true for iron powders. Figures 1 and 2 show particles of various types of iron powders. As a general rule, the shape of various iron powder particles can be characterized as follows:

Manufacturing Process	Particle Shape
Carbon monoxide reduced ore	Spongy, of irregular shape
Hydrogen reduced mill scale	Spongy, of irregular shape
Electrolytic	Dendritic irregular
Atomized	Spherical hollow or spherical spongy
Carbonyl	Spherical regular
Pulverized	Angular

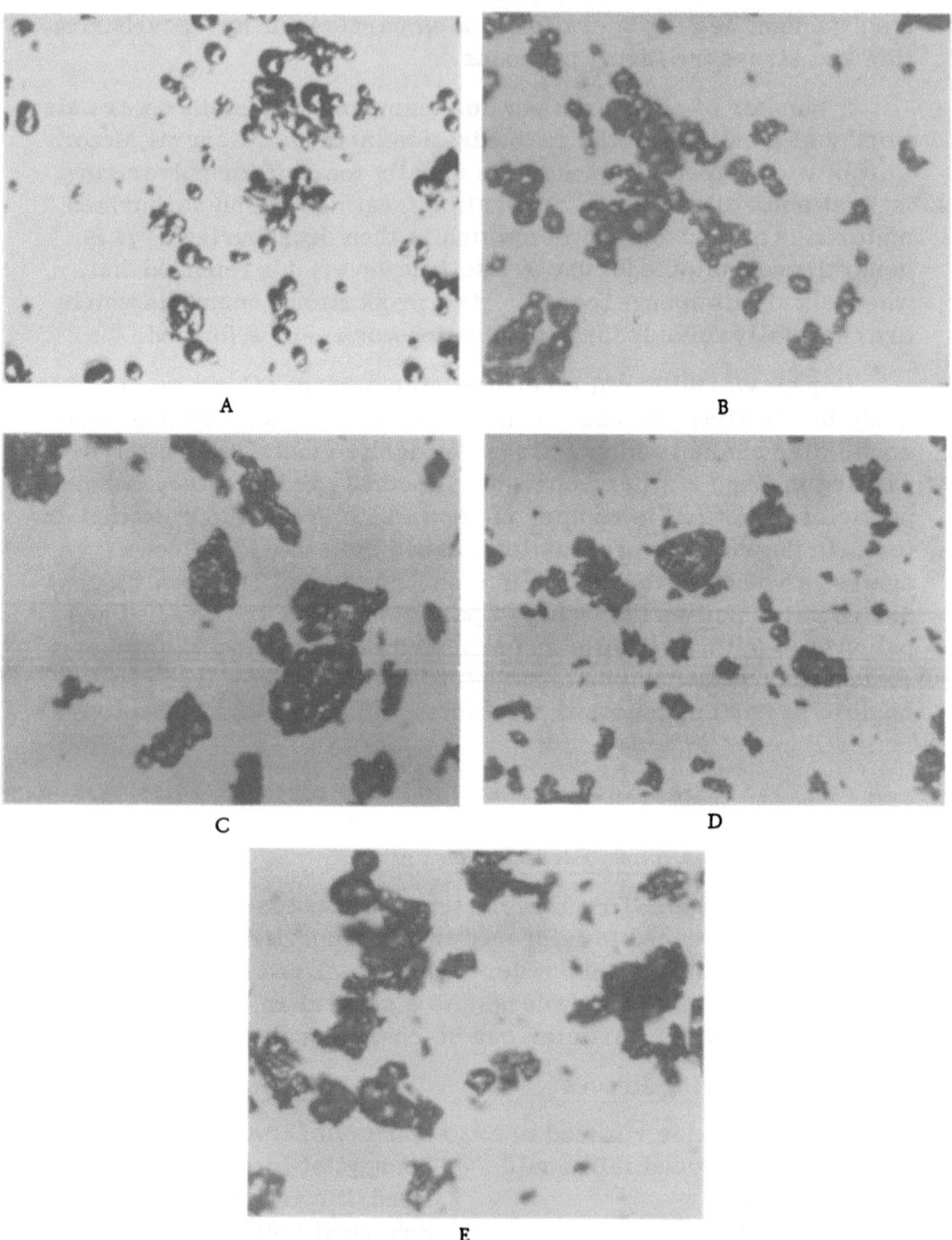

Fig. 1. Photomicrographs showing the surface of iron powder particles manufactured by various methods. A) Carbonyl process, 600 ×; B) atomization process, 100 ×; C) electrolytic process, 100 ×; D) hydrogen reduction of mill scale, 100 ×; E) carbon monoxide reduction of iron ore concentrates, 100 ×.

The spongy irregular appearance of powders is a characteristic of the reduction method.

Characteristically electrolytic iron particles have a dendritic appearance and have been termed irregular dendritic, dendritic globular, irregular flaky nodular, irregular globular, and dendritic globular. This illustrates the wide diversity of particle shape which can be obtained by the electrolytic method.

Atomized particles are generally considered to have a spherical regular shape. In the Mannesmann Process, the heating step after atomizing contributes significantly to the particle shape, giving them a hollow, spherical appearance of irregular size and somewhat spongy appearance.

The carbonyl particles are spherical, and the pulverized particles are irregular and angular in shape.

Particle density: There is scant information in the literature on particle density for iron powders. Leadbeater et al. give some interesting figures for electrolytic, reduced oxide and carbonyl iron particles, and, in particular, particle densities for a number of different particle sizes.

Although the values given vary over a wide range, a summary of the data is possible and this is shown in Table 1.

It can generally be stated that carbonyl iron particles are most dense with figures close to theoretical density being given. Pulverized iron particles are also of extremely high density since they are made from fully dense iron scrap. The least dense are materials derived by reduction methods.

Properties of the Powder

The standard properties which are used by the manufacturers to describe iron powders are given in Tables 2 through 6 for six carbonyl, three atomized, ten carbon monoxide reduced from ore, ten hydrogen reduced mill scale, and ten electrolytic powders. The powders which were chosen for these tables are representative of the production for the various types.

An examination of these tables quickly shows that it is not possible, with any degree of certainty, to determine the method of manufacture of a given powder from these standard powder properties. The bulk properties such as iron content, impurities, ap-

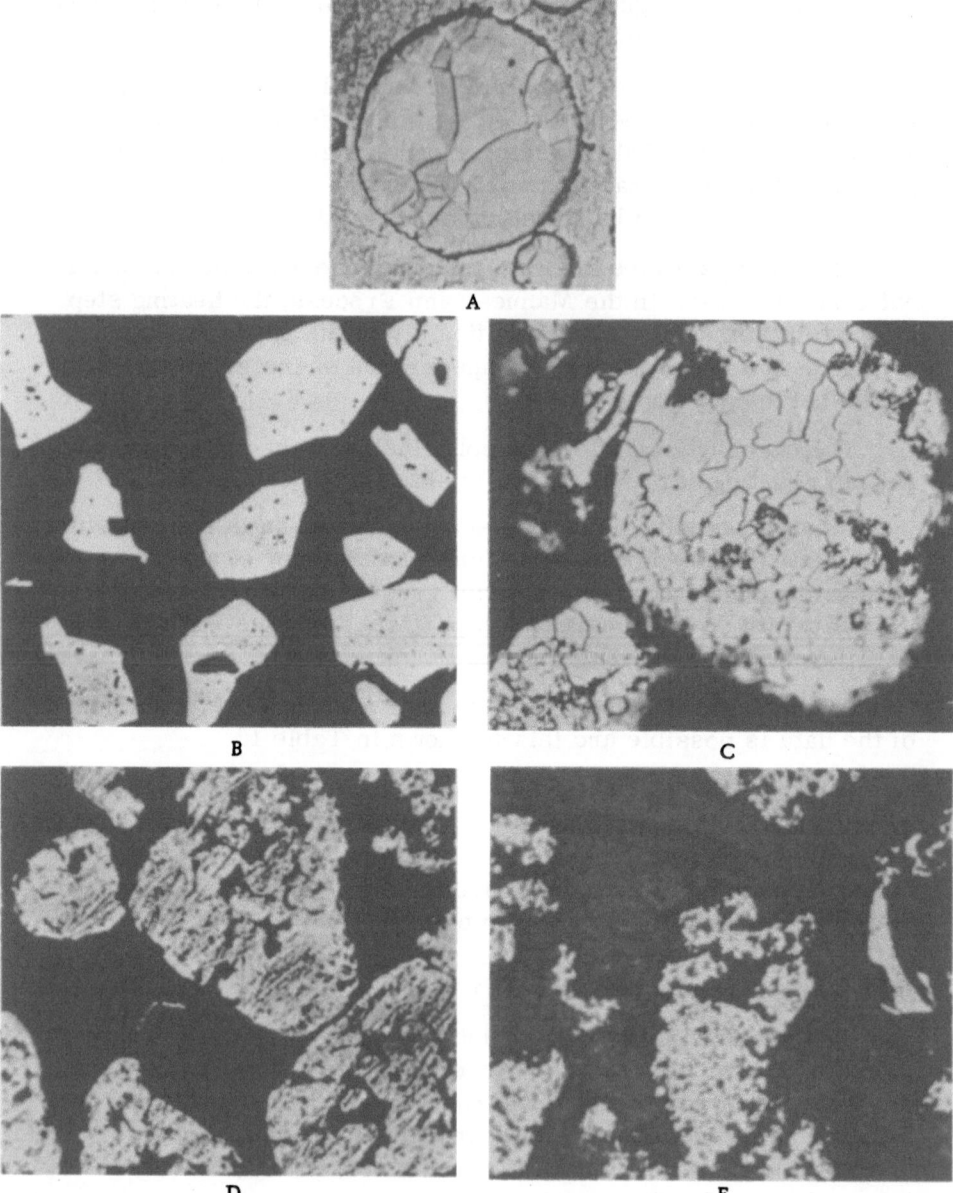

Fig. 2. Photomicrographs showing the cross section of iron powder particles manufactured by various methods. A) Carbonyl process, 2500 ×; B) crushed steel, 200 ×; C) electrolytic process, 500 ×; D) hydrogen reduction of mill scale, 200 ×; E) carbon monoxide reduction of iron ore concentrates, 200 ×.

TABLE 1. Particle Density for Three Types of Iron Powders (Summary of Leadbeater Data)

Powder Type	Particle Density (g/cc)
Hydrogen Reduced Oxide	7.5—7.7
Electrolytic	7.6—7.85
Carbonyl	7.84

parent density, particle density, powder flow, or sieve analysis do not serve to characterize the material.

In numerous instances in the literature authors have compared bulk properties such as powder flow, apparent density, or sieve analysis for a number of materials simply designating the various powders by the terms "sponge iron," "reduced iron," etc. Their comparisons are not really valid since no one powder is represent-

TABLE 2. Specifications of Various Carbonyl Iron Powders*

Powder No.	1	2	3	4	5	6
% Fe (min.)	98	98	98	99.6	99.5	99.5
% C (max.)	0.8	0.8	0.9	0.1	0.1	0.075
% O (max.)	0.3	0.3	0.6	0.2	0.3	0.3
% N (max.)	0.9	0.9	0.7	0.02	0.1	0.05
% Wt. Loss in Hydrogen (max.)	0.3	0.3	0.6	0.2	0.3	0.3
Apparent Density (g/cc)	2.2—3.2	2.0—3.0	2.5—3.5	2.5—3.0	2.2—3.2	2.2—3.2
Tap Density (g/cc)	3.5—4.5	3.0—4.0	—	—	3.0—4.0	3.0—4.0
Particle Density (g/cc)	—	—	7.8	7.8	—	—
Average Particle Size (Microns)	3—5	3—4	8.0	10.0	6—8	6—9

*Manufacturer's specifications.

TABLE 3. Specifications of Various Atomized
and Heated Iron Powders (Mannesmann Process)*

Powder	1	2	3
% Fe	98.9	98.7	98.7
% C	0.05	0.06	0.09
% P	0.011	0.011	0.011
% S	0.022	0.022	0.022
% Mn	0.15	0.15	0.15
% Si	0.04	0.04	0.04
% Wt. Loss in Hydrogen	0.38	0.52	0.43
% Acid Insolubles	0.12	0.12	0.12
Apparent Density (g/cc)	2.60	2.45	2.52
Hall Flow Rate (sec)	27	28	28
Sieve Analysis:			
+80 mesh	2.5 max.	2.5	2.5
−80 +100 mesh	5 max.	5 max.	5 max.
−100 + 200 mesh	35−50	35−50	35−50
−200 + 325 mesh	25−35	25−35	25−35
−325 mesh	20−30	20−30	20−30

*According to the manufacturer.

ative of a given type and, as the tables show, there is no real dis-
tinction in any of the properties with regard to a given manufacturing method. In many instances valid comparisons are made by noting the specific grades or designation for a powder type versus
other powder types also of specific grade.

It can be seen from the tables that electrolytic iron powder,
as an example, is made to consist of very coarse particles, 89%
+100 mesh, or very fine particles, 85% -325 mesh, or any one of a
number of distributions between these extremes. Flow rates may
vary from 23 to 38 seconds, apparent densities from 2.20 g/cc to
3.32 g/cc and weight loss in hydrogen from 0.06% to 0.75%! Equal
degrees of variation occur within the other powder types.

If one is well acquainted with the iron powder industry, it is
possible by noting the parameters used to describe a powder to
make a reasonable guess as to the manufacturing company and the
method of manufacture. For example, SiO_2 is usually reported only
for one type of iron powder made by carbon monoxide reduction of
ore concentrates, and nitrogen content is reported only for carbonyl
iron powders. This is not to say that SiO_2 is reported for all sponge
iron powders or nitrogen for all carbonyl iron powders.

TABLE 4. Specifications of Various Iron Powders Made by Carbon Monoxide Reduction of Ore Concentrates*

Powder No.	1	2	3	4	5	6	7	8	9	10
% Fe	98.0	98.0	98.0	98.0	98.5	98.5	98.6	98.6	98.6	98.8
% C	0.06	0.1	0.06	0.06	0.06	0.06	0.08	0.10	0.09	0.03
% P	0.02	—	0.03	0.03	0.014	0.01	0.01	0.01	0.01	0.01
% S	0.02	—	0.04	0.04	0.015	0.015	0.02	0.02	0.02	0.01
% Mn	0.35	—	0.15	0.15	0.03	—	—	—	—	—
% Si	0.25	—	0.1	0.1	—	—	—	—	—	—
% SiO$_2$	—	—	—	—	0.3	0.2	0.10	0.10	0.10	0.15
% Wt. Loss in Hydrogen	0.8	1.0	0.8	0.8	0.4	0.5	0.60	0.65	0.65	0.45
Apparent Density (g/cc)	2.30	2.95	2.50	3.0	2.40	2.80	1.70	2.35	1.55	2.45
Hall Flow Rate (sec)	—	24	30	24	—	30	—	31	—	30
Sieve Analysis:										
+ 48 mesh	—	—	—	—	12	—	—	—	—	—
−48 +65 mesh	—	—	—	—	22	—	—	—	—	—
−65 +80 mesh	—	—	—	—	—	tr	0.5	0.5	0.5	tr
−80 +100 mesh	10	2	1	1	23	1	11.0	5.0	3.5	1
−100 +150 mesh	—	27	14	14	—	20	26.0	17.5	20.0	18
−150 +200 mesh	47	18	21	20	37	27	21.5	19.0	26.0	26
−200 mesh	—	—	—	—	3	—	—	—	—	—
−200 +250 mesh	—	—	—	—	—	8	8.0	8.0	10.0	9
−250 +325 mesh	25	32	34	30	—	21	16.0	20.0	20.0	24
−325 mesh	18	22	30	35	—	23	17.0	30.0	20.0	22

*According to the manufacturer.

TABLE 5. Specifications of Various Hydrogen Reduced Mill Scale Iron Powders*

Powder No.	1	2	3	4	5	6	7	8	9	10
% Fe	98.25	98.35	98.50	99.30	97.85	96.0-98.25	97.18	98.25	96.0-98.25	96.0-98.25
% C	0.10	0.15	0.07	0.05	0.15	0.015-0.022	0.018	0.015-0.022	0.015-0.022	0.015-0.022
% P	—	—	—	—	—	0.012	0.012	0.012	0.012	0.012
% S	—	—	—	—	—	0.005	0.005	0.005	0.005	0.005
% Mn	—	—	—	0.01	—	0.3-0.6	0.4	0.45-0.65	0.3-0.6	0.3-0.6
% Wt. Loss in Hydrogen	0.65	0.70	0.60	0.30	1.20	0.95	0.95	0.70-1.20	0.7-1.5	0.7-1.2
% Acid Insolubles	—	0.25	—	0.30	0.25	0.20-0.50	0.55	0.20-0.45	0.2-0.7	0.2-0.5
% Other Impurities	1.00	0.55	0.83	0.55						
Apparent Density (g/cc)	2.55	2.80	2.90	2.75	2.65	1.4-1.8	2.40	2.30-2.50	2.3-2.4	2.25-2.35
Hall Flow Rate (sec)	28.5	27	24	27	—	37 (poor)	—	27-34	Poor	33-39
Sieve Analysis:										
+100 mesh	0.3	tr	0.5	1.0	—	2 max.	—	1 max.	—	—
−100 +150 mesh	33.0	12.0	20.0	25.0	—		—	9-14	—	2 max.
−150 +200 mesh	26.0	22.0	21.0	25.0	5.0		—	19-23	—	—
−200 +250 mesh	6.0	6.0	6.0	7.0	5.0		—	6-9	2 max.	—
−250 +325 mesh	18.0	24.0	22.5	20.0	30.0		5	20-28	10-30	—
−325 mesh	18.0	36.0	30.0	22.0	60.0	20-35	95	28-42	70-90	50-75

*According to the manufacturer.

TABLE 6. Specifications of Various Electrolytic Iron Powders*

Powder No.	1	2	3	4	5	6	7	8	9	10
% Fe	99.65	99.65	99.47	99.45	99.40	99.40	99.40	99.17	99.00 min.	99.00 min.
% C	0.02	0.03	0.03	0.04	0.20	0.20	0.03	0.03	0.02 max.	0.02 max.
% P	–	–	–	–	0.01	0.01	–	–	0.015 max.	0.015 max.
% S	0.007	–	–	–	0.008	0.008	–	–	0.02 max.	0.02 max.
% Mn	–	–	–	–	0.07	0.07	–	–	0.03 max.	0.03 max.
% Ni	–	–	–	–	0.01	0.01	–	–	–	–
% Cu	–	–	–	–	0.01	0.01	–	–	–	–
% Si	–	–	.	0.06	0.02	0.02	–	–	0.01 max.	0.01 max.
% Other Impurities	0.05	0.07	0.05	0.06	0.05 chlorine	0.05 chlorine	0.06	0.05	Tr. ea.: Ni, Cr, Cu	Tr. ea.: Ni, Cr, Cu
% Wt. Loss in Hydrogen	0.06	0.25	0.45	0.45	0.30	0.30	0.50	0.75	0.40	0.40
Apparent Density (g/cc)	2.65	2.33	2.20	2.40	3.32	2.72	2.45	2.48	3.0–3.3	2.7–2.9
Hall Flow Rate (sec)	–	–	–	–	23	29	35	38	25–30	31–24
Sieve Analysis: +28 mesh	13.20	–	–	–	–	–	–	–	–	–
–28+48	35.70	tr.	–	–	–	–	–	–	–	–
–48+65	20.15	15.00	–	–	–	–	–	–	–	–
–65+100	19.75	23.00	–	–	3	4	tr.	tr.	–	0–2
–100	11.20	–	–	–	–	–	–	–	–	–
–100+150	–	23.00	tr.	–	26	11	10	0.20	0	11–17
–150+200	–	13.00	1.50	–	18	18	20	12.00	2–10	9–15
–200+250	–	–	2.50	–	6	16	4	7.00	10–10	5–11
–250+325	–	10.00	15.00	15.00	16	18	18	25.80	40–50	23–32
–325	–	16.00	81.00	85.00	31	33	48	55.00	32–42	32–42

* According to manufacturer.

Because of the wide variation in properties for various grades
of the same powder type, it is not possible to compare the bulk
properties of the various types of iron powders.

Fabrication

Compaction

It is not feasible to generalize to any extent on the ability of
various types of iron powders to be compared to specific density
levels. Because of the many variables affecting the compaction
process and the different ways in which these variables may apply,
it is also not meaningful to present single figures or single line
curves as representative of the situation.

The many variables which affect the compaction of iron powder include:

1. Basic ductility of the particle as determined by purity,
   stress condition, and structure;
2. Particle size, size distribution, shape, and shape distribution;
3. Moisture content of the powder;
4. Condition and type of surface oxides;
5. Compact shape and size;
6. Pressing speed.

The extent to which a powder manufacturing process determines the properties of the compact is determined by the extent
to which the process controls one or many of the factors listed
above.

In order to achieve a comparison of the overall effect of powder processing on the compact density for iron powders, the percent of theoretical density for many different grades of each powder type was plotted as a function of compacting pressure for pressures from 20 to 60 psi. All of the data were taken from samples
compacted in standard MPIF tooling for a density bar from powders
containing either 0.75 or 1.00% zinc stearate lubricant. Data were
gathered from more than one source for each type of iron powder
and in many instances more than one set of data was available for
a given grade within a type.

For the atomized, ore reduced, oxide reduced, and electrolytic powders data were taken for those commonly used materials

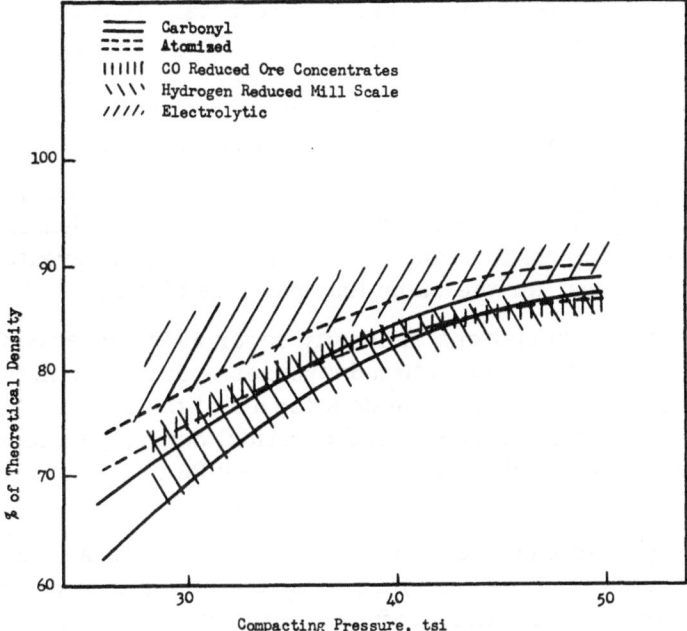

Fig. 3. Compacting pressure vs. compact density expressed as percent of theoretical for iron powders made by various methods. Variation in density for numerous powders of each type is shown by the range for the type.

whose particle size centered in the -60 mesh size range with 35% max. in the -325 mesh range. The data for carbonyl iron powder were taken for materials all of 10-micron Fisher subsieve size, since carbonyl iron powders are available only in this fine size.

Figure 3 shows the ranges over which the data for the various types of iron powder fell. The ranges are indicated either by crosshatch markings or by solid or dashed lines.

A number of conclusions can be drawn from the data shown in Figure 3.

1. Density increases with compaction pressure, as can be expected, the greatest increase being with carbonyl iron powder. Electrolytic iron powder shows the least density change with increasing compaction pressure.

2. The highest density over the compaction pressure range of 20 to 60 tsi is obtained with electrolytic powders. The density range for electrolytic iron is from approximately 77-85% of theo-

retical density at 20 tsi and from approximately 89-93% of theoretical density at 60 tsi. There is some overlapping of the results for electrolytic and atomized powders. Over the entire pressure range the lowest density figures for the electrolytic powder are still higher than for the highest density atomized iron.

3. Iron powder derived by carbon monoxide reduction of ore concentrates gives the smallest variation in compact density data over the entire pressure range. The variation is not greater than ±1% of theoretical for the many iron powders plotted.

4. Greater variations in density occur at the lower compacting pressures and the variation decreases with increasing pressure. Variation over all of the data for all of the types at 20 tsi is from approximately 62 to 85% of theoretical density, while at 60 tsi the variation is only from approximately 86 to 92% of theoretical density.

5. There is a considerable overlapping of values in the ranges for the various iron powder types. For the majority it can be said that the process of powder production does not determine to any large degree the ability of the material to form compacts of characteristic density. The notable exception to this is the overall higher density of electrolytic powder.

6. The fact that carbonyl iron powder compacts to a density equal with that of oxide reduced iron at low pressures, and better than sponge iron, and equal to reduced oxide and atomized iron at high pressures, appears to eliminate the consideration of particle size or particle-size distribution as the determining factor. Carbonyl iron powder not only has a very fine average particle size but the size distribution of 0.5 to 20.0 microns is narrow, compared with the other materials.

Sintered Properties

It is well known that when sintered samples of equal density are compared, all processed under identical conditions from various types of iron powder, their properties when sintered depend mainly upon the powder particle size and not upon the method of powder production. In this respect it can generally be said that the properties when sintered increase with decrease in powder particle size.

Investigations concerning the properties of sintered samples made from various types of iron powders have been sparse in recent years. However, during the years just after World War II considerable effort was made to determine the properties of various iron powders and a number of very excellent reports appear in the literature. The following discussion is drawn mainly from the work of Squire and Kuzmick of approximately twenty years ago and from work reported within the last five years.

The properties of sintered iron powders appear to be consistent in their relationship to density. Regardless of the method of powder manufacture, each material has its own particular density curve, which is determined to the largest extent by the green density, and along this curve properties after sintering increase with increase in density. Density, therefore, is an important consideration in this discussion.

Work of Squire (as reported by Goetzel)

Squire established density as the primary criterion of sintered mechanical properties. The powders used by Squire are listed in Table 7 and the resulting data in Fig. 4. As shown in the figure, the effect of increasing density for samples from all of the powders is to increase tensile strength, elongation, hardness, and impact strength as sintered. The relation, however, is not always linear, especially for the measurement of ductility, impact strength, and elongation.

Squire also fractionated an electrolytic and a hydrogen reduced iron powder into –140 +200 mesh and –325 mesh fractions and tested each fraction for both materials. Following is a summary of the result of this work.

a) Impact Resistance: The –325 mesh fraction produced samples having greater impact resistance than the coarser fractions. Typical figures are 10 ft-lb for the –140 +200 mesh fraction and 22 ft-lb for the –325 mesh fraction when comparison is made for 7.0 g/cc dense samples.

b, c) Elongation and Hardness: Neither the difference in material nor size fractions had any effect on these properties.

d) Tensile Strength: The finer-sized fraction for each material produced sinterings which were approximately 25% stronger than

PROPERTIES OF SINTERED IRON

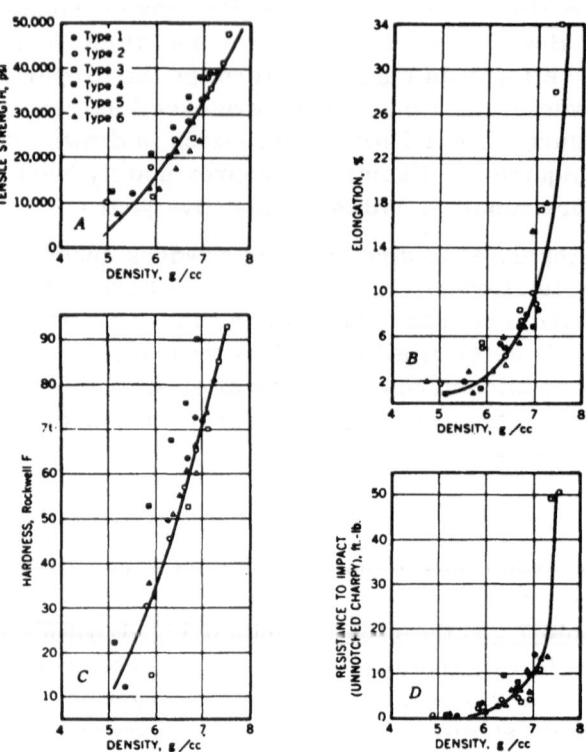

Fig. 4. Influence of the type of iron powder used on the physical properties
of compacts after sintering in hydrogen for 1 hour at 1100 °C (2010 °F) (according to Squire). A. relation between tensile strength and density; B. relation between elongation and density; C. relation between hardness and
density; D. relation between impact strength and density. Points designate
the following types of iron powder: 1.carbon monoxide reduced mill scale;
2. hydrogen reduced iron oxide; 3. hydrogen reduced mill scale; 4-6. electrolytic iron.

those produced from the coarser fractions. Typical figures at 6.8
g/cc density are 25,000 psi for the coarse fraction and 30,000 psi
for the fine fraction.

From both of these investigations it is concluded that the exact relationship between density and respective physical properties
is dependent only upon the particle size and independent of the type
or history of the powder used.

Work of Kuzmick

The effect of particle-size distribution on properties after sintering was studied by Kuzmick et al. for two oxide reduced and two electrolytic-type iron powders. In this study each powder was separated into various fractions and each fraction, along with the unfractionated powder, was used to form pressed and sintered test bars. The properties of the powders used by Kuzmick are given in Table 8 and Table 9 shows the fractions for each powder. Samples from the unfractionated as well as the fractionated powders were

TABLE 7. Characteristics of Iron Powders Used
by Squire (from Goetzel)

Powder No.	1	2	3	4	5	6
Powder Type	CO red. mill scale	H₂ red. oxide	H₂ red. mill scale	Elec.	Elec.	Elec.
Chemical Analysis, %						
Carbon	0.19-0.18	0.02	0.04	0.04	0.08	0.02
Manganese	0.18	0.34	0.0	0.14	0.25	0
Silicon	0.075	0.20	0.01	0.17	0.10	0
Sulfur	0.022	0.009	0.012	0.009	0.23	0.008
Phosphorus	0.005	0.007	0.003	0.019	0.007	0.002
Nickel	0.09	0.06	0	0	0.045	Tr
Chromium	0.03	0.04	0	0.03	0.03	0.03
Screen Analysis, %						
+100	0.05	0.35	1.2	0	1.8	1.5
−100+140	7.40	4.2	15.4	0	5.9	8.5
−140+200	23.25	29.6	39.5	0	34.25	24.8
−200+270	9.75	9.9	10.4	0.05	10.65	9.8
−270+325	12.35	13.5	11.8	0.15	14.15	14.0
−325	46.80	43.0	22.0	99.45	32.75	41.2
Apparent Density, g/cc	1.71	2.27	2.88	2.79	2.67	2.52
Tap Density, g/cc	2.55	2.75	3.30	4.28	3.05	3.52
Specific Surface, cm²/g	1059	522	235	1460	488	530
Flow Rate, g/sec	0	0	1.50	0	1.86	1.06

TABLE 8. Chemical Analysis of Powders Used by Kuzmick, %

Element	Powder 1 Reduced	Powder 2 Electrolytic	Powder 3 Reduced	Powder 4 Electrolytic
Total Fe	97.55	99.67	97.55	99.12
Metallic Fe	90.97	95.29	93.95	95.20
Total C	0.29	0.04	0.24	0.09
SiO_2	0.15	0.02	0.22	0.04
S	0.007	0.024	0.028	0.011
P	0.023	0.005	0.017	0.005
Mn	0.17	0.08	0.20	0.03
Other Elements	—	—	0.1-0.15	—

compacted at 25 and 50 tsi followed by sintering for 30 min at a temperature of 2012°F in a deoxidized and dried hydrogen atmosphere.

The results of this study clearly demonstrate the following with regard to properties after sintering:

1. Shrinkage is mainly determined not by the type of powder but by the amount of fine material present. In general, radial and axial shrinkages during sintering increase with increasing amount of fine particles.

TABLE 9. Particle–Size Distribution
of Powders Used by Kuzmick

Fraction, μ	Mesh	Powder 1 Reduced	Powder 2 Electrolytic	Powder 3 Reduced	Powder 4 Electrolytic
Screen Test					
Oversize	+100	0.05%	1.18%	1.23%	Tr
B (105-149)	−100+150	9.57	11.09	15.35	2.70%
C (74-150)	−150+200	19.78	24.36	31.28	32.27
D (44-74)	−200+325	31.93	22.25	27.88	32.70
Subsieve	−325	38.67	41.12	24.26	32.33
Subsieve Distribution					
E (30-44)		13.72%	10.05%	7.09%	10.31%
F (20-30)		13.38	11.02	8.26	12.88
G (10-20)		11.39	19.79	8.76	8.89
H (0-10)		0.18	0.26	0.15	0.25

TABLE 10. Tensile Strength and Elongation of Sintered Samples in Kuzmick Study

Fraction	Powder 1 Reduced		Powder 2 Electrolytic		Powder 3 Reduced		Powder 4 Electrolytic	
	Tensile Strength, psi	Elongation, %	Tensile Strength, psi	Elongation, %	Tensile Strength, psi	Elongation, %	Tensile Strength, psi	Elongation, %
Compacting Pressure, 25 tsi								
A (original)	12,900	6	17,000	9	7,200	2	15,200	7
B	11,100	4	13,100	6	7,200	2	14,400	10
C	10,800	5	14,800	8	5,800	2	13,000	9
D	12,200	4	15,900	9	8,800	1	15,500	1C
E	13,100	5	17,700	10	15,200	5	18,300	7
F	14,700	4	18,300	11	15,400	5	18,100	6
G	14,400	4	19,000	12	10,700	3	17,100	4
Compacting Pressure, 50 tsi								
A (original)	22,500	8	25,000	14	13,700	2	28,100	12
B	20,400	6	21,400	13	12,300	3	23,400	16
C	20,700	6	22,800	15	11,800	1	22,800	14
D	21,800	8	24,600	14	16,200	3	27,700	14
E	24,100	8	24,600	11	24,500	8	24,000	10
F	24,100	8	26,200	14	23,600	8	30,800	11
G	22,000	8	26,300	15	18,000	5	27,700	7

2. Hardness is little affected by the initial powder particle size. Samples made from the various fractions and from the un-fractionated powder gave essentially the same results.

Hardness did vary between the two types of materials since they were of different densities.

The hardness of samples made from unfractionated reduced powders are R_H 74 and 70, for unfractionated electrolytic powders, R_H 78 and 81. For the -150 +200 mesh fraction, the hardness after sintering is R_H 72 and 71 for reduced powder, and R_H 79 and 80 for electrolytic powder.

3. Tensile Strength and Elongation: Table 10, from Kuzmick, shows that the tensile strength and elongation after sintering are both higher for samples made from electrolytic powder and that there is little effect on these properties due to particle size regard-less whether the unfractionated or fractionated powder is consid-ered.

The Kuzmick results appear to be in direct conflict with statements previously made to the effect that higher properties are obtained with finer powders after sintering. However, a notable difference between the work of Squire and Kuzmick was the sinter-ing time. Although the sintering temperature was essentially the same, 2010°F for the work of Squire and 2012°F for the work of Kuzmick, and the sintering atmosphere was hydrogen in both in-stances, the time of sintering varied considerably, being 60 min for the Squire work and 30 min for the Kuzmick work.

The question of comparing properties of equal density sam-ples or of samples made from equal fractions after sintering is, in a way, somewhat academic. In most instances powders are chosen for their ability to perform within the confines of a specific manu-facturing process for a specific purpose. For example, in most in-stances it is not feasible to compact sponge, reduced oxide, or atomized powder to densities above 7.0 g/cc since the pressures to reach this density with this material are excessive. So the high-er-compact-ability electrolytic powder is used. Likewise, it is not reasonable to use the electrolytic powder at lower densities for most applications since lower-priced sponge, oxide reduced, and atomized powder will do the same job. In many applications there are equipment limitations which restrict or dictate one type of pow-der rather than another type. A frequently employed method to compensate for the lower property values is to make additions to iron powders which normally form lower density sinterings. In this way copper or carbon additions will increase properties of lower density compacts to well above the properties of the higher density compacts.

Because powder flow is such an important property for proper filling of the die cavity, the use of finer iron powders is restricted. As previously discussed, finer electrolytic powders produce sintered samples of superior properties. However, they are not in common use because they do not readily flow.

It becomes important then to secure a knowledge of the properties for sintered samples made from standard materials as they are produced.

Properties for Sintered Standard Iron Powders

An interesting comparison of the properties for various types of sintered iron powders is shown in Figs. 5, 6, and 7. In this work standard carbonyl, electrolytic, and carbon monoxide reduced ore powder types were used. The properties of the powders used for

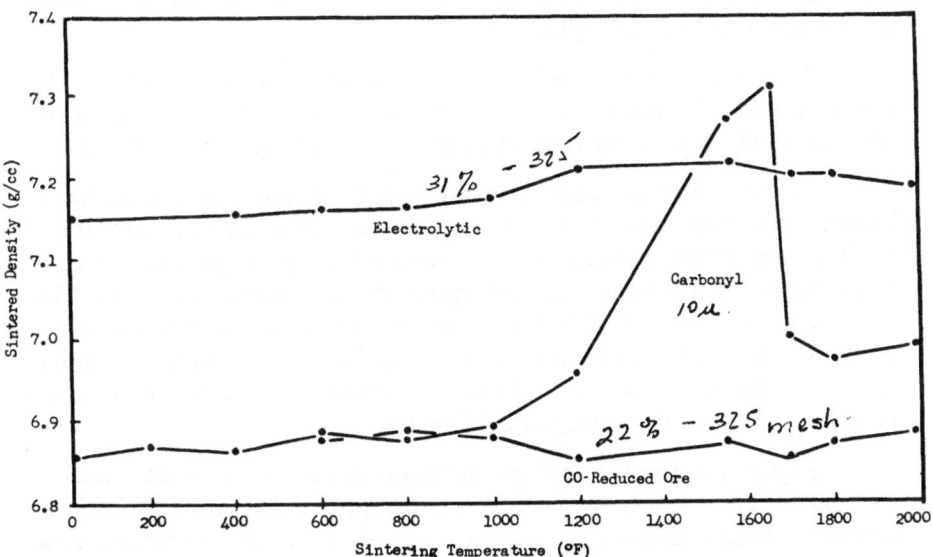

Fig. 5. Density vs. sintering temperature for sintered samples made from various iron powders. (Compaction at 50 tsi using die wall lubrication. Sintered for 1 hour in dissociated ammonium atmosphere.)

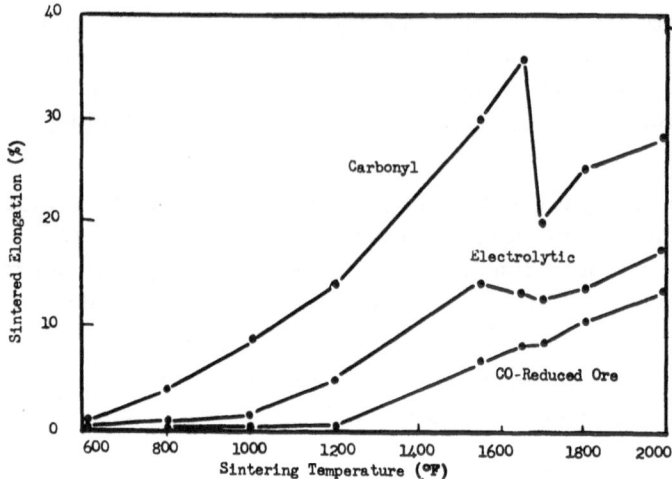

Fig. 6. Elongation vs. sintering temperature for sintered samples made from various iron powders. (Compaction at 50 tsi using die wall lubrication. Sintered for 1 hour in dissociated ammonia atmosphere.)

the results in Figs. 5, 6, and 7 and also Table 11 are given in the following tables: Table 2, powder 4 for carbonyl; Table 3, powder 2 for atomized; Table 4, powder 10 for CO reduced ore; and Table 6, powder 5 for electrolytic.

Compacts were pressed at 50 tsi using a die-wall lubricant consisting of 1% stearic acid in carbon tetrachloride, and sintered for one hour at various temperatures in a hydrogen atmosphere.

Density changes are given in Fig. 5 for the three materials. Electrolytic and carbon monoxide reduced iron powder samples change very little in density over the sintering temperature range while carbonyl iron powder undergoes considerable changes. The density curves in Fig. 5 for the sintered electrolytic and carbon monoxide reduced materials correspond to the compacted density curves. The curve for the sintered carbonyl iron powder is entirely independent of the compacted density.

Figures 6 and 7 show the changes in tensile strength and elongation as a function of sintering temperature. Elongation and tensile strength increase as sintering temperature increases in a somewhat regular way for the electrolytic and carbon monoxide reduced materials. However, the carbonyl iron powder samples

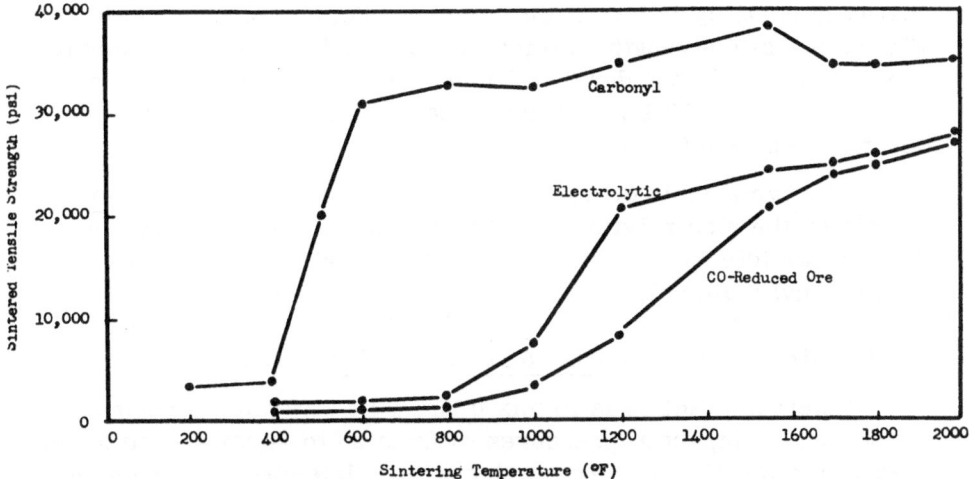

Fig. 7. Tensile strength vs. sintering temperature for sintered samples made from various iron powders. (Compaction at 50 tsi using die wall lubrication. Sintered for 1 hour in dissociated ammonia atmosphere.)

TABLE 11. Sintered Properties of Carbonyl, Electrolytic Reduced and Atomized Iron Powder Samples

Com-pacting Pressure	Iron Powder Type*			
	Atomized	Electrolytic	Carbonyl	H_2 Reduced Mill Scale
Tensile Strength (psi)				
20 tsi	14,600	15,400	24,500	—
30	24,200	20,000	33,000	20,000
40	26,200	21,500	37,500	—
Elongation (%)				
20 tsi	6.5	6.4	9.5	—
30	7.2	7.0	15.5	9
40	8.2	7.8	24.0	—
Density (g/cc)				
20 tsi	5.96	6.39	5.86	—
30	6.64	6.95	6.55	5.95
40	6.98	7.12	6.90	—

*0.75% zinc stearate added to each powder. Sintered for 1 hour at 2050°F under hydrogen pressure.

change in ductility in a way similar to the density change shown in
Fig. 5. Tensile strength changes for carbonyl iron powder samples
are different from the density changes, increasing considerably between 400°F and 600°F and undergoing only minor changes for temperatures up to 2000°F.

The two notable differences between carbonyl iron powder
and either the electrolytic or carbon monoxide reduced powders is
the fine particle size (20 micron), and spherical particle nature of
the carbonyl material.

With Addition of Compacting Lubricant

Common practice is to mix 0.75% zinc stearate lubricant to
the powder, compact at pressures of from 25 to 50 tsi and sinter at
2050°F for 30-60 min. Properties after sintering are given in
Table 11 for a number of iron powders after undergoing this treatment. This table shows quite conclusively that the property increase as a function of increasing density cannot apply to all types
of iron powders.

In these data the lower density carbonyl iron powder produced
samples having the highest tensile strength and elongation. The
atomized powder samples were second in tensile strength and the

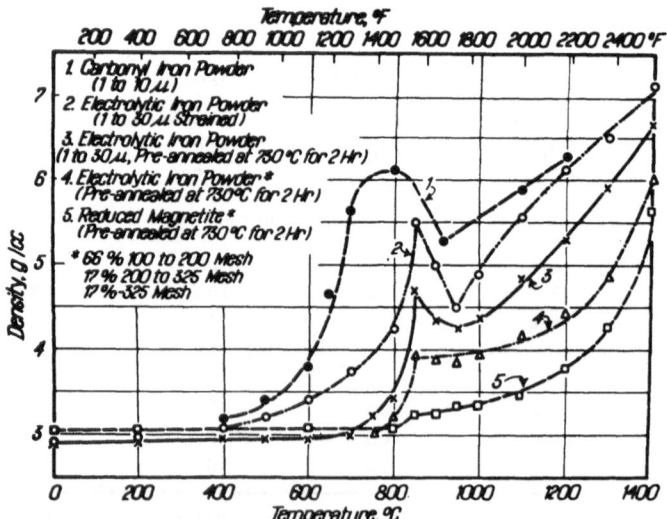

Fig. 8. Density changes of loose iron powders of various origin
and particle-size distribution when annealed 24 hours in hydrogen at temperatures up to 1400°C (2550°F).

electrolytic powder samples lowest. The dependency of properties on powder type, in this comparison, is also shown when equal density sample, produced at different compacting pressures, are compared. For example, the tensile strength and elongation figures are 20,000 psi and 7.0%, respectively, for electrolytic iron powder samples at 6.95 g/cc density. At 6.95 g/cc density for carbonyl powder samples these figures are 37,500 psi and 24%.

Loose Powder Sintering

Figure 8 shows the work of Schlecht et al. on the sintering of various iron powders without first forming a compact. The sinterability of the finer material is obvious from this figure.

Sintered Structure

The sintered structures of samples made from carbon monoxide reduced ore, electrolytic, and carbonyl iron powders compacted at 50 tsi and sintered for one hour at 1800°F in hydrogen are shown in Fig. 9. There are notable differences in these microstructures in both grain size and the configuration and distribution of the porosity. The sample made from carbon monoxide reduced material shows the finest grains with the electrolytic somewhat coarser and the carbonyl the coarsest.

Grain size as shown in Fig. 9 for the three materials is not consistent with the results of Forss shown in Fig. 10. In this latter figure the grain size change is given as a function of sintering temperature for a number of types of iron powder. In this work, the grain size for the carbonyl sample is considerably smaller than for the other materials and electrolytic iron powder is shown to produce the largest sintered grains. However, the results shown in Fig. 10 are for samples sintered for 30 min in dissociated ammonia while the results shown in Fig. 9 are for samples sintered for 60 min in hydrogen.

A second difference between the microstructures shown in Fig. 9 is the form and distribution of the porosity. The porosity in the sample prepared from carbonyl iron powder is more uniform in size, is generally spherical, and also well distributed. Although the total porosity is less for the sample made from electrolytic powder than for the carbonyl sample, the porosity is distributed in a much different way. The porosity in the microstructure of the electrolytic sample and the carbon monoxide reduced sample is

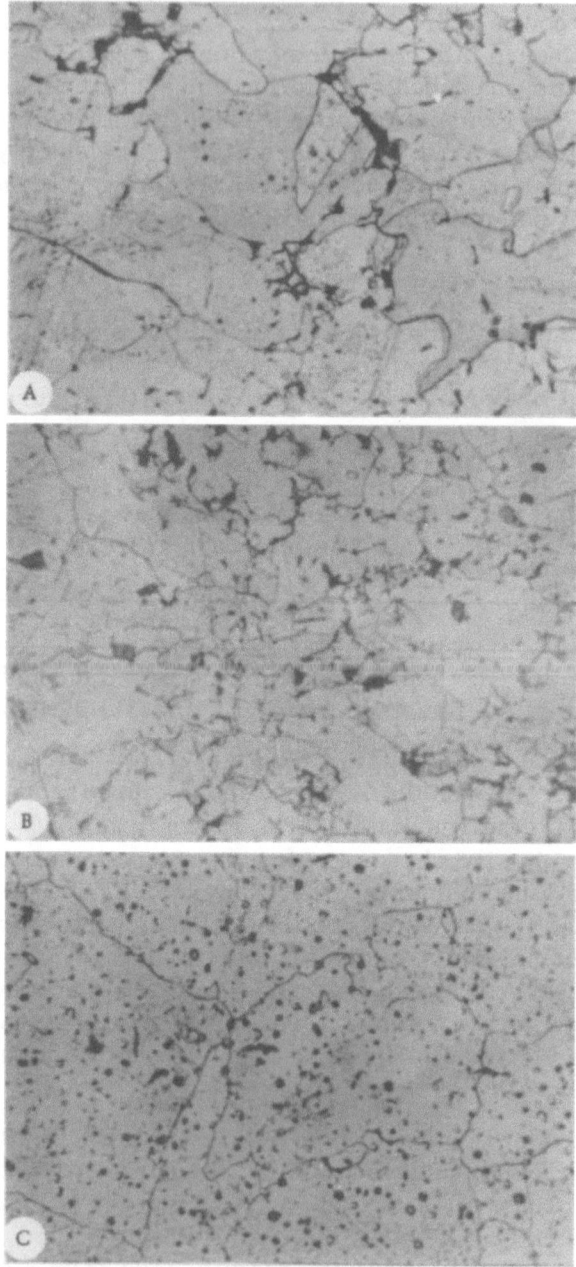

Fig. 9. Photomicrographs at 300× of the structure of sintered samples made from various iron powders after compaction at 50 tsi and sintering at 1800°F for 1 hour in hydrogen. A) From electrolytic iron powder; B) from iron powder made by carbon monoxide reduction of ore concentration; C) from carbonyl iron powder.

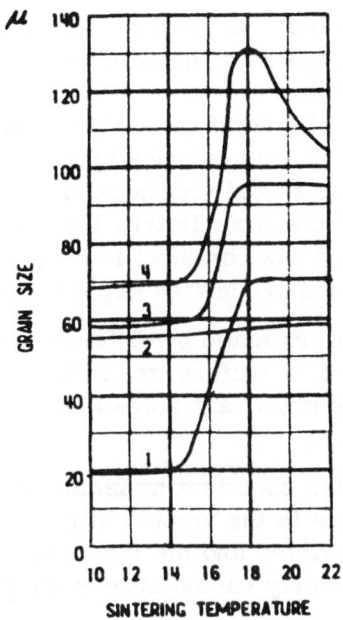

Fig. 10. Approximate average grain size after 30 min at indicated
temperature for samples made from various iron powders. 1. Carbonyl
iron; 2. hydrogen reduced iron; 3. sponge iron; 4. electrolytic iron
(after Forss).

distributed mainly at the grain boundaries in a nonuniform manner,
while the porosity in the carbonyl structure is distributed very
uniformly.

Discussion

In this chapter the methods utilized in the manufacture of
iron powder, the characteristics of the powder particle and the
mass of particles, and the properties of samples made from the
various powders when compacted and sintered have been discussed.

In a general way it can be said that the particular method of
powder manufacture appears to have very little responsibility for
either pressed or sintered properties, although the characteristics
of the various powder types differ considerably. The literature
strongly indicates that a number of iron powders of equal properties made by different methods would form compacts and sinterings
of equal properties.

From the preceding discussion it appears that there are two dominating factors concerning the powder which determine sintered properties. The first factor is the complex of characteristics which determine the ability of a powder to be formed into compacts of a given density. These characteristics may be plasticity of the particle as affected by purity and stress conditions, particle density, particle size and particle-size distribution, interparticle contact, surface conditions, and gas content. For a given set of powder characteristics sintered properties depend upon density.

In practice powders made by different methods do not have similar properties and, therefore, the individual characteristics of each powder type play important roles in determining the properties after sintering.

The second major contributing factor in determining pressed and sintered properties is the amount of fine particles: the greater the content of fine particles and the finer the particle size, the higher will be the property values after sintering.

References

L. S. Forss, Raw Materials and Properties in Ferrous Powder Metallurgy, American Society for Metals Technical Report No. 14.4 (1963).

C. G. Goetzel, Treatise on Powder Metallurgy, Vols. I and II, Interscience Publishers, Inc., New York, N. Y. (1949).

W. D. Jones, Metal Powders for Engineering Purposes: A Review, "Symposium on Powder Metallurgy, Special Report No. 58," Iron Steel (London), 1954, pp. 1-7.

W. D. Jones, Principles of Powder Metallurgy, Edward Arnold, London (1937).

J. F. Kuzmick, J. D. Shaw, C. L. Clark, T. W. Frank, W. V. Knapp, and A. S. Margolies, "Effect of Particle Size on Iron Powder Properties," The Iron Age, Dec. 5, 1946, pp. 72-80.

C. J. Leadbeater, F. Hargreaves, and L. Northcott, Some Properties of Engineering Iron Powders, "Symposium on Powder Metallurgy, Special Report No. 38," Iron Steel (London), Dec. 1947, pp. 15-34. (See also pp. 72-121 of this book.)

J. Libsch, R. Volterra, and J. Wulff, "The Sintering of Iron Powder," in: Powder Metallurgy, by J. Wulff, Am. Soc. Metals (1942).

G. L. Miller, Production of Ferrous and Nonferrous Metal Powders, "Symposium on Powder Metallurgy, Special Report No. 38," Iron Steel (London), Dec. 1947, pp. 8-14.

A. R. Poster, Handbook of Metal Powders, Reinhold Publishing Corp., New York, N. Y. (1966).

A. R. Poster, "Alpha and Gamma Phase Sintering of Carbonyl and Other Iron Powders," in: Modern Developments in Powder Metallurgy, Vol. 2, 1966, pp. 26-44.

Powder Metallurgy; Report of Technical Assistance Mission No. 141 of the Organization for European Economic Co-operation, June, 1955, pp. 145-163.

Relations between Pore Structure and Densification Mechanism in the Compacting of Iron Powders

Gerhard Bockstiegel

Höganäs AB
Höganäs, Sweden

Compacting Properties in Relation to the Pore Structure Inside and in between Powder Particles

Introduction

In compacting metal powders, the total porosity of the compact decreases rapidly at first and then more and more slowly with increasing compacting pressure. During the last three or four decades numerous attempts have been made to find an adequate mathematical description of this empirically well–known relationship. However, none of the many models of powder compacting so far proposed has proved to be a satisfactory description of reality. From a recent critical review of the relevant literature [1] it emerges that most of these models fail either due to oversimplification of the involved mechanism, or simply because they are derived from curve-fitting rather than considerations about the underlying physical principles.

Until recently, little or no attention had been paid to the fact that the total porosity of a powder compact cannot be changed without affecting its pore size distribution. This very circumstance, however, offers an ideal opportunity to gain deeper insights into the mechanism of powder compacting. In a recent paper this author [2] has shown that with increasing compacting pressure the pore size distribution in iron powder compacts changes in a way as if the individual pores were eliminated from the compact strictly in order of size, progressing from larger to smaller ones.

This observation has been related to the yielding behavior of a hollow sphere under increasing external pressure. From the theory of elasticity it can be derived that a hollow metal sphere cannot yield unless the external pressure exceeds a certain threshold, P_{min}, which is the higher, the smaller the volume of the hole is compared to the volume of the solid metal. In the formula [2]

$$P_{min} = \frac{2}{3}\,\sigma_o\,\frac{r_a^3 - r_i^3}{r_i^3} \tag{1}$$

where σ_0 is the upper yield stress of the metal in tension, r_a is the outer radius and r_i the inner radius of the hollow sphere.*

---

*In this connection it may be of interest to mention that in 1948 Torre [3] had tried to calculate a yielding criterion for the hollow metal sphere in order to lend theoretical support to an empirical formula which earlier the same year K. Konopicky [4] had derived from curve fitting of experimental results over a limited range of compacting pressures. Torre derived the theoretical formula

$$P = \frac{2}{3}\,\sigma_o\,\ln\left(\frac{r_a}{r_i}\right)^3$$

where σ_0 = yield stress in tension, r_i = inner radius, r_a = outer radius, and $(r_i/r_a)^3$ = porosity of the hollow sphere.

Konopicky proposed the empirical formula

$$P = A\,\ln\frac{U_o}{U}$$

where A and U_0 are constant factors and U = porosity of the powder compact. Although this seemingly good agreement between theory and experiment had elicited much attention since the formulas first were published, a critical analysis made recently [1, 2] revealed that neither is Torre's formula the correct yielding criterion for a hollow metal sphere under uniform external pressure, nor is Konopicky's formula a satisfactory description of reality, when applied to a wider range of compacting pressures.

If, in rough approximation, a porous compact is anticipated as being composed of a great number of hollow spheres of different sizes, each being subjected to roughly the same external pressure, the above yielding criterion can be used to explain qualitatively why, on the average, larger pores disappear at lower pressures than smaller ones.

The value of this rather primitive kind of explanation can, of course, be questioned, because the geometry of pore structure in real powder compacts is much too involved to permit a straightforward application of the simple mathematics valid for the stress distribution in a hollow metal sphere.

However, as will be shown later on in this paper, it is possible, on the grounds of certain kinetic and statistical considerations, to derive a mathematical model which, in much more detail than the simple model of the hollow sphere, can account for the experimentally observed changes of the pore size distributions in metal powder compacts.

An important prerequisite for the success of theoretical speculations is, of course, to have a sound basis of experimental facts. It is, therefore, the purpose of the present paper to broaden this basis by collecting more experimental information about the changes of pore structure under the influence of increasing compacting pressure.

Since metal powders, and in particular iron powders, depending upon their mode of manufacturing, can have widely different particle structures, varying from rather compact to extremely porous particles, it is a natural proposition to compare their compacting behavior on the basis of varying particle porosity.

An interesting aspect in this connection is the circumstance that there actually can be two different kinds of pores in a powder compact: those inside the powder particles, which are inherited from the powder manufacturing process, and those which occur in between the particles in the bulk of the powder. These two kinds of pores can, of course, be either of comparable or of quite unequal sizes, depending upon the particle structure and the granulometry of the powder in question. Especially interesting from the theoretical point of view is the case where the pores inside the powder particles are much smaller than those in between them. One can ask, for instance, whether even for such extreme combination of pore sizes the previously reported observation [2] holds true that with increasing compacting pressure the pores disappear in order of size, i.e., first the large ones and then the small ones.

The experimental results discussed below will throw light upon this and other relevant questions.

Experimental Procedures

Raw Materials and Preparation of Iron Powders

A convenient way of producing iron powders of different particle porosity is the reduction of pure magnetite (Fe_3O_4) at different temperatures in hydrogen or dissociated ammonia.

As a raw material, a fine-grained natural magnetite, type MK_{10}, of high purity from the Gellivare mines of Sweden was chosen. The chemical analysis and the granulometry of this material can be seen from Table 1.

The reduction was carried out in a streaming atmosphere of dissociated ammonia (75% H_2 + 25% N_2) in a laboratory tube furnace, at the following temperatures: 500, 700, 850, 1000, 1150, and 1250°C. The reducing material was charged, approximately 1000 grams at a time, in trays of iron plate with a bottom area 10 cm × 20 cm and a height of 3 cm. The time intervals to achieve complete reduction to iron powder were: 170, 250, 62, 25, 3, and 1.5 hours, respectively. Since the iron powder reduced at 500°C was expected to be highly pyrophoric, it was subjected to a short temperature shock of 15 min at 800°C before taking it out of the reduction fur-

TABLE 1. Chemical Analysis and Granulometry of Gellivari
Magnetite, MK_{10}, as Used in This Investigation

Chemical analysis, wt%:

Fe-tot	FeO	Fe_2O_3	SiO_2	P	S	HCl insol.
71.6	31.1	68.4	0.47	0.009	0.003	0.51

Sieve analysis:

Tyler mesh	− 35 +100	− 100 +150	− 150 +200	− 200 +250	− 250 +325	− 325
wt%	12.4	16.0	15.9	6.6	12.1	37.0

TABLE 2. Granulometry and Oxygen Content (H_2-loss)
of Hydrogen-Reduced Iron Powders As Used
in This Investigation

Red. temp., °C	H_2-loss* wt%	Sieve analysis, Tyler mesh, wt. %						Apparent density,† g/cc	Flow-ability,‡ sec/50 g
		+65	−65 +100	−100 +150	−150 +200	−200 +250	−250 +325	−325		
500	0.60	0	9.0	15.0	16.5	3.6	12.1	41.8	1.82	∞
700	0.45	0	5.5	21.0	21.5	6.8	14.5	30.7	1.58	59.6
850	0.48	0	4.8	22.4	24.0	8.0	15.6	25.2	1.74	49.5
1000	0.29	0	5.0	25.8	28.0	8.2	15.4	17.6	1.77	44.4
1150	0.22	0	11.3	35.3	25.7	5.3	10.3	12.1	1.77	45.6
1250	0.18	0	2.4	33.4	31.6	7.4	12.8	12.4	2.08	37.7

*H_2-loss determined according to MPI-Standard No. 2-48.
†Apparent density determined according to MPI-Standard No. 4-45.
‡Flowability determined according to MPI-Standard No. 3-45.

nace. This simple procedure proved sufficient to render the powder nonpyrophoric. Reduction temperatures higher than 850°C yielded a reduced iron in the form of spongy cakes, which were carefully disintegrated in a mortar and subsequently soft-annealed for 15 min at 850°C in dissociated ammonia. All powders were screened on a 65-mesh Tyler screen.

The granulometry and the residual oxygen contents (H_2-loss) of the so-obtained iron powders are presented in Table 2, the pore structure of their particles emerges from the photomicrographs of Fig. 1.

For one particular series of experiments a very close size fraction, i.e., between 24 and 100 Tyler-mesh, was isolated from the iron powder reduced at 500°C. For comparison, a similar size fraction, i.e., between 40 and 150 Tyler-mesh, of electrolytic iron powder, type "HVA-Standard," consisting of practically nonporous particles (manufacturer: Husqvarna Vapenfabriks AB of Sweden) was also used.

Testing Procedures

The above-described iron powders were investigated with regard to the following properties: compacting pressure required to achieve a green density of 6.0 g/cc, green strength, particle porosity, pore size distribution including most frequent pore size inside particles, and specific surface area.

The compacting pressure required to achieve a green density of 6.0 g/cc was determined on cylindrical compacts of 10 mm diameter and approximately 7 mm height, made in a hardened steel die.

The green strength was determined as transverse rupture strength of rectangular compacts (unsintered) with the dimension $1^1/_4 \times ^1/_2 \times ^1/_2$ in. and having a green density of 6.0 g/cc according to MPI-Standard No. 15-51 T. In all compacting experiments 1% zinc stearate was admixed to the iron powders as a lubricant.

The specific surface area of the powders was determined by nitrogen absorption due to the well-known BET-method, using the Ströhlein-Aerometer.

The total porosity of the powder compacts was determined both with the hydrostatic balance and by means of microscopic

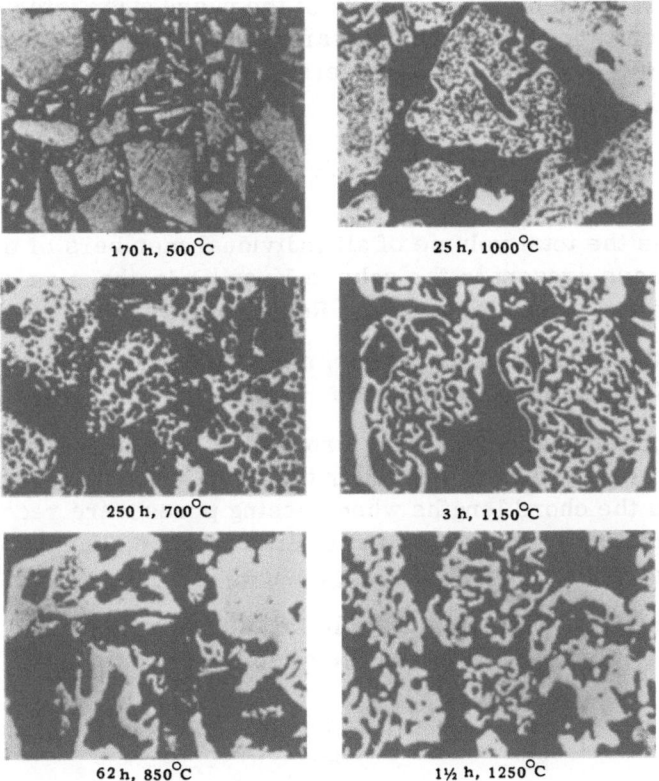

170 h, 500°C   25 h, 1000°C

250 h, 700°C   3 h, 1150°C

62 h, 850°C   1½ h, 1250°C

Fig. 1. Photomicrographs showing pore structure in iron powder particles obtained from pure magnetite reduced in hydrogen at temperatures and time intervals as indicated below pictures. Unetched. 400 ×.

lineal analysis, the principle of which is described in some detail below. Satisfactory agreement was found between the two methods. The porosity inside individual powder particles as well as the porosity arising from pores in between the particles was determined by lineal analysis only.

The pore size distribution, including the most frequent pore size, both inside and in between the powder particles was also determined by linear analysis.

Lineal Analysis

Lineal analysis (also called line-scanning) is based on a mathematical principle discovered by A. Rosiwal [5] according to which the volume proportion of a dispersed phase in a body of volume V is equal to the linear proportion of the sum of the chord lengths produced by the members of the phase when intersected by a (sufficiently long but not necessarily straight) measuring line of length L (so-called Rosiwal-traverse), which is laid arbitrarily through the body. In the formula

$$\frac{v}{V} = \frac{\lambda}{L} \tag{2}$$

where v is the total volume of all individual members of the dispersed phase present in the volume V, and λ is the sum of all chord lengths encountered on the Rosiwal-traverse of length L.

In the present investigation, microscopic lineal analysis has been applied in the following way.

A polished cross section through the carefully prepared powder compact [2] was passed under the cross-hair center of a microscope and the chord lengths when passing pores were recorded and automatically classified in 16 different size classes by means of an electronic impulse counting device.

Applying Rosiwal's principle [5] the total porosity U of the compact is then obtained from the simple expression

$$U = \sum_{j=1}^{k} n_j \, l_j \tag{3}$$

where n_j is the number of chords with lengths between l_{j-1} and l_j which are encountered per unit length of the Rosiwal-traverse, and where $j = 1$ and $j = k$ are the numbers of the size classes containing the smallest and the largest chord length, respectively.

Pore Size Distribution

Before discussing pore size distribution, it is necessary to define the term "pore size." Since pores as they occur in powder compacts are of rather irregular shape and partly interconnected, it may seem difficult at first glance to find a reasonable and unambiguous parameter which could be used for quantitative mathematical treatment.

Fortunately, as mentioned above, lineal analysis delivers a clearly defined size distribution of chord lengths, and it is possible to define an equivalent size distribution of spherical pores that not only yields the same chord length distribution, but also the same specific surface and the same total porosity, as the irregular pores of the real powder compact [2, 6]. If n_j is the number of chords with lengths between l_{j-1} and l_j, which occur per unit-length of the Rosiwal-traverse, one can calculate the number per unit-volume, N_j, of pores with equivalent sphere diameters between $D_{j-1} = l_{j-1}$ and $D_j = l_j$ from the relationship

$$N_{j+\frac{1}{2}} = \left(\frac{4}{\pi} \; \frac{n_j}{l_j^2 - l_{j-1}^2} - \frac{n_{j+1}}{l_{j+1}^2 - l_j^2} \right) \tag{4}$$

where the index $j + \frac{1}{2}$ means that in graphical presentation, the value of $N_j + \frac{1}{2}$ is to be plotted over the middle between l_j and l_{j+1} while n_j and n_{j+1} are to be plotted over l_j and l_{j+1}, respectively.

A pore size distribution is most conveniently represented by the accumulated porosity $u(D_i)$ arising from pores having equivalent sphere diameters $\leq D_i$, which is defined as follows:

$$u(D_i) = \frac{\pi}{6} \sum_{j=1}^{i} D_j^3 \, N_j \tag{5}$$

where N_j and D_j have the same meaning as explained in connection with (4). If D_k designates the diameter of the largest pores present in the compact, the total porosity is given by $U = u(D_k)$.

Another important characteristic of a pore size distribution is the most frequent poor size, D_m, which can be determined graphically by plotting the so-called accumulated frequency function

$$H(D_i) = \sum_{j=1}^{i} D_j^3 N_j \bigg/ \sum_{j=1}^{k} D_j^3 N_j \qquad (6)$$

over D_i and reading the value of D_m where the curve representing $H(D_i) = 50\%$. More information about the determination of pore size distributions by means of linear analysis is available [2, 6].

Experimental Results

In the graphs of Fig. 2 the structural and the mechanical powder properties as described above are plotted as functions of the reduction temperature. A comparison of the trends of the respective curves reveals the following interesting features.

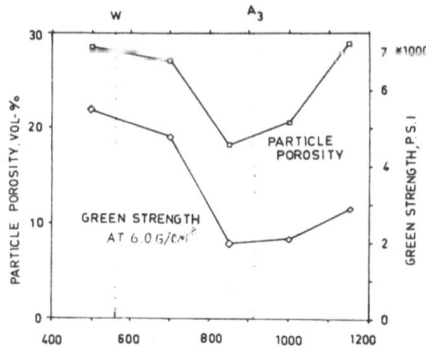

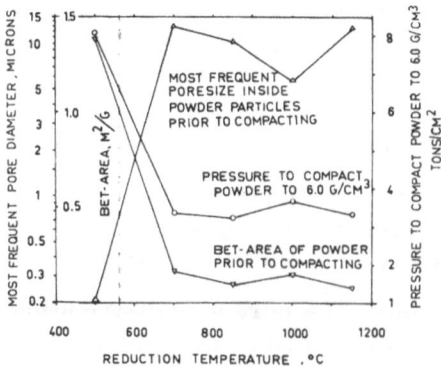

Fig. 2. Influence of reduction temperature upon particle porosity and other green properties of iron powder obtained from hydrogen-reduced pure magnetite. W = Lower temperature limit for existence of wüstite (= $Fe_{1-x}O$). Below W, reduction proceeds according to: $Fe_3O_4 \rightarrow Fe$, above W, according to: $Fe_3O_4 \rightarrow Fe_{1-x}O \rightarrow Fe$. $A_3 = \alpha/\gamma$—transition temperature.

Relations between Reduction Temperature and Pore Structure

Looking first at the structural properties, we find that the particle porosity on one hand and the most frequent pore size inside the powder particles, and the specific surface area (BET) of the powder on the other hand, are affected by the reduction temperature in a remarkably different way. When passing the so-called Wüstite temperature at 560°C, both the specific surface area and the most frequent pore size change abruptly (the first one dropping by a factor of ~ 60); the particle porosity, however, varies only little.

As a further contrast, we notice that for reduction temperatures in the neighborhood of the α/γ-transition temperature (910°C), particle porosity adopts a pronounced minimum, while the most frequent pore size and the specific surface area change comparatively little or hardly at all.

Although it is not the object of the present paper to get more deeply involved in speculations about how different reduction mechanisms influence the pore structure of reduced iron, it may well be of interest to briefly mention the following two associated aspects: (1) Below the Wüstite temperature of 560°C the reduction of magnetite proceeds according to the one-step reaction $Fe_3O_4 \rightarrow$ Fe, i.e., directly from the magnetite to metallic iron, while above this temperature it proceeds according to the two-step reaction $Fe_3O_4 \rightarrow Fe_{1-x}O \rightarrow Fe$, i.e., via an intermediate phase of Wüstite. (2) When the crystal structure of iron at 910°C changes from the body centered cubic α-lattice to the face centered cubic γ-lattice, the self-diffusion coefficient of iron drops by a factor of ~ 400. A corresponding, though smaller, drop is observed also in the shrinkage rate of iron powders, when sintered just below and just above this transition temperature.

After these brief hints, we have to leave further speculations about how reducing and sintering mechanisms in combination produce the observed pore structures here in question, and revert to the real object of this investigation, i.e., how the pore structure in the particle influences the compacting properties of the iron powder. More information about the influence of different reduction conditions on the structure of the reduced iron can be found in papers by J. O. Edström [7, 8] and by K. Kohl and B. Marincek [9].

Relations between Pore Structure

and Compacting Properties

Taking now another look at the graphs in Fig. 2, it appears that the pressure required to compact the various iron powders to a density of 6.0 g/cc closely follows the same pattern as the specific surface area of the powders, to which the curve of the most frequent pore size inside the particles (prior to compacting) forms a mirror image.

From this correlation we can draw the conclusion that an iron powder is the more easy to compact the smaller its specific surface area is, or the larger the pores inside its particles are, which in principle agrees well with the above-mentioned yielding criterion equation (1), due to which large pores should require less high external pressure for yielding than small pores.

In contrast, green strength, which is a measure for the strength of cohesion between particles in the unsintered compact, closely follows the same pattern as particle porosity, which is quite different from that of specific surface area. At this point, it may be interesting to ponder about why green strength should depend upon the porosity of the powder particles but not upon the size of their pores. The following speculations, however, are not meant to be an accurate theory of green strength, but rather a simple model to explain the main features of the mechanism behind the observed experimental facts.

Green strength, σ_g, is, as already mentioned, a measure for the strength of bonding between the particles in an unsintered powder compact. Mainly responsible for the bonding between two adjacent particles is a kind of mechanical interlocking which occurs when, under the applied pressure, metal of one particle is forced into pores of the other one.

It is reasonable to assume that the strength of this interlocking is proportional mainly to the total cross-sectional area of the interlocking protrusions formed that way. This cross-sectional area, in turn, is proportional to the probability, W_{mp}, that metal area in the periphery of one particle faces pore area in the periphery of the adjacent particle. In short, one can assume that

$$\sigma_g \sim W_{mp} = \left(\frac{a_p}{a_p + a_m}\right)_1 \cdot \left(\frac{a_m}{a_p + a_m}\right)_2 , \qquad (7)$$

where a_p is the total pore area and a_m is the total metal area in the periphery of one particle. The subscripts 1 and 2 indicate that the first quotient is to be taken for one particle, and the second quotient for the adjacent particle.

From a geometrical principle discovered by A. Delesse [10] it follows that the area proportion of pores occurring in the periphery of a porous body is equal to the volume proportion of pores contained in the volume of the body. Applying this principle to our case, and assuming that the two adjacent particles both have the same porosity U, equation (7) takes the form

$$\sigma_g \sim W_{pm} = U\,(1-U) \qquad\qquad (8)$$

Evidently, this function has a maximum for U = 0.5, and disappears for both U = 0 and U = 1. Particle porosities greater than 50% (U > 0.5), however, hardly ever occur in metal powder particles because such particles would already in the manufacturing process either disintegrate into smaller ones, or become densified.

Thus, one can conclude that in most practical cases (U < 0.5) green strength should increase steadily with increasing porosity of the powder particles. Since, on the other hand, particle porosity is not necessarily correlated with pore size or specific surface area, it is not surprising to find by experimental evidence that green strength, though being closely related to the first mentioned variable, does not show much dependence on the two latter ones.

The Changes of Particle Porosity
and Interparticle Porosity during Compacting

Interesting features of the mechanism of powder compacting are revealed when studying porosity changes in a powder where the pores inside the particles are much smaller than those in between the particles.

For this purpose, the iron powder obtained by hydrogen reduction at 500°C (compare Fig. 1) was screened between 24 and 100 mesh (Tyler) in order to achieve as large pores as possible in between the powder particles. In this way it was possible to achieve a maximum pore size between particles of the uncompacted powder in the order of 100 microns, while the maximum pore size inside the particles was in the order of 1-2 microns.

If the model of the hollow sphere, equation (1), due to which large pores should yield at lower pressures than smaller ones,

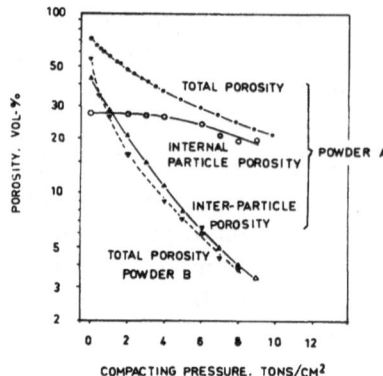

Fig. 3. Change of total porosity, porosity inside powder particles and porosity in between powder particles with increasing compacting pressure. Powder A = iron powder obtained from pure magnetite reduced at 500°C in hydrogen, particle size 24—100 mesh, high degree of internal particle porosity. Powder B = electrolytic iron powder, particle size 40—150 mesh, practically no internal particle porosity.

would be applicable in principle even to this extreme combination of pore sizes, one would expect the porosity arising from the large pores in between the particles to decrease much faster with increasing compacting pressure than the porosity arising from the very small pores inside the particles. As can be seen from the graph in Fig. 3, this is indeed what happens.

One might argue that the great difference in shrinkage rate between large and small pores, observed in this case, could be partly due to the circumstance that the large pores between the particles could shrink not only through deformation but also through particle rearrangement, while the small pores inside the particles could shrink through deformation only.

Experimental evidence from previous investigations [2] proves, however, that particle rearrangement in iron powders ceases already at compacting pressures as low as 0.25 tons/cm^2. It can, therefore, be assumed that the large pores between the particles and the small pores inside the particles both shrink due to the same mechanisms, i.e., through deformation.

Another interesting feature, emerging from the graph, Fig. 3, is that the decrease of interparticle porosity of the powder with highly porous particles (A) follows almost exactly the same curve as the decrease of total porosity of a comparative powder with practically pore-free particles (B). Since in the latter case, total porosity is identical with interparticle porosity, one can draw the conclusion that the yielding conditions for the large pores between the particles were the same in both cases. In other words, the ductility of the powder particles must have been approximately the same regardless of whether they contained a great number of very small pores or not. In this connection it must be remembered, of course, that ductility is not the same as compressibility. So there is no contradiction between highly porous powders particles being easily deformable and at the same time difficult to densify.

The photomicrographs in Fig. 4 clearly illustrate how this unique combination of properties works when the powder is compacted.

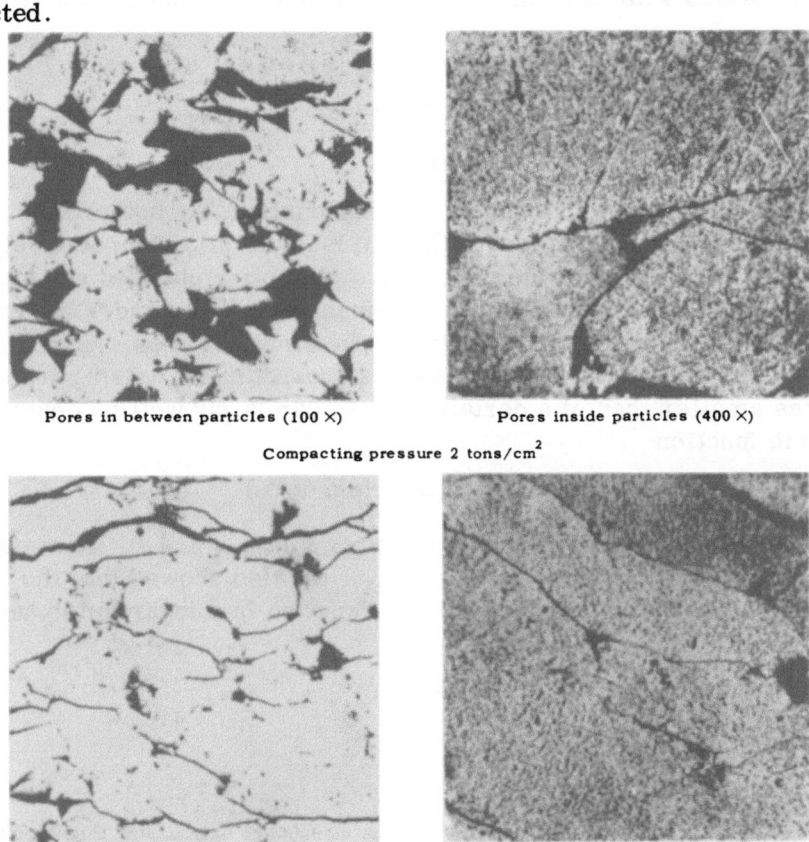

Pores in between particles (100 ×)   Pores inside particles (400 ×)

Compacting pressure 2 tons/cm^2

Pores in between particle (100 ×)   Pores inside particle (400 ×)

Compacting pressure 9 tons/cm^2

Fig. 4. Photomicrographs showing pore structure of iron powder compacts after compacting at 2 tons/cm^2 (top row) and 9 tons/cm^2 (bottom row), respectively. Pictures to the left show pores in between powder particles (at 100 diameters); pictures to the right show pores inside powder particles (at 400 diameters). (All pictures taken from specimens sectioned parallel to pressing direction, ground, polished with diamond paste and etched electrolytically for 1 sec at 0.01 Amp/cm^2.)

Summarizing, we have learned three things from the graph, Fig. 3, and from the photomicrographs, Fig. 4: (1) Large pores shrink much faster with compacting increasing pressure than small pores. (2) The yielding condition for the large pores between the powder particles is approximately the same whether the particles themselves contain many small pores or are practically nonporous. (3) Powder particles containing a great number of small pores can be deformed without greatly densifying their internal pore structure

The Effect of Compacting Pressure upon the Size Distribution of Pores Inside and in between Powder Particles

As already mentioned in the introduction, the author has shown in a previous paper [2] that, in isostatically compacted electrolytic iron powders, the pore size distribution changed as if the pores were eliminated from the compact strictly in order of size, i.e., first the largest and eventually the smallest ones. In particular, he showed that the accumulated porosity, u(D), arising from pores smaller than of a certain size D, can fairly well be described by the function

$$u(D) = u(D_R)\, \text{erfc}\, ln\, (D/D_R)$$

where erfc is the complementary error function and Dr a constant but arbitrary reference value $< D_{max}$. He also showed that the straight line, which represents this function in a logarithmic-nor-

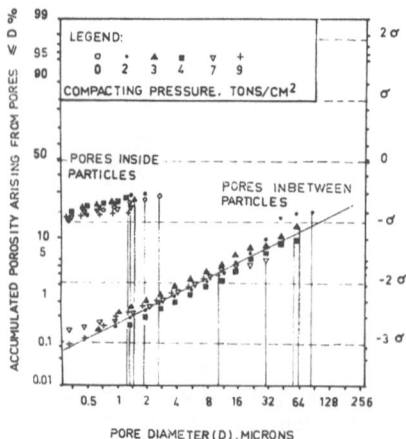

Fig. 5. Change of pore size distribution inside and in between powder particles with increasing compacting pressure, for iron powder type A (compare Fig. 3). Logarithmic-normal plot, i.e. pore diameter, D, on logarithmic scale, and accumulated porosity arising from pores $\leq$ D on Gaussian scale.

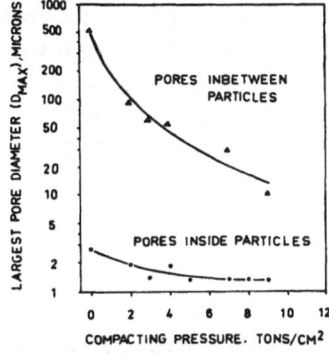

Fig. 6. Change the largest pore diameter, D_{max}, inside and in between powder particles with increasing compacting pressure, for iron powder type A (compare Fig. 3). Pore diameters on logarithmic scale, compacting pressure on linear scale.

mal coordinate system, neither substantially changes its position nor its slope if the compacting pressure is gradually increased, but only creases at its upper end at smaller and smaller values of the pore size D.

As can be seen from the graph, Fig. 5, this interesting feature is very well confirmed by the results obtained with the type of reduced iron powder discussed in the preceding paragraph. Again, the small pores inside the powder particles, and the large ones between them, have been studied separately. As can be seen from the graph, two distinct pore size distributions are obtained, one for the very small pores inside and one for the large ones between the particles. Both have in common that the relevant measuring data, irrespective of the compacting pressure at which they were obtained, fall closely into one and the same narrow stroke of measuring dots. In the case of the pores between particles, the pattern of measuring dots can be represented in fairly good approximation by a straight line. In the case of the small pores inside particles, the range of investigated pore sizes is too small to tell exactly whether the respective pattern of measuring dots resembles a straight line or not. It can be seen, however, that it follows the same general pattern as in the case of the large pores between particles. In both cases, the curves cease at smaller and smaller pore sizes, as the compacting pressure increases.

The graph, Fig. 6, shows how with increasing compacting pressure the size of the largest pores between the particles shrinks in comparison to the size of the largest pores present inside the particles. This graph, in which the largest pore diameters, D_{max}, are plotted on a logarithmic scale and the compacting pressure, P, on a linear scale, reveals the following two interesting features:

(1) The slope of the curves, dlnD/dP, becomes flatter with increasing compacting pressure; (2) it is not only the largest pores (between the particles) that shrink under pressure, but simultaneously also the smaller pores (inside the particles), though to a much minor degree. In general mathematical terms, this important observation can be expressed as follows:

$$-\frac{d\ln D}{dP} = F(D, P)$$

where F(D, P) is a positive function that decreases with increasing compacting pressure, P, but increases with increasing pore size, D.

References

1. Bockstiegel, G., and J. Hewing, "Kritische Betrachtungen des Schrifttums über den Verdichtungsvorgang beim Kaltpressen von Pulvern in Starren Pressformen," Archiv Eisenhüttenw. 36(10):751-57 (1965).
2. Bockstiegel, G., "The Porosity-Pressure Curve and its Relation to the Pore Size Distribution in Iron Powder Compacts," in: Modern Developments in Powder Metallurgy, Vol. 1, Fundamentals and Methods, Plenum Press, New York (1966), pp. 155-87.
3. Torre, C., "Theorie und Verhalten der zusammengepressten Pulver," Berg-Hüttenmänn., Monatsh. montan. Hochschule Leoben 93:62-67 (1948).
4. Konopicky, K., "Parallelität der Gesetzmässigkeiten in Keramik and Pulver-metallurgie, "Radex-Rundschau (1948), pp. 141-48.
5. Rosiwal, A., "Über geometrische Gesteinsanalysen," Verhandl. K.K. Geol. Reichsanstalt (Austria) (516):143-75 (1898).
6. Bockstiegel, G., "Eine einfache Formel zur Berechnung räumlicher Grössen-verteilungen aus durch Linearanalyse erhaltenen Daten," Z. Metallk. (August, 1966).
7. Edström, J. O., "Reduktionsstrukturer hos järnpulver," Jernkontorets Ann. 140:116-29 (1956).
8. Edström, J.O., "Solid State Diffusion in the Reduction of Hematite," Jernkontorets Ann. 141:809-36 (1957).

9.   Kohl, K., and B. Marincek, "Uber die Reduktion der Eisenoxyde mit Graphit," Archiv Eisenhüttenw. 36(12):851-59 (1965).

10.   Delesse, A., "Procédé méchanique pour déterminer la composition des roches," Ann. Mines 13:379-88 (1848).

Some Properties of Engineering Iron Powders

C.J. Leadbeater, L. Northcott,
and F. Hargreaves

Armament Research Department
Ministry of Supply, Royal Arsenal
Woolwich, England

Introduction

Increasing interest is now being taken in the technique of powder metallurgy for the manufacture of small engineering components that are generally made by machining from bar stock. Since the properties of the powder used as raw material have a considerable influence on the properties of the sintered compact, as well as on the ease of making it, close control of the powder properties is essential for successful operation.

At the outset of the present work, information on the apparatus and procedure for testing the various properties of powders was found to be incomplete and it became necessary to investigate and standardize testing apparatus and conditions in order that a true comparison could be made between the properties of various powders; this work is described in Part I of the paper. This work has been confined to iron powders, but much of the information concerning methods of testing will undoubtedly apply to other metallic powders; in view of the disparities in present practice the need for standardizing the apparatus and procedure for testing both metallic powders and sintered compacts is also indicated.

Using the technique described in Part I, a study was made of twenty-eight iron powders, thirteen of which were obtained from American and the remainder from British manufacturers; the results are described in Part II. The properties of sintered compacts prepared from these powders under standard conditions of pressing and sintering are given in Part III; and Part IV comprises a discussion of the results obtained.

Part I – Methods of Testing

The examination of an iron powder may entail (1) the determination of its physical properties such as particle, apparent, and tap densities; flow behavior; shape factor; specific surface and the chemical composition, including the oxygen, hydrogen, and nitrogen contents; and (2) the properties of the pressed and sintered compact after treatment under standard conditions of pressure and sintering temperature, namely, ultimate tensile strength, yield point, elongation, hardness, resistance to impact, density, linear and volume changes, and residual gas contents.

Such a range of properties is probably wider than will be applied in the routine examination of powders in commercial practice, but was considered necessary for the present purpose and formed the basis on which the following methods were examined.

Density

In considering the density properties in powder metallurgy, there are four types involved: (1) The particle density, which refers to the mass of the actual volume of the individual particles; (2) the apparent density, obtained from the mass of the apparent or bulk volume of the powder under conditions of loose packing; (3) the tap density, sometimes referred to as the load factor, which is obtained from the mass of the apparent volume after tapping to ensure packing or consolidation; and (4) the density of the compact before and after sintering. The methods for determining these different densities are given below.

(1) Particle Density

The individual particles of some grades of powder may be porous, and information about this porosity is useful both in the manufacture of powder and in its possible influence on the pressing operation. The specific-gravity-bottle method is convenient for determining this property. The complete removal of the entrapped air in the powder sample is necessary in order to determine the correct volume of the powder, and this is accomplished by immersing the sample in a suitable liquid, which is then transferred to a container, the atmosphere of which is reduced to a pressure of approximately 3 mm of mercury. Xylene has been found suitable for use with iron powders owing to its nonreactivity, low vapor pressure, low surface tension, and low viscosity. Before use the liquid is placed in a container which is then evacuated to remove dissolved gases and impure liquids which boil at this reduced pressure.

The following precautions should be observed:

(a) In order to minimize experimental error the volume of the powder should be large compared with small volume changes produced by temperature variations and by incorrect determination of the volume of the specific-gravity bottle and the density of the xylene; a satisfactory volume is provided by 20 g of metal. Increasing the weight of sample was found to have an effect on the density values; thus with one powder (No. 9) samples of 5, 10, 20, and 50 g gave densities of 5.00, 7.22, 7.69, and 7.70 g/cc, respectively.

(b) The density of the xylene at the operating temperature should be known to ± 0.0003 g/cc.

(c) During the period of evacuation the bottle plus contents should be removed from the container and thoroughly shaken in order to aid the removal of air, after which the bottle is replaced for further evacuation.

(d) The volumes of the density bottles should be determined, since the actual capacities of bottles used in the present work differed from their nominal values by as much as 0.3%.

(2) Apparent Density

The apparent density of a powder is the mass of a bulk volume of the powder. The property is important from the standpoint of die-fill. To obtain a compact of a given weight a greater volume is required of a bulky powder, i.e., one of low apparent density, than of a powder having a high apparent density. The die design or plunger stroke suitable for one grade of powder would probably require modification for another grade having a different apparent density.

Preliminary trials were made to determine the effect of the conditions of testing on the apparent density (a) by filling the cup from bronze cones of different radii to provide a continuous stream of powder; (b) by adding small quantities of powder (e.g., 0.5 g) at a time to a cone having an orifice of 2.29 mm radius at a standard height above a cup of 25 cc capacity into which the powder would fall; and (c) by direct additions into the cup by means of a spatula. Excess powder forming a heap on the cup is removed by a scraper so that the surface of the powder in the cup is flush with its rim. The powder filling the cup is weighed, allowing the density to be calculated. The results of these tests are given in Table 1, which shows that significant differences may be obtained.

On the basis of these results the following method was standardized: A glass funnel, the stem of which was 2 cm long and 2 mm in radius — large enough to permit the flow of the coarsest powder — was placed 10 cm above a glass cylinder of 25 cc capacity; small quantities of powder were added to the sloping surface of the cone until the cylinder was full, after which the free surface was leveled by means of a scraper.

During the test the powder must not be induced to "pack" or shrink in volume by disturbing the apparatus by tapping either the container or the stand on which the apparatus is placed. The ap-

TABLE 1. Variation of Apparent Density (g/cc)
with Method of Determination

Method of Filling 25-c.c. Container	Powder No.	Apparent Density, g /c.c.
Stream falling from orifice 2·29 mm in radius	8 20	2·766 2·382
Stream falling from orifice 1·27 mm in radius	8 20	2·745 2·342
Small quantity falling from orifice 2·29 mm in radius	8 20	2·882 2·44, 2·47
Direct addition with spatula	8 20	2·87 2·40

parent densities of a large number of iron powders have been found to lie within the range 1-4 g/cc, which is equivalent to porosities of approximately 87-50%.

(3) Tap Density or Load Factor

The tap density or load factor is the apparent density of a metal powder after it has been "shaken down" or "packed" by tapping or vibration and is important in questions of packing, storage, or transport of commercial powders, and also in pressing practice where dies are vibrated to consolidate the die-fill. Five methods for the determination of tap density were investigated. Two of these embody the tapping produced by allowing a container to fall freely for a short distance and stop instantaneously, and the remaining three the shaking down produced by horizontal or vertical vibrations; these methods may be briefly described as follows:

1. Free Fall, Operated Manually. A graduated cylinder containing a known weight of powder was lifted a distance of 1.5 in. above a rigid table and allowed to fall freely and normally onto the table; this operation was repeated until the minimum volume was obtained.

2. Free Fall, Operated Mechanically. A graduated cylinder containing powder was fixed on a table mounted on a vertical rod fixed on a table mounted on a vertical rod which could be raised by means of a cam connected to a motor-driven reduction gear; the cylinder was allowed to fall freely at the rate of 80 falls/min through a distance which was controllable up to a maximum of 1.35 in.

3. Vertical Vibration Produced Electromagnetically. A steel reed, acting as the support for the cylinder, was vibrated in a fluctuating magnetic field produced by passing alternating current (50 cycles/sec) through a coil mounted on a suitable former.

4. Horizontal Vibration Produced Mechanically. The cylinder containing the powder was placed on a table which was vibrated by means of two rotating off-center balance weights which were self-compensating except in the direction of the vibration. Test were carried out over a range of frequencies of 900−1500 cycles/min.

5. Horizontal Vibration Produced Pneumatically. A pattern vibrator as employed in foundry work was used, and the general testing procedure was as follows: A known weight of powder was

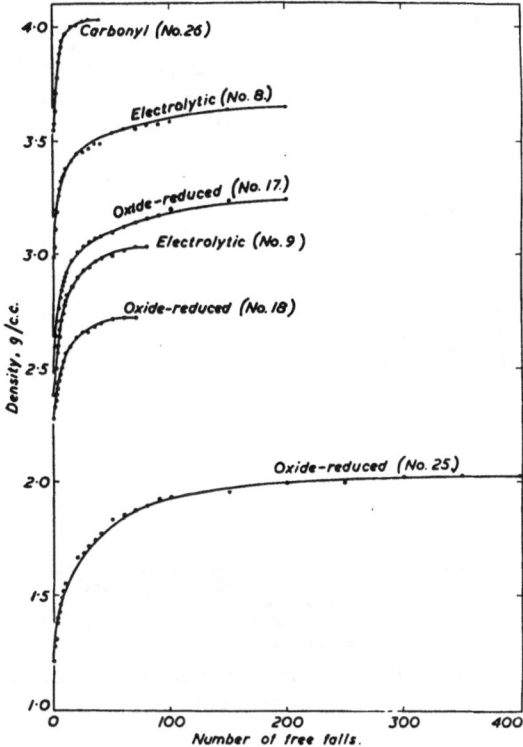

Fig. 1. Effect of number of free falls on the tap densities of different powders.

placed in a graduated glass cylinder of specified size. The first surface of the powder was leveled by tapping gently with the finger so that the minimum packing was produced; some leveling was produced by inclining the container to the vertical direction and rotating the cylinder slowly, and the volume of the powder was determined by direct observation. The loaded cylinder was then placed upon the vibrating mechanism to cause packing of the powder. Vibration was discontinued when no further reduction in volume occurred. The free surface of the powder was again leveled if necessary and the volume of the powder noted.

Six powders were tested by each of the five methods in each of three glass containers of different sizes. If is noteworthy that in the free-vertical-fall method (mechanical) the shrinkage of the powder was visible after each of the first 2–4 falls, so that enough time should elapse between two consecutive falls to avoid possible interference with the packing mechanism. The methods involving

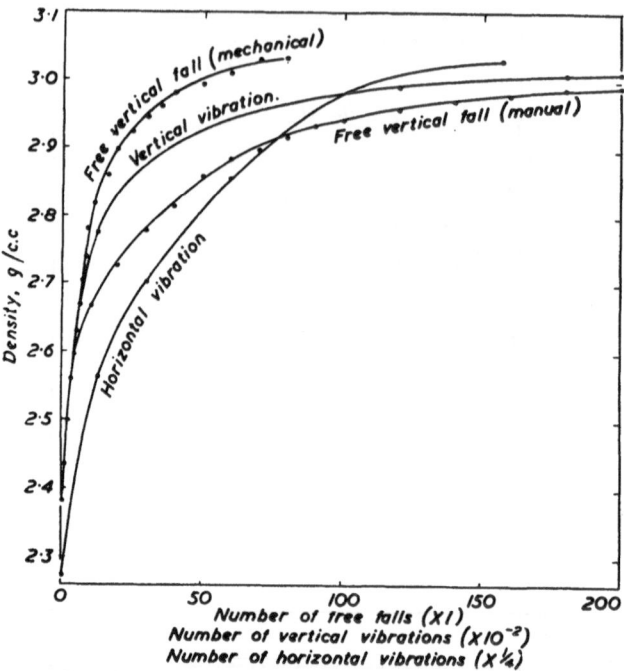

Fig. 2. Effect of the method of vibration on tap densities of powder No. 9.

horizontal and vertical vibrations were characterized by swirling of the powder, which was pronounced in the upper half of the sample, a condition that is not conducive to packing, since a slight increase in density could be obtained by applying the free-fall method after the vibration test. The results are given in Table 2, and the following conclusions may be drawn:

(1) The tap density of a powder may be determined by any method which embodies vibration, provided that the number of vibrations is sufficient, this number varying for different powders and different methods of applying the vibration (Figs. 1 and 2).

(2) The average of the density values obtained by using three cylinders of different sizes showed that the maximum tap densities are generally produced by the pneumatic-vibrator method, which is recommended for routine examination, particularly as the type of vibration is practiced.

(3) The size of cylinder producing the maximum density varies with the method employed; cylinders of 50 and 25 cc capa-

TABLE 2. Effect of Method of Determination on Tap Density (g/cc)

| | | Vertical vibration | | | | | | | | | | | | Horizontal vibration | | | | | | | |
| | | Manual (1·50-in. drop) | | | | Mechanical (1·35-in. drop) | | | | Electromagnetic (50 cycles/sec) | | | | Mechanical (25 cycles/sec) | | | | Pneumatic | | | |
Powder No.	Size of Graduated Cylinder, c.c. :	100	50	25	Max. Diff.	100	50	25	Max. Diff.	100	50	25	Max. Diff.	100	50	25	Max. Diff.	100	50	25	Max. Diff.
9	Density :	2·99	3·01	2·99	0·02	3·03 / 3·03* / 3·03†	2·99	3·09	0·10	3·01	3·01	3·05	0·04	3·03 / 3·03§ / 2·94‖	2·99	3·05	0·06	3·18	3·28	3·21	0·10
	No. of cycles :	180	180	80		70 / 90 / 150	150	150		18,000	12,000	4500		625 / 800 / 1875	525	600					
8	Density :	3·45	3·52	3·52	0·07	3·64	3·56	3·60	0·08	3·33	3·5	3·52	0·19	3·57	3·53	3·57	0·04	3·64	3·82	3·70	0·18
	No. of cycles :	200	120	120		150	90	50		12,000	18,000	6000		400	1300	250					
17	Density :	3·21	3·23	3·20	0·03	3·23	3·22	3·23	0·01	3·00	3·21	3·23	0·23	3·08	3·12	3·13	0·05	3·31	3·30	3·40	0·10
	No. of cycles :	70	80	80		150	150	60		18,000	12,000	9000		1125	900	825					
18	Density :	2·69	2·73	2·72	0·04	2·72	2·73	2·72	0·01	2·68	2·69	2·70	0·02	2·72	2·63	2·75	0·12	2·86	2·81¶	2·83	0·05
	No. of cycles :	80	120	120		60	50	60		12,000	9000	9000		725	875	1000					
25	Density :	2·05	2·05	2·00	0·05	2·03	2·01	1·92	0·11	1·61	1·68	1·89‡	0·28	1·79	1·56	1·58	0·23	1·77	1·69¶	1·66	0·11
	No. of cycles :	180	180	120		350	250	80		36,000	18,000	21,000		3900	1575	1750					
26	Density :	4·03	4·17	4·17	0·14	4·02	4·08	4·03	0·06	3·85	4·17‡	4·03	0·32	4·00	4·00	4·07	0·07	4·02	4·26¶	4·03	0·24
	No. of cycles :	30	40	30		25	20	10		18,000	12,000	3000		1050	525	825					

* 1-in. drop.

† 0.5-in. drop.

‡ Large amplitude owing to resonance.

§ 20 cycles/sec.

‖ 15 cycles/sec.

¶ Violet swirling and slight loss of powder.

city developed maximum densities in four methods, whereas the 100-cc cylinder produced the highest value in one method. Emphasis must be placed on the greater accuracy of measuring volumes in the smaller containers in view of the calibration of smaller fractions of volume.

(4) An increase of 100% in the weight of the sample caused an inconsistent change in the tap density of -0.06 to $+0.22$ g/cc as determined by the free-vertical-fall method (mechanical) with powder No. 9 in three different containers.

(4) Density of Pressed and Sintered Compacts

This property provides information on the porosity of the finished component. Since most sintered iron compacts are porous and the mechanical properties in general increase with decrease in porosity, the value of the final density may give some indication of the properties of the material which in turn reflect the pressing and sintering technique.

Two methods of determining this property, depending on the shape of the compact, are available. When a compact is regular in form, e.g., an oblong, cube, or cylinder, the dimensions from which the volume may be computed can readily be measured. However, compacts having irregular contours owing to design, fracture, or uneven shrinkage, present some difficulty; the normal method of water displacement is unsatisfactory since liquids of low surface tension become absorbed in the component; mercury, however, has been found to be a suitable medium.

The procedure consisted in weighing the specimen in air and also completely immersed in mercury. In the latter operation, the specimen was attached to the cross-wires of a weighted stirrup by means of a thin wire; the weighted stirrup being at least twice the weight of the compact. In this procedure the specimen and the mercury should be at the same temperature to avoid convection currents, and the mercury should not be contaminated. After treating under the following conditions:

Compacting pressure . . . . 30 tons/sq in.
Sintering temperature . . . . 1050°C
Period of sintering at temperature . . . . 1 hr

the range of densities of pressed and sintered iron powders was found to be $4.60 - 7.05$ g/cc.

Flow Behavior

A method of determining the flow behavior of metal powders has been described by Hardy [1] and by Greenwood [2] whose apparatus consisted of a metal cone with an internal angle of 60°, and a parallel stem ending with an orifice from which the powder issued in the course of the test.

Different sizes of orifice and also different methods of assessing the flow behavior were used by these authors and consequently investigation of the effect of different factors on the period of flow in the above type of apparatus was considered desirable. The following factors have been studied: variations in weight of sample, radius of the orifice, length of the parallel stem of the cone, and the effect of tapping the cone, and the effect of tapping the cone during a test.

For the present work the testing procedure was as follows: (1) The sample of powder was weighed, 50 g being adopted as a minimum weight. (2) The orifice of the cone was sealed by means of a metal disc; the finger of an operator should not be used, since moisture evaporating from moist fingers has been found to prevent flow. (3) The powder was transferred to the cone by pouring it onto the inclined surface from a minimum height. (4) Care was taken

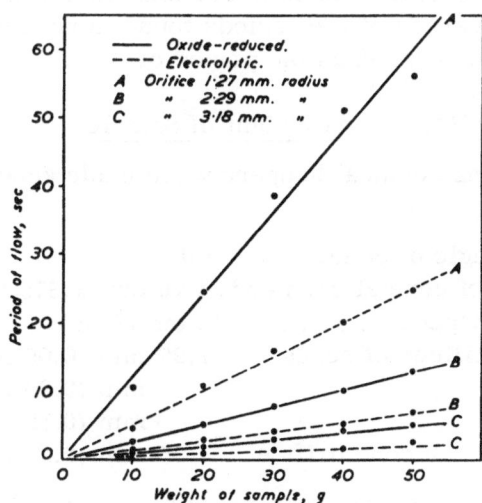

Fig. 3. Effect of radius of cone orifice and weight of powder sample on flow periods for two powders.

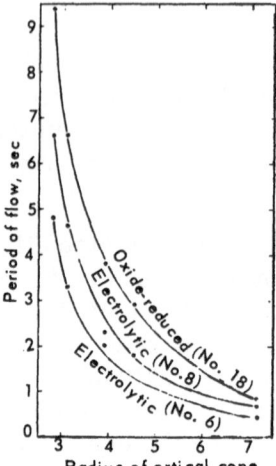

Fig. 4. Effect of radius of
cone orifice on flow periods
for three powders.

not to move or disturb the cone in any manner which would lead to
greater packing of the particles. (5) The metal disc was removed
from the orifice to permit flow, the time being noted on a stop-
watch. (6) The "end-point" of the test was taken as the time at
which the last portion of the powder flowed through the orifice; this
was best observed from a point above and almost directly over the
cone. (7) The average of the periods found in not less than three
determinations was taken as the period of flow.

Effect of Orifice Radius and Weight of Sample

Three bronze conical hoppers were made according to the
following design:

Internal angle of cone . . . . . . 60°
Diameter of cone at open end . 4.76 cm (1.875 in.)
Height of hopper . . . . 5 cm (2 in.)
Radii of orifices of cones . . . 1.27 mm (0.05 in.), 2.29
                                 mm (0.09 in.), 3.18
                                 mm (0.125 in.)
Length of each cone orifice . . 6.35 mm (0.25 in.)

Samples of different powders, of progressively increasing
weights, were tested in each of the cones. The results are illustrated in Figs. 3 and 4. The period of flow was found to be directly

proportional to the weight of the sample and inversely proportional to the n-th powder of the radius, where n, calculated from the experimental data, was found to be approximately 2.5. In the case of some powders it was necessary to tap the cone with the finger in order to maintain the flow. The effect of tapping the cone during the flow of a normally free-flowing powder is shown in Table 3 and is critical for the smallest orifice only (1.27 mm in radius).

Under the above conditions of testing, the powders examined can be classified into (a) free-flowing powders, (b) those which flow only when the cone is tapped, and (c) those which do not flow even under conditions of tapping.

Effect of Stem Length

Two glass cones were prepared to which were attached parallel stems of varying lengths and 1.5 and 2 mm in radius, respectively. Powders 8, 10, and 18 were tested in the stem of 1.5 mm in radius and powder No. 8 in the stem of 2 mm in radius. The following procedure was adopted: A standard weight of 50 g of powder was placed in the cone; in all tests involving any parallel length of stem, sufficient powder was poured in before the flow test to fill the stem to the level of the intersection of the stem and the cone. The observation of flow was made on the powder flowing from the cone and not on that issuing from the free orifice of the long stem. The results are illustrated in Fig. 5. Thus a variation in the length of the stem produces an appreciable change in the period of flow, which becomes critical for lengths of stem of the order of 0-7 cm. It is desirable, however, in any flowmeter, that specific points of design should maintain a constant effect on the

TABLE 3. Effect of Tapping on Period of
Flow of Free-Flowing Powders

Powder No.	Period of Flow, sec					
	Radius of Orifice, 1·27 mm		Radius of Orifice, 2·29 mm		Radius of Orifice, 3·18 mm	
	Un-tapped	Tapped	Un-tapped	Tapped	Un-tapped	Tapped
8	24	55	9·7	9·7	2·6	2·6
20	62	76	13·2	15·4	5·4	5·4

period of flow for different powders. A strict comparison of the experimental data obtained from cones of different designs should, therefore, be made on the basis of equivalent ratios of length of stem of the cone to radius of orifice, but the development of such equivalent cones would be unnecessarily complicated for practical adoption and could be replaced by a series of truncated cones.

A New Method for Assessing Flow Behavior

The necessity of tapping the cone to initate and maintain the flow of some powders indicated that a more rational basis of comparison of the flowing properties of a powder might be available if the conditions of test permitted any powder under examination to flow freely; this was achieved by a series of cones possessing a sufficient wide range of orifice sizes.

On this basis eight glass cones, truncated to eliminate parallel stems, were prepared to the following design:

Internal angle..........................60°
Radius at open end...................6.5 cm
Height ....~5 cm
Radii of orifices ....0.936 – 7.16 mm

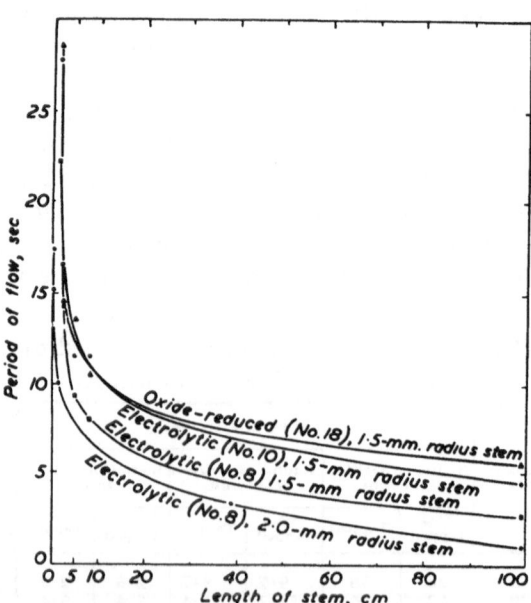

Fig. 5. Effect of length and radius of the cone stem
on flow periods for three powders.

The cones were mounted conveniently in a wooden support. A large
number of metal powders were tested to determine the relationship
between the period of flow, the weight of the sample, and the radius
of the orifice. The testing procedure described above was used,
commencing with cone 1. If free flow was obtained, further testing
in cones of larger radius would normally be unnecessary, but for
the present purpose it was desirable to obtain as much data as possible on the effects of various testing conditions. A powder which
did not flow in cone 1 was further tested in successive cones until
the sample flowed freely. From these results it was possible to
obtain a theoretical comparison (flow coefficients) and also a
practical scale of differences (cone numbers) of the relative flowing
behavior of different powders. The results indicated that the period of flow in a truncated cone is inversely proportional to (radius)^n,
n having an average value of 2.58. It is interesting to note that values of 2.5 and 2.63 have been quoted [3, 4] for n in tests on ferti-
lizers and sand. The period of flow in truncated cones was proportional to the weight of the sample in agreement with the results illustrated in Fig. 3.

A Coefficient of Flow

Since the period of flow is directly proportional to the weight
of the sample, and inversely to the radius expanded to a powder n,
the following expression may be defined:

$$t = \frac{kw}{r^n}; \text{ hence } k = \frac{tr^n}{w},$$

where t = period of flow (sec), r = radius of orifice (mm), w = weight
of sample (g), and k = proposed coefficient of flow.

Values of n and k were computed for the various powders; n
ranged from 2.43 to 2.85, giving an average value of 2.58 and values
of k were found from the actual and average values of n. The values of k for different powders increased from approximately 1.24 to
greater than 6.4 for free and nonflowing behavior, respectively; a
good correlation was found to exist between the k value and the
minimum size of the orifice permitting free flow, a summary of
which is given in Table 4; either of these criteria could therefore
be used to discriminate between the different flow behavior of powders.

In order to minimize the experimental error of timing the
period of flow, further tests, using samples weighing 1000 g, were

TABLE 4. Summary of Flow Coefficients and Cone Numbers

Number of Powders	Percentage	Range of k Values	Average k Value	Minimum Cone Size for Free Flow	Range of Free Flow
15	71·4	1·24–2·59	1·92	1 or 2	Unrestricted
2	9·5	2·64–2·72	2·68	5 or 6	Restricted
4	19·0	6·4 or greater	>6·4	8	Very restricted or none

carried out in the series of glass cones used in the previous tests, but fitted with a tower to contain the greater bulk of the powder. The results showed that (a) the relative order of k between different powders is maintained; (b) the variation in the value of k is still present; and (c) the value of n should be calculated from data obtained from two orifices differing widely in size, that is, cones 1 and 8; the average value of n is 2.43, as compared with 2.56 obtained in tests using 50 g of powder; the average k value is up to 9% higher when a small sample is used.

The Physical Basis of the Flow Coefficient

The coefficient k may be regarded as a general summation of the effects of those factors controlling the flowing mechanism of metal powders; such factors, some of which supplement or vitiate others, are conveniently classified as follows:

(a) The conditions of the testing apparatus, which involve (1) the internal angle of the conical hopper, (2) the smoothness of the internal surface of the cone, resulting in possible variations of friction between the surface and the moving powder, (3) the radius of the orifice, and (4) the presence and the length of the parallel stem leading to the orifice.

(b) The bulk properties of the powders, for example, (1) particle size, shape, and size distribution, (2) the segregation of coarse and fine particles within any one sample, (3) the method of determining the flow coefficient, that is, whether dense packing is induced by accidental vibration of pressing of the powder, or (4) the formation of bridging as a result of different shape factors for any given average size or cohesive attractions caused by absorbed liquid films, and (5) the presence of pores within single particles.

(c) The surface properties of the powders affecting the friction of particles; this is governed by (1) the type of contact between the surfaces in the shear plane, and (2) the chemical constitution

TABLE 5. Effect of Additions of a Fine to a
Coarse Powder on the Flow Behavior

Fine $= 0 - 10 \mu$
Coarse $= + 52 \mu$

Test No.	Additions, %	Period of Flow, sec		n	k
		Cone 1	Cone 8		
1	0	79·35	0·70	2·33	1·37
2	1	72·91	0·59	2·37	1·25
3	2	75·36	0·54	2·43	1·29
4	3	77·70	0·49	2·49	1·32
5	4	81·98	0·48	2·53	1·40
6	5	86·45	0·48	2·55	1·45
7	6	91·91	0·42	2·65	1·55
8	7	96·63	0·42	2·67	1·61
9	8	104·17	0·46	2·67	1·76
10	9	No flow	No flow	...	...

of the surface, that is, whether or not it is partly or entirely covered with a film of oxide, basic hydroxide, or adsorbed moisture or oil.

A low degree of graphical correlation was found between k and some of the more likely factors indicated above.

Miscellaneous Tests

In the course of the above work, various tests were carried out to determine the effect of drying and also of demagnetization of the sample powder, but no changes in the period of flow were observed.

The results of an addition of a $-10-\mu$ fraction to the $+52-\mu$ fraction of powder No. 9 are shown in Table 5. A sample of 50 g, to which additions of the $-10-\mu$ fraction were made at increments of 1 wt%, was used; the original and subsequent samples were tested in cones 1 and 8, having radii of 0.936 and 7.16 mm, respectively.

Thus a small proportion of fines improves flow but an excess is detrimental to it. It is conceivable that some fines break down bridging tendencies by reducing the kinetic friction obtaining during movement by acting as localized ball bearings for the larger particles.

The following notes on a possible standard test of flow and for the determination of k are suggested:

(a) The design of the individual container should consist of a cone, or a cylinder tapered at one end in the form of a cone, and should be large enough to contain an appreciable quantity of powder in order to minimize the experimental error.

(b) The cones should be truncated so as to eliminate the parallel stems.

(c) A suitable number of cones (e.g., four) possessing radii of the order of 1.27 mm (0.05 in.), 3.18 mm (0.125 in.), 6.35 mm (0.25 in.), and 12.70 mm (0.50 in.) would permit discrimination between different varieties and compositions of powders; 72% of all the powders tested in the present work flowed freely in the two cones with radii of 0.936 and 2.78 mm.

(d) An appreciable difference should exist between the radii of the smallest and largest cones in order to obtain a value of n which will produce nearly equal values of k for any one powder tested in different cones.

Compression Ratio

The height of the die required for any component is partly governed by the degree of compaction produced at a given pressure, which is equivalent to a reduction in height and an increase in density; this may be conveniently found from the following ratios and is described as the compression ratio (C.R.):

$$C.R. = \frac{\text{Height of uncompressed powder in the die}}{\text{Height of the pressed compact}}$$

$$= \frac{\text{Density of pressed compact}}{\text{Apparent density of powder}}.$$

Determination of the height of the powder in the die is difficult and the compression ratio (for a specific pressure) is more accurately determined from the ratio of densities (see above).

Size Distribution

For engineering purposes, commercial iron powders range in size from less than 1 to 200 μ (1 μ = 0.001 mm). The complete size determination involves two stages: (1) the use of wire-mesh sieves or screens of different mesh sizes to sieve out all material coarser than 52 μ; this mesh corresponds to a 300-mesh sieve which is the smallest standardized by the British Standards Institution; and (2) the use of one of the methods based on sedimentation, elutriation, or microscopical counts, for the distribution of fractions less than 52 μ in size. As the median size of the majority of iron powders examined has been mostly above 52 μ, sieving has been adopted throughout and has been supplemented by microscopical and elutriation methods only for those powders having an appreciable proportion finer than 52 μ.

Sieves 8 in. in diameter and corresponding to B. S. sieves of 120, 150, 170, 200, and 240 mesh were used; an additional sieve with an aperture width of 44 μ (325 mesh) was also used, although this is not included in the B.S. range. Sieving was carried out in a machine in a frame mounted on three flexible bearings, vibrated by means of a rotating, off-center, balance wheel. The movement of the three vibrating supports for the sieve frame was an out-of-phase, inclined, elliptical movement which resulted in a clogging of the meshes of the sieve.

As a result of preliminary tests the general procedure of sieving was standardized as follows: (1) A 200-g sample was placed in the sieve of coarsest mesh, fixed in the machine, and covered with a lid. (2) The sample was sieved for 15 min, after which the frame and lid were removed integrally from the machine, held above a sheet of paper, and shaken vigorously for 15 sec; if the fines passing through the sieve exceeded 0.2 g the sieve was replaced in the machine and shaken for a further 15 min; this process was repeated until the fines shaken through by hand weighed less than 0.2 g. (3) The fines collected in the pan and on the test paper were transferred to the next finest sieve, i.e., 150 mesh, and the sieving procedure was repeated for all the range of available meshes. (4) All fractions were collected, weighed, and expressed as a percentage of the original. (5) If dusting or other losses exceeded 0.5% of the sample weight, the test was repeated with a new sample.

Median Size

The median size is the particle size above or below which 50 wt% of the powder occurs. It provides a useful measure of the relative fineness of different size distributions, and is best determined from a graphical correlation of cumulative percentage weight greater or less than a stated size against size of particle.

Specific Surface

The specific surface considered for the present work is defined as the surface area per unit weight of powder. The compacting and sintering operations are considerably influenced by the contact between the metal particles and "sintering" may, in fact, be looked upon as a surface reaction. The actual surface area involved is therefore an important factor in behavior of the material on heating and this applies whether or not the compacts contain a liquid phase at the sintering temperature. If all powder particles were symmetrical and of the same shape it would be possible to estimate the total surface area from particle-size distribution, but the shapes of powder particles are irregular and vary with the method of manufacture and subsequent treatment.

There are several methods for determining specific surface, and much literature exists on the subject; a useful review has been published recently [5, 6, 7]. The air-permeability method developed by Lea and Nurse [8] was selected for an investigation of its applicability to iron powders.

The following conclusions may be drawn from this work: (1) The specific surface of iron powders is not independent of the porosity of the bed, owing to the presence of a double system of porosity, i.e., one system around the powder particles and the other inside them. (2) For powders possessing these characteristics only an arbitrary value of specific surface may be obtained at a predetermined value of porosity for the beds. (3) The values of specific surface obtained using the burette method are of the order of 1.14 times those obtained in the normal apparatus. (4) The modified air-permeability method provides a useful order of differentiation for different varieties of powders and is easily carried out.

Microscopical Examination

For the microscopical study of particle size, shape factor, and surface texture, the sample should satisfy the following conditions: (1) There should be a suitable dispersion of particles so that each particle lies in the field of view independently of others, thus permitting easy and precise examination, and (2) the actual size distribution must be truly represented by the dispersion selected for final examination.

Two methods for preparing suitable dispersions of powders have been used. Firstly, the dry powders are dusted onto a suitable surface; if the median size is small, gentle grinding by means of a slip assists in distintegrating groups of particles. Secondly, a wet suspension is prepared by shaking 0.1 g of powder in 10 cc of dilute Necol varnish, from which a sample is quickly taken and placed on a suitable surface; if the median size is small, gentle grinding by means of the cover slip may be used. For the purpose of photographing individual particles of a powder, a stainless-steel mirror used as a mounting slide has given the best results.

Particle Size and Its Determination

The sizes and shapes of individual particles of any one metal powder vary appreciably and their measurement offers some difficulty [9, 14].

The size or diameter of a particle has a definite physical meaning only when the object measured has a regular geometrical form. Hence, in practice, the measurement of the diameter of an irregularly shaped particle is related to one or more ideal standards, e.g., the diameter of (a) the circle of equivalent projected area, (b) the sphere of equivalent surface area, or (c) the sphere of equivalent volume. A technique suitable for obtaining the diameter of the circle of equivalent projected area has been established by Martin [15], who proposed the equivalent diameter as the distance between opposite sides of the particles, measured crosswise on the field of view of the microscope and on a line bisecting the projected area of the particle.

Heywood [16] measured 100 particles of irregular shape, but of similar size, and showed that the mean Martin diameter is within $1\frac{1}{2}\%$ of the mean diameter of the circles of equivalent area, and the agreement should be closer for a greater number of particles. The Martin diameter has been adopted for the present microscopical investigations on particle size and shape.

Shape Factors

The subject of the shape factor of particles has been considered by many investigators in the fields of fuels and refractories, with a view to relating the varying dimensions, surface, and shape of any one particle or group of particles to similar properties of a symmetrical solid body, for which the sphere has been universally adopted. Very little use appears to have been made of the statistical ratio of length to breadth of a particle, but it is considered to be a useful value, complementary to other shape-factor constants such as the coefficient of rugosity [17], which is the ratio of the measured specific surface to the hypothetical surface area which the particles would possess if they were all spherical.

The ratio of length to breadth was calculated for each particle in a sample of 200 particles from each powder, after which the statistical average ratio was calculated. The results can be conveniently arranged as a percentage frequency of particles of different ranges of ratios occurring within different ranges of particle sizes.

Preparation of Sections for Metallographic Examination

A sample of metal powder was mixed with bakelite mounting powder, preferably of an average size less than 100 mesh, in the proportion of 1 to 8, respectively, by volume, the mixture being heated and pressed by the method in general use for mounting photographic specimens; the mounting of a sintered specimen is necessary only when the dimensions do not permit easy polishing or when an edge requires support.

Grinding and polishing procedures included those normally used for soft metals; it was found that light pressure in all stages of polishing, together with continuous rotation on the polishing wheel, is advantageous.

The etching and drying of polished surfaces was found to be the most difficult stage in the preparation; grain boundaries and oxide impurities are highly reactive to the etching medium which individual pores tend to retain. Less etching solution is absorbed when the specimen, except the surface to be examined, is embedded in bakelite, and also when the surface is immersed in the etching solution facing downwards, enabling the entrapped gases in the pores to reduce the diffusion of liquid into the porous metal. No set procedure has been standardized for washing and drying the specimens, but repeated rinsings in hot water and alcohol, followed by mopping with a dry absorbent cloth and holding in a stream of warm air, are generally found necessary.

Surface Texture

In addition to the determination of numerical shape factors, qualitative information may be obtained on the projected shape and form, and the surface properties of a particle from a microscopical examination. The results obtained have suggested a descriptive system which is based upon an arbitrary selection of geometrical shapes and forms, together with various types of rough and smooth surface. Examples of the classification are given in Part II, when the properties of individual iron powders are dealt with.

Chemical Composition

Impurities were determined by the standard method of chemical analysis; the only feature requiring comment is that, owing to their relative high specific surfaces, metallic powders are prone to considerable gas adsorption. Particular attention has, therefore, been paid to oxygen, hydrogen, and nitrogen contents, and these have been determined by the vacuum-fusion method described by Sloman [18-20].

General Properties after Pressing and Sintering

The standard die forms a pressed compact approximately 6.45 cm long, 1 cm wide, and 1 cm high. After sintering, the compact is tested for density and shrinkage; it is then machined into a round tensile specimen, 0.25 in. in diameter and with a 1-in. parallel length, so that the ultimate tensile strength, yield point, and elongation may be determined. Since the tests involved in this section are so well known and standardized, further description is unnecessary.

Part II – The Properties of Some Commercial Iron Powders

Several manufacturers supply powders possessing different size distributions and made by one or more methods, and they are prepared to meet any particular specification as regards size distribution. It is considered that the twenty-eight iron powders examined (thirteen were obtained from American manufactures and the remainder from British) represented a fair average of the types available. Thirteen powders, Nos. 1-13, were produced

C. J. LEADBEATER, L. NORTHCOTT, AND F. HARGREAVES

TABLE 6. Visual and Microscopical Appearance of Iron Powders (The code numbers used in the tables are defined in the classification given in Table 7)

Type of Powder	Powder No.	General Visual Appearance	Oxygen, %	Microscopical Appearance			
				Perimeter (A)	Shape (B)	Smooth Surface (C)	Rough Surface (D)
Electrolytic	1	Grey	...	1, 3, 4	3, 4	3	2
,,	2	Grey	...	1, 3	3, 4	3	2
,,	3	Brown	3·0*	4	5	...	4
,,	4	Brown	...	4	5	...	4
,,	5	Grey	0·10	4, some 3	3, 4, 5	2	4
,,	6	Light grey, some bright particles	...	3	6	...	3
,,	7	Slightly dark grey	...	1, 3	1, 5	...	4
,,	8	Medium grey	...	4, some 1	3, 4, 5	Some 2	4
,,	9	Slightly dark grey	0·41	4	3	2	...
,,	10	Slightly dark grey	0·45	2, 3	4	...	3
,,	11	Dark grey	...	1, 2, 3	4	...	3
,,	12	Grey, some bright particles	0·56	1, 2, 3	4	...	3
,,	13	Very dark, slightly brown, gritty	...	Some 5	5	...	3
Oxide-reduced	14	Light grey	0·71	1, 2	5	...	3
,,	15	Slightly dark grey	0·72	1, 3, 4	5	...	3
,,	16	Grey	1·00	3	4, 5	...	3
,,	17	Slightly dark grey	...	3	4, 5	...	3
,,	18	Slightly dark grey, bright particles	...	3	4, 5	...	2
,,	19	Grey	2·09	2, 3	5	...	...
,,	20	Slightly dark grey	1·22	2, 3, 4	5	...	2
,,	21	Light grey	...	1, 3	5	...	2
,,	22	Dark grey	1·35	1, 3	5	6	3
,,	23	Dark grey	1·03	1, 3	4	...	...
,,	24	Dark, some bright particles	1·18	3	4	...	3
,,	25	Very dark	...	3, 4	4	...	3
Carbonyl	26	Grey	0·26	1	1, 2	6	...
Abrasion	27	Grey	1·47	1, 3	4, 5	...	2, 3
Chloride-reduced	28	Slightly dark, granular	...	3	5	...	3

* The maximum oxygen content found in grey powders was 2·49%, so that 3% for a brown powder is likely to be a minimum estimate.

electrolytically; twelve, Nos. 14-25, by the oxide-reduction process; one, No. 26, by the decomposition of iron carbonyl; one, No. 27, by mechanical comminution; and one, No. 28, by the reduction of iron chloride in hydrogen. Ten American and six British manufacturers are represented by these various powders.

The powders have been examined according to the technique described in Part I, the results of which are now presented.

Appearance

In general, the iron powders varied in color from a silvery to a dark gray, but three were brown or reddish-brown. There was close correlation between color shade and oxygen content in the electrolytic-iron powders (Table 6), a darkening in color occurring with an increase in the oxygen content and, in some cases, with the carbon content. A reddish or brown appearance may indicate the presence of rust or excessive oxide. Direct correlation was impossible with the oxide-reduced powders, since the oxide was not confined to the surface of the powder.

Microscopical examination, based on the tentative classification given in Table 7, showed that individual particles of different varieties of powders exhibit differences of appearance and shape, details of which are given in Table 6. An analysis of this data revealed that electrolytic particles possessed perimeters Nos. 1 (15.4%), 2 (9%), 3 (34.6%), and 4 (42.0%), whereas 64% of the oxide-reduced particles had perimeters No. 3, with a corresponding diminution in the other numbers. The shapes of electrolytic particles were mainly No. 3 (21%), No. 4 (37%), and No. 5 (33%), but of oxide-reduced particles the shapes were No. 4 (30%) and No. 5 (70%). As to surface texture, 33% of the electrolytic-iron particles exhibited relatively smooth surfaces, compared with 8% for those oxide-reduced. Typical photographs of individual particles of the powders are reproduced in Figs. 6(a) to 11(a).

Shape Factor

The shape factor determined in the present work was the ratio of length to breadth, both dimensions being based on Martin's diameter and measured on a microscopical micrometer scale. A factor of 1 indicated perfect symmetry for a circle and a square. Two hundred particles for each of seven typical powders were analyzed into frequency distributions for ranges of shape factors and

TABLE 7. Tentative Classification of the Shape and Surface Texture of Powder Particles

Types of Perimeters and Shapes		Types of Surfaces on Individual Particles	
(A) Perimeters as Projected under the Microscope	(B) Shape as Observed under the Microscope	(C) Smooth	(D) Rough
1. Approximately circular	1. Spherical	1. Horizontal or inclined to the horizontal axis of the particle when lying in its stable position under the microscope	
2. Square	2. Conical	2. One or more smooth plane surfaces at different levels forming part or the whole of the surface of the particle	2. Uniformly dark and scaly
3. Approximately rectangular	3. Prismatic	3. One or more plane surfaces at different levels, showing network of cracks or strain lines, forming part or the whole of the surface of the particle	3. Reticulation resulting from upper edges of numerous pitted or crater-like depressions
4. Lenticular	4. Lamellar	4. One or more smooth isolated plane faces	4. Dark, with axial or transverse dendrites
5. V-shaped	5. Ellipsoidal	5. One or more isolated plane faces showing strain lines or cracks	
6. Annular (genorally circular)	6. Complex forms formed by undulating surfaces	6. Smooth, curved, or undulating surfaces	

particle sizes. Table 8 shows that the powders possessed a wide range of shape factors, namely, 1-5, although powder No. 9 possessed a much wider range, namely, 1 to greater than 10. At least 50% (in number) of the particles of each powder fell within the ranges 0-40 μ for size and 1-2.49 for shape factors, but powders prepared by reduction of oxide possessed at least 50% (in number) within the shape-factor range 1-1.49. Electrolytic powders had the greatest, and carbonyl powders the least, shape factors of all the powders.

Microstructures

Polished and etched sections of individual powder particles were examined after mounting the powder in bakelite. Photographs of representative structures are reproduced in Figs. 6(b) to 11(b).

Electrolytic powders were nonporous and mostly showed a polycrystalline appearance, Fig. 6(b), but one of the powders, No. 13, showed a periodic, laminated structure typical of highly stressed, brittle, hard, and un-heat-treated electrodeposits, Fig. 7(b).

All oxide-reduced powders, Figs. 8(b), 9(b), and 10(b), were characterized by a number of pores, mostly interconnected and uniformly distributed throughout the particle; some oxide may have been removed from these cavities during the preparation of the specimens for microscopical examination. The oxide-reduced powders also contained unreduced oxide, usually irregularly distributed and sometimes segregated at one end of the sectional axis of the particle. The crystal structure of the ferrite in the oxide-reduced powders was detected only with difficulty. Powder No. 25, Fig. 10, was different from the other oxide-reduced powders in having a greater dispersion of the oxide phase and also pores which appeared more interconnected.

Particle Density and Porosity

Density values ranging from 7.26 to 7.89 were obtained in testing twenty-one powders (see Table 9) of which those notably different from the density of iron contained oxygen as oxide and porosity. The calculation of porosity should be based on the density value, which takes into account the oxide content of the particle; for this purpose densities of iron and iron-oxide mixtures (FeO)

TABLE 8. Percentage Frequency Distribution for Various Ranges of Shape Factors and Particle Sizes

Range of Sizes, μ	Percentage Frequency Distribution									
	Ranges of Shape Factors									
	1·0–1·49	1·5–1·99	2·0–2·49	2·5–2·99	3·0–3·49	3·5–3·99	4·0–4·99	5·0–5·99	6–10	10 +
Electrolytic, No. 9, Mean 1·76										
0– 10	13·0	4·0	2·5	1·5	...	...	0·5	...	...	...
11– 20	11·0	5·0	4·5	1·5	1·0	...	1·5	...	0·5	...
21– 40	8·0	5·0	4·5	1·5	2·5	2·0	3·0	1·0	0·5	0·5
41– 60	3·0	2·0	2·0	1·0	0·5	0·5	1·0	0·5	1·5	0·5
61–100	0·5	...	1·0	1·5	0·5	...	2·0	1·0	1·0	...
101–200	...	1·0	0·5	0·5	...	0·5	1·0	0·5	0·5	0·5
Electrolytic, No. 13, Mean 1·72										
0– 10	19·5	12·5	7·0	0·5	...	0·5	0·5	0·5*	...	...
11– 20	7·0	7·0	6·0	1·0	0·5	...	0·5	...	...	...
21– 40	9·5	4·5	3·0	1·0	1·0	0·5	1·0	0·5*	...	...
41– 60	2·5	2·5	2·0	...	...	...	0·5	...	...	...
61–100	1·5	3·5	1·5	...	...	0·5	...	0·5*	...	...
101–300	0·5	0·5	...	...	...	...	...	...	...	...
Oxide-Reduced, No. 17, Mean 1·63										
0– 10	15·0	8·0	3·0	1·5	...	...	...	...	...	...
11– 20	15·5	4·5	4·5	1·0	1·5	0·5	...	1·5*	...	...
21– 40	10·5	8·5	1·0	1·5	...	0·5	0·5	0·5*	...	...
41– 60	7·5	4·5	1·0	1·0	...	...	...	...	...	...
61–100	3·5	0·5	0·5	...	...	...	0·5	...	...	...
101–200	0·5	...	0·5	...	0·5	...	...	...	...	...
Oxide-Reduced, No. 24, Mean 1·47										
0– 10	32·5	12·0	5·0	0·5	1·5	...	...	...	...	...
11– 20	15·5	8·0	1·5	1·0	...	0·5	...	...	...	...
21– 40	7·0	4·5	0·5	2·0	0·5	...	...	...	...	...
41– 60	2·0	0·5	0·5	...	...	...	...	...	...	...
61–100	0·5	2·5	0·5	...	...	0·5	...	...	...	...
100 +	...	0·5	...	...	...	...	...	...	...	...
Oxide-Reduced, No. 25, Mean 1·38										
0– 10	61·0	17·5	14·0	1·5	2·5	...	0·5	...	...	...
11– 20	2·0	1·0	...	...	...	...	...	...	...	...
Carbonyl, No. 26, Mean 1·06										
0– 10	99·0	1·0	...	...	...	...	...	...	...	...
Abrasion, No. 27, Mean 1·50										
0– 10	4·0	0·5	0·5	...	...	...	...	...	...	...
11– 20	17·5	8·0	1·0	...	0·5	...	...	...	...	...
21– 40	30·5	13·0	3·0	1·5	0·5	...	...	...	...	...
41– 60	3·5	6·0	1·5	0·5	...	...	0·5	...	...	...
61–100	3·0	1·0	1·0	...	...	...	0·5	1·0*	...	...
101–200	1·0	...	...	...	...	...	...	...	...	...

*Shape factor greater than 5.

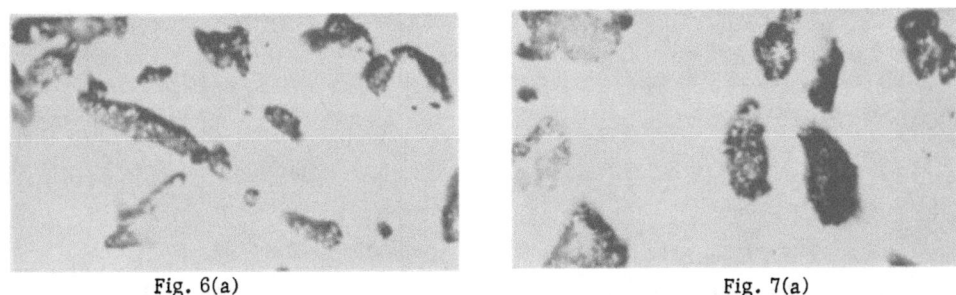

Fig. 6(a)   Fig. 7(a)

Figs. 6(a) and 7(a)—Size and shape of individual particles. ×500.

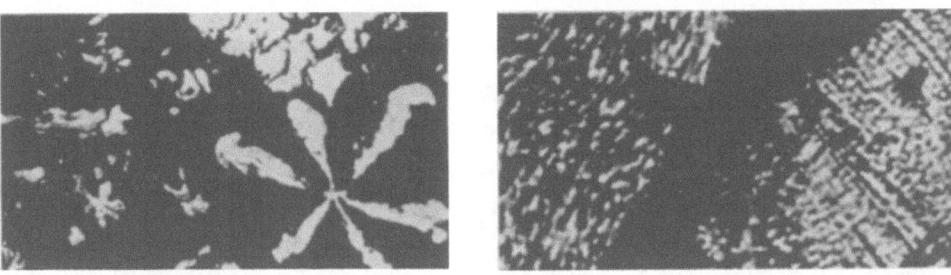

Fig. 6(b)   Fig. 7(b)

Figs. 6(b) and 7(b)—Microstructure of individual particles. ×1000.

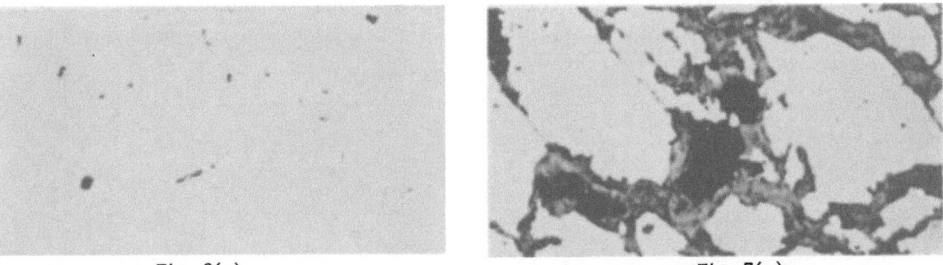

Fig. 6(c)   Fig. 7(c)

Figs. 6(c) and 7(c)—Polished sections of sintered compacts. ×500.

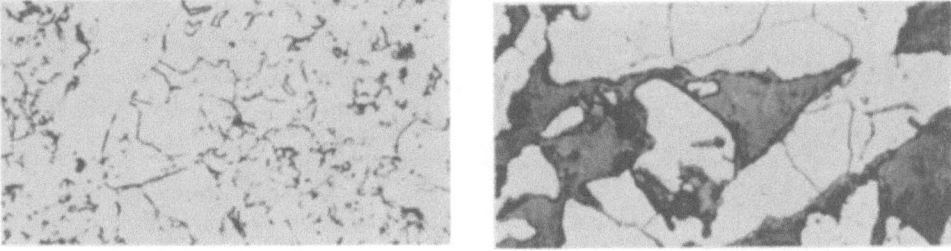

Fig. 6(d)   Fig. 7(d)

Figs. 6(d) and 7(d)—Etched sections of sintered compacts. ×500.

Fig. 6. Electrolytic-iron powder (No. 9).   Fig. 7. Electrolytic-iron powder (No. 13

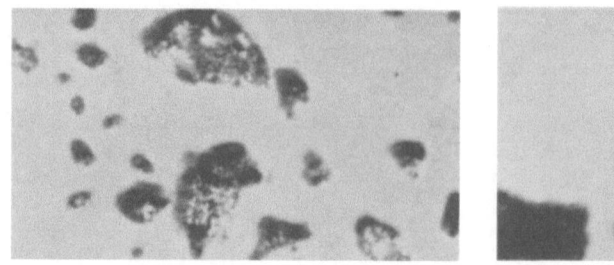 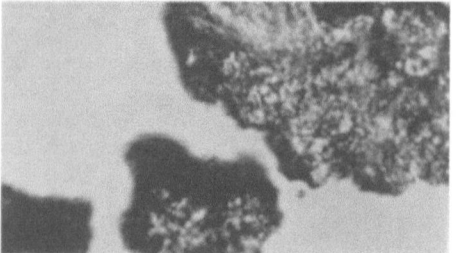

Fig. 8(a)   Fig. 9(a)

Figs. 8(a) and 9(a)—Size and shape of individual particles. × 500.

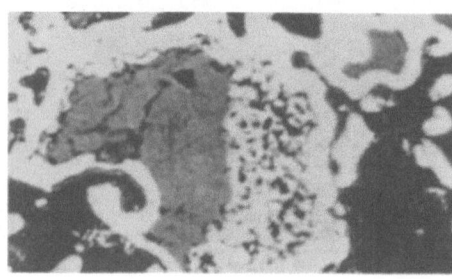 

Fig. 8(b)   Fig. 9(b)

Figs. 8(b) and 9(b)—Microstructure of individual particles. × 1000.

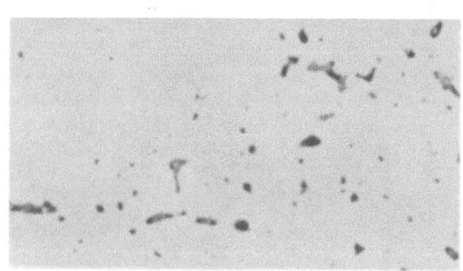 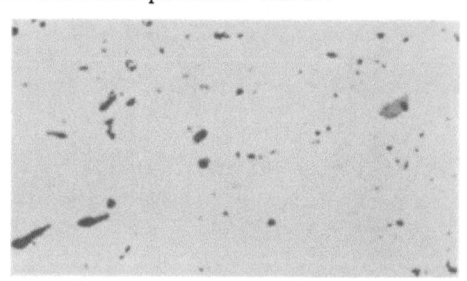

Fig. 8(c)   Fig. 9(c)

Figs. 8(c) and 9(c)—Polished sections of sintered compacts. × 500.

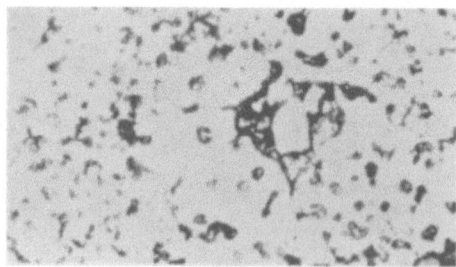 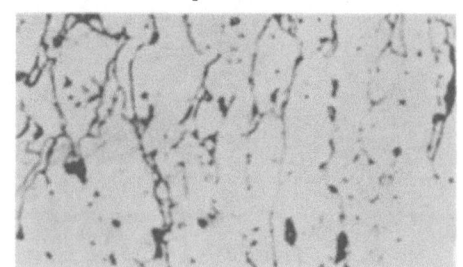

Fig. 8(d)   Fig. 9(d)

Figs. 8(d) and 9(d)—Etched sections of sintered compacts. × 500.

Fig. 8. Oxide-reduced powder (No. 17).   Fig. 9. Oxide-reduced powder (No. 24).

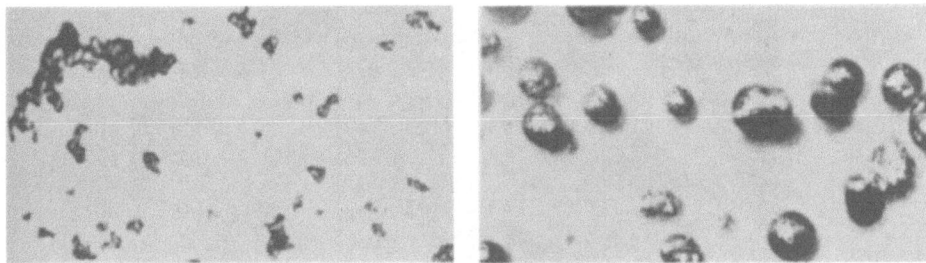

Fig. 10(a)   Fig. 11(a)
Figs. 10(a) and 11(a)—Size and shape of individual particles. × 1000.

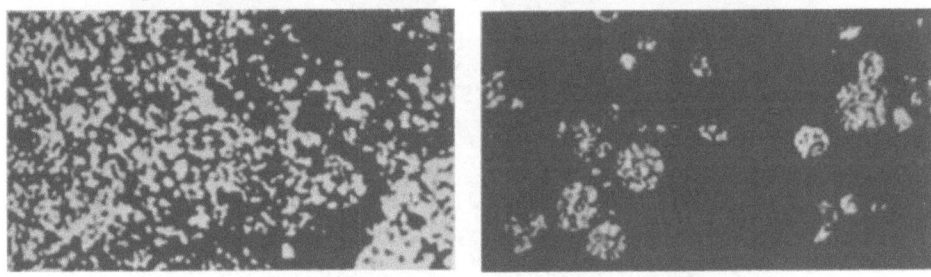

Fig. 10(b)   Fig. 11(b)
Figs. 10(b) and 11(b)—Microstructure of individual particles. × 1000.

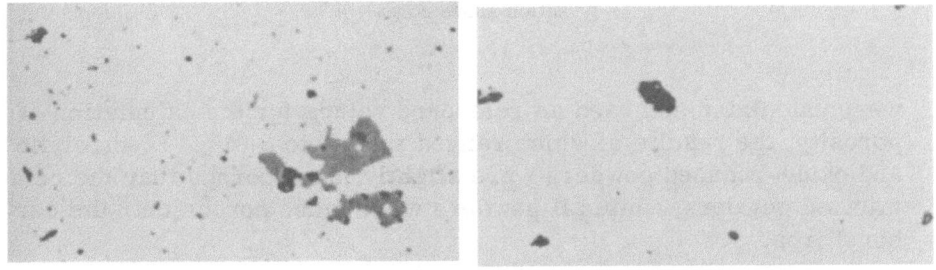

Fig. 10(c)   Fig. 11(c)
Figs. 10(c) and 11(c)—Polished sections of sintered compacts. × 500.

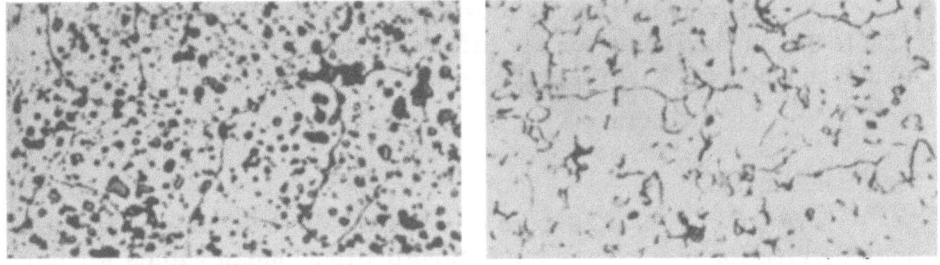

Fig. 10(d)   Fig. 11(d)
Figs. 10(d) and 11(d)—Etched sections of sintered compacts. × 500.
Fig. 10. Oxide-reduced powder (No. 25).   Fig. 11. Carbonyl powder (No. 26).

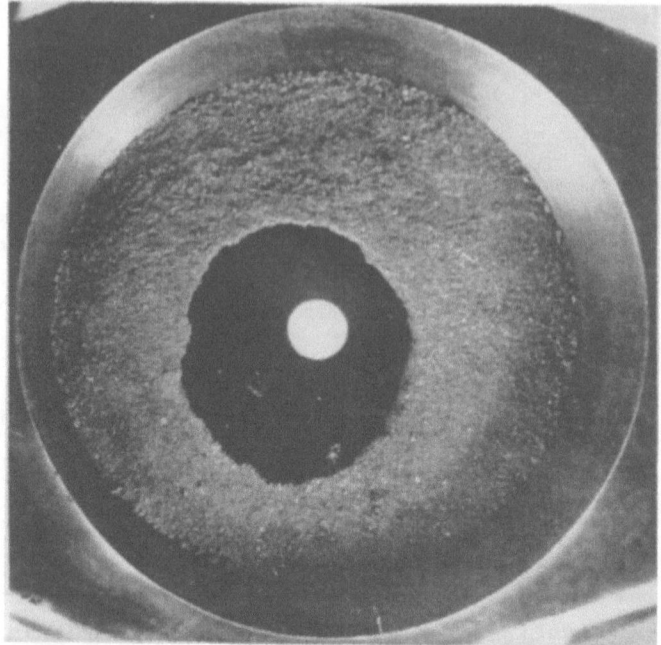

Fig. 12. Example of bridging in powder No. 27, viewed from position above cone.

were calculated and used as reference values for the calculation of
porosity, the results of which ranged from 0 to 3.81%. Electrolytic
and oxide-reduced powders were slightly more porous than the comminuted powders, while all powders were more porous than the carbonyl iron.

Apparent Density

The apparent-density values ranged from 0.97 to 3.40 g/cc
(Table 9). Mean values for the different types of powders were:
oxide-reduced 2.17, abrasion 2.49, electrolytic 2.77, and carbonyl
3.40 g/cc.

Tap Density

A range of 1.09-4.29 g/cc was found (Table 9), using the pneu-
matic-vibration method for the determination. Mean values for the
types of powders examined were: chloride-reduced 1.09, oxide-reduced 3.12, electrolytic 3.35, abrasion 3.52, and carbonyl 4.02 g/cc.

TABLE 9. Physical Properties of Different Types of Powders

Powder No.	Method of Manufacture	Densities, g/c.c.			Flow			Size Distribution, wt.-%							Median Size, μ	Properties after Pressing at 30 tons/sq. in.			Specific Surface, sq. cm/g
								Nominal Mesh Sizes											
		Apparent	Tap	Particle	Period, sec. r = 2·78 mm	Cone No.	Coefficient	+120	120-150	150-170	170-200	200-240	240-325	-325		Apparent Density, g/c.c.	Porosity, %	Compression Ratio	
								Actual Micron Sizes											
								+139	139-102	102-86	86-82	82-60	60-44	-44					
1	Electrolytic	3·32	4·05	7·88	4·6	2	1·24	17·41	16·75	12·77	4·86	15·11	28·27	4·43	84	5·60	28·88	1·69	....
2	,,	3·04	3·92	7·87	7·0	2	2·10	3·58	3·79	3·48	2·53	11·77	71·72	2·83	52	5·70	27·64	1·88	....
3	,,	..	2·22	..	..	..	..	2·64	16·18	7·25	5·03	14·64	46·89	7·26	57	..	..	..	....
4	,,	..	2·37	..	..	..	2·58	6·00	11·39	6·32	2·88	11·16	57·49	4·74	114	..	..	..	....
5	,,	..	2·63	7·89	..	2	1·90	29·18	30·59	15·97	9·74	5·12	4·10	5·06	53	..	..	..	....
6	,,	3·32	4·02	7·64	4·8	2	1·30	15·87	13·39	6·01	10·38	13·28	40·88	(−240)	78	6·19	21·31	1·86	265
7	,,	2·05	3·18	7·85	No flow	>8	>6·40	0·71	1·67	1·64	19·38	6·75	67·15	2·34	53	5·64	28·41	2·75	1149
8	,,	2·76	3·64	7·69	6·6	1	1·99	0·03	0·33	0·93	3·15	5·45	77·26	11·90	47	6·05	23·11	2·19	385
9	,,	2·08	3·18	7·89	No flow	8	6·40	1·08	..	..	2·77	9·70	81·04	5·53	51	6·11	22·42	2·94	733
10	,,	2·56	3·46	..	6·4	1	1·63	1·18	20·90	12·55	3·56	14·79	36·22	10·33	63	6·06	22·09	2·37	452
11	,,	2·43	3·44	..	6·5	1	1·82	7·45	19·93	9·96	10·43	12·36	32·39	13·45	66	6·32	19·76	2·60	478
12	,,	..	3·03	..	..	..	..	..	12·46	10·35	6·82	6·20	45·91	5·52	59	..	..	..	....
13	,,	3·37	4·29	7·45	5·5	2	1·48	11·31	25·34	..	..	10·96	23·96	10·96	84	4·54	42·36	1·34	275
14	Oxide-reduced	3·03	3·98	7·86	7·4	2	1·87	12·14	16·41	10·87	5·60	9·13	45·44	0·87	68	5·84	25·83	1·93	516
15	,,	..	4·00	..	..	..	1·42	22·03	25·75	7·94	3·09	4·50	19·25	15·36	95	..	..	..	....
16	,,	..	3·82	..	No flow	8	6·40	0·02	0·02	1·04	4·82	9·45	57·61	28·77	48	..	..	..	....
17	,,	2·19	3·31	7·64	9·4	8	6·40	26·30	23·51	1·51	7·79	13·66	67·80	12·14	51	5·64	28·36	2·58	945
18	,,	2·17	2·86	7·26	7·5	2	2·59	0·51	7·73	6·14	8·95	3·39	33·06	3·48	95	5·36	31·91	2·47	684
19	,,	2·18	2·91	7·72	8·2	1	2·29	3·25	20·11	10·40	8·47	14·87	62·72	1·94	55	5·38	29·40	2·55	513
20	,,	2·38	2·81	7·40	7·8	1	2·45	0·90	23·47	11·37	9·95	8·50	46·31	4·17	62	5·62	28·56	2·26	532
21	,,	2·09	2·61	7·78	7·0	1	2·04	3·91	16·08	10·63	9·42	12·21	35·27	8·46	70	5·45	30·71	2·69	562
22	,,	2·08	3·16	7·56	6·2	1	1·91	2·47	11·44	17·74	9·55	17·49	37·60	2·72	69	5·45	30·73	2·62	506
23	,,	2·53	3·19	7·72	7·5	2	1·73	0·21	30·92	1·18	1·04	17·40	47·02	1·46	62	5·43	30·97	2·16	448
24	,,	2·16	1·77	7·55	..	..	2·32	1·05	2·44	..	..	13·61	27·06	0·87	85	5·15	34·48	2·51	488
25	,,	0·97	..	7·58	No flow	>8	>6·40	..	..	..	..	2·44	10·93	78·41	6	..	..	5·31	5161
26	Carbonyl	3·40	4·02	7·84	No flow	5	2·72	2·56	0·04	0·03	0·06	0·66	36·46	95·30	7	5·69	27·73	1·67	3459
27	Abrasion	2·49	3·52	7·60	..	6	2·64	1·00	0·50	0·40	0·78	2·57	31·18	58·00	41	5·79	26·44	2·33	585
28	Chloride-reduced	..	1·09	..	..	..	..	30·97	16·32	6·64	5·56	9·0	..	−240	97	..	..	..	....

TABLE 10. Composition of-Iron Powders

Pow-der No.	Type	General Appearance	C, %	Si, %	S, %	P, %	Mn, %	Ni, %	Cr, %	Al, %	Pb, %	Cu, %	O_2, %	H_2, %	N_2, %	Loss on Heating in H_2 at 1000°C, %	Total Impur-ities, %	Fe (by differ-ence), %
6	Electrolytic	Light grey, some bright particles	0·060	0·040	0·025	0·015	0·040	0·050	0·007	<0·005	<0·010		0·100	0·001	0·010		0·363	99·637
7	,,	Slightly dark grey	0·060	0·005	0·140	0·018	0·110	0·170	0·080	0·080	0·013	0·054	(Insols. 0·180; Ti <0·005)				0·915	99·085
8	,,	Medium grey	0·020	0·060	0·049	0·018	0·130	0·110	<0·010	0·076	0·015	0·023	(Insols. 0·176; Ti <0·005)				0·592	99·408
9	,,	Slightly dark grey	0·080	0·050	0·020	0·006	0·010	0·110	0·006	<0·010	<0·010	0·140	0·410	0·004	<0·010		0·866	99·134
10	,,	Slightly dark grey	0·040	0·030	0·010	0·006	<0·010	<0·010	Nil	<0·005	<0·010		0·450	<0·001	<0·010	0·460	0·582	99·418
11	,,	Dark grey	0·030	<0·010	0·012	<0·005	<0·010	<0·010	Nil	<0·005	0·010	Nil	0·560	0·001	0·020	0·700	0·673	99·327
12	,,	Grey, with bright particles														1·100		
13	,,	Very dark, slightly brown, gritty	0·050	0·010	0·029	<0·005	<0·010	<0·010	Nil	<0·005	0·040	0·030	1·230	0·025	0·020	1·440	1·464	98·536
15	Oxide-reduced	Slightly dark grey	0·060	0·130	0·013	0·012	0·150	0·710	0·002	0·041	<0·010		0·710	0·002	0·010		1·750	98·250
16	,,	Grey	0·060	0·200	0·011	0·030	0·690	0·030	0·024	0·029	0·020		0·720	0·002	0·030		1·846	98·154
17	,,	Slightly dark	0·070	0·180	0·110	0·030	0·660	0·100	0·017	0·031	<0·010	(Ti 0·100)	1·000	0·004	0·020		2·232	97·768
18	,,	Slightly dark grey, bright particles	0·070	0·460	0·230	0·025	0·350	0·310	<0·010	0·020							>1·575	<98·425
19	,,	Grey	0·080	0·130	0·010	0·019	0·750	0·020	0·030	0·029	0·010	0·060	2·090	0·007	0·070	0·230	3·305	96·695
20	,,	Slightly dark	0·090	0·020	0·008	0·017	0·330	0·030	0·018	0·021	<0·010	0·070	1·220	0·001	0·040	1·350	1·875	98·125
21	,,	Light grey														1·060		
22	,,	Dark grey														2·250		
23	,,	Dark grey	0·120	0·040	0·015	0·007	0·230	0·060	0·014	0·011	<0·010	0·080	1·350	0·005	0·020	1·450	1·962	98·038
24	,,	Dark, bright parti-cles	0·360	0·060	0·119	0·045	0·300	<0·010	<0·002	0·031	<0·010	0·040	1·030	0·005	0·050	1·000	2·062	97·938
25	Carbonyl	Very dark	0·050	<0·010	0·014	<0·005	<0·010	<0·010	Nil	<0·005	0·010	Nil	1·180	0·004	0·010	1·300	1·308	98·692
26	Abrasion	Grey	0·120	<0·010	0·008	<0·005	<0·010	<0·010	<0·002	<0·005	0·010	0·060	0·260	0·004	0·030	0·310	0·474	99·526
27	Chloride-reduced	Grey	0·060	<0·020	0·016	0·013	0·360	0·040	0·032	0·009	<0·010	0·060	1·470	<0·005	0·020	1·300	2·115	97·885
28		Slightly dark, granular	0·050	0·040	0·005	0·010	0·260	0·220	0·020	0·086	0·009	0·025	(Insols. 0·106)				0·831	99·169

The increase in density of a powder that is consequent of consolidation by vibration has been expressed by the ratio of tap density to apparent density and for all the powders tested a range of
ratios of 1.18-1.82 was found; values for the various types of powder increased in the following order: carbonyl, electrolytic, oxide-
reduced, and abrasion, so that the greatest degree of "packing" was
produced in the abrasion powder.

Size Distribution

Details of the percentage-weight distribution in seven size
ranges are given in Table 9. All the powders are distributed in the
range 0-150 μ, with most of them in the range 0-100 μ; fourteen
powders possessed 50-97 wt% in the 0-60 -μ size. The percentage-
weight distribution on a size basis was plotted graphically; of ten
electrolytic powders, eight possessed a large size dispersion and
as many as three modes, while the remaining two exhibited a normal or Gaussian distribution; three oxide-reduced powders out of
a total of eleven possessed normal distributions, while the remainder were widely dispersed, with two modes in the distribution. The
carbonyl and abrasion powders possessed normal distributions.
Tentative median sizes, based on the actual dimensions of the
apertures of a sieve and determined from cumulative-percentage
curves are also given in Table 9, and have a range of 6-114 μ, the
average values for electrolytic and oxide-reduced powders being
66 and 64 μ, respectively. Only two powders, Nos. 25 and 26 are of
exceptionally small median size.

Several interesting points about sieving behavior were observed: A wide range of sieving times was found between methods;
some powders sieved the coarse fractions at a slow rate, others the
fine. The formation of dust was a feature of a number of samples
and this necessitated precautions in handling and testing.

Specific Surface

The values obtained in the modified air-permeability method
are given in Table 9. Exceptionally high values are found for Nos.
25 and 26, viz., 5161 and 3459 sq cm/g, respectively. Comparing
the oxide-reduced with the electrolytic-iron powders, the average
specific surface for the former was 1036 sq cm/g, as against 534
sq cm/g for the electrolytic powders.

TABLE 11. Properties of Sintered Components

Powder No.	Type	Mechanical Properties						Shrinkage, %				Gas Content, %		
		Ultimate Tensile Strength, tons/sq. in.	Yield Point, tons/sq. in.	Elongation, %	Vickers Diamond Hardness	Density, g/cc.	Porosity, %	Length	Breadth	Height	Volume	Oxygen	Hydrogen	Nitrogen
1	Electrolytic	4·99	5·75	4·4	25·8	5·71	27·45	0·65	0·81	0·90	2·33	…	…	…
2	,,	6·37		5·6	36·5	5·79	26·41	0·71	0·31	1·03	2·12	…	…	…
3	,,	4·25		2·2	23·5	5·04	35·96	5·99	7·06	7·36	18·37	…	…	…
4	,,	4·09	7·04	2·2	22·5	6·07	32·88	1·41	1·62	2·86	5·79	…	…	…
5	,,	8·01	4·40	6·8	32·5	6·23	20·21	0·00	0·30	2·13	−1·56	…	…	…
6	,,	7·50		7·5	28·7	5·91	24·91	0·29	0·30	0·97	1·57	0·06	0·015	<0·01
7	,,	6·87		3·3	33·7	6·23	21·22	1·57	1·27	1·77	4·07			
8	,,	7·85		3·3	37·7	6·13	21·48	0·28	0·20	0·55	0·96			
9	,,	10·31	7·30	8·3	36·2	6·17	21·60	0·35	0·30	1·20	1·84	0·14	0·016	0·02
10	,,	8·30			43·6	6·42	18·43	0·31	0·13	1·83	2·27	0·06	0·004	<0·01
11	,,	9·02		6·8	31·6			0·29	0·58	1·32	2·16			
12	,,	7·70			61·2							0·19	0·002	0·015
13	Oxide-reduced	7·41		2·2	42·2	4·63	41·55	1·13	2·07	1·55	4·71			
14	,,	4·19			41·5	6·09	22·62	1·53	1·59	2·01	5·04	0·39	0·020	0·01
15	,,	6·80	6·50	3·0	32·9	6·11	21·99	1·22	1·38	2·15	4·63	0·53	0·074	<0·01
16	,,	8·90	6·80	3·0	33·4	5·94	30·00	0·55	0·85	2·57	2·07	0·43	0·013	0·01
17	,,	6·30		1·1	41·8	5·71	24·53	1·69	1·80	2·74	7·29			
18	,,	6·33		2·2	32·9	5·71	27·45	2·44	2·42	1·24	2·92			
19	,,	6·48	5·70	2·2	35·0	5·47	30·50	0·96	0·73	1·56	3·10	0·97	0·073	0·01
20	,,	7·60	6·00	4·4	30·2	5·28	27·58	0·87	0·55	1·23	2·53	0·37	0·083	0·05
21	,,	4·46	5·90	1·65	26·9	5·47	33·42	0·79	0·78	0·20	1·66			
22	,,	4·66	2·00	1·65	32·2	5·47	29·23	0·71	1·00	1·54	2·17	0·29	0·021	0·01
23	,,						30·50	0·78	0·57	1·02	2·17	0·32	0·089	0·03
25	Carbonyl	20·07	13·10	7·15	113	7·05	10·42	10·10	10·30	9·28	27·00	1·18	0·003	<0·01
26	Abrasion	8·08	7·68	3·35	49·8	5·98	24·02	1·87	2·22	1·54	5·52	1·05	0·008	<0·01
27	,,			1·1	32·2	5·79	26·43	0·57	0·72	0·91	2·22			
28	Chloride-reduced	11·5		2·2	39·5	6·38	18·94	0·98	0·00	0·70	1·20	0·48	0·031	0·02

Flow Behavior

The powders were tested in a series of glass cones, the orifices of which formed a series of progressively increasing radii; the cones were without stems. Comparisons of the behavior of different powders were made on the basis of cone numbers previously described, and also on the basis of flow coefficients from the expression $k = tr^n/w$, in which k represents the flow coefficient, t the period of flow (sec) of w g of powder, and r the radius (mm) expanded to the power of n, which has an average value of 2.58.

The results of these tests are given in Table 9, which shows that 72% of all the powders flowed freely in cones 1 and 2; two of the powders (8%) flowed freely in cones 5 and 6, respectively, while 20% possessed cone numbers of 8 or higher. The mean values of the flow coefficient were 1.93 for the free-flowing powders and greater than 6.40 for nonflowing powders, corresponding to cones 1 and 8.

Three types of flow were observed. (a) The powder issuing from the orifice resulted in a general setting of the mass in the cone (plug flow); (b) the mass near the axis of the cone moved toward the orifice, the remainder being stationary until two-thirds of the mass had issued from the cone when general setting occurred (axial flow); and (c) some powders showed no flowing tendency at all, while others flowed for a short period, and when an axial capillary had been formed in the mass, flow suddenly ceased, indicating a strong bridging mechanism, which is possible in certain types of powders. An example of bridging formed in powder No. 27 is shown in Fig. 12. It is interesting to note that axial flow could be converted to plug flow in a given powder by increasing the radius of the orifice.

Certain powders which possessed high flow factors outside the range of experimental interest could be caused to flow by tapping the cone.

Compressibility and Compression Ratio

The relationship between compacting pressure and density was examined, and the increase in density was found to be rapid and continuous up to a compacting pressure in the neighborhood of 50 tons/sq in., whereas, above this, the increase in density was less marked and attained an asymptotic value near 70 tons/sq in. Differences in density between the powder compacts prepared at any fixed pressure indicated general differences in the compressibility of powders.

The compression ratios which are important from the standpoint of die design were determined after pressing at 30 tons/sq in. and fell within the range of 1.34-5.31 (Table 9). An increasing order of values was given by the following powders: carbonyl, electrolytic, abrasion, and oxide-reduced.

Chemical Composition

The chemical compositions of twenty powders are given in Table 10. The total amounts of impurities are also given, from which it may be seen that the purity of the electrolytic and carbonyl powders is much higher than that of the oxide-reduced and abrasion powders. Electrolytic powders contained 98.536-99.637% of iron; in contrast to this, oxide-reduced powders contained 96.-695-98.692% of iron; the carbonyl iron, 99.526%; chloride-reduced, 99.169%; and abrasion, 97.885%.

All powders contained little carbon, with the exception of No. 24 which contained 0.36%, as compared with a range of 0.02-0.12% for the other powders. The silicon content for twelve powders was less than 0.06%, but for others was in the range 0.06-0.46%. The sulfur content was low for all powders, except Nos. 7, 17, 18, and 24, which had sulfur contents ranging from 0.11 to 0.23%. Appreciable amounts of manganese were present in three powders, and of nickel in one powder; other elements were present only in small quantities.

Oxygen, hydrogen, and nitrogen contents varied over the ranges 0.1-2.09 wt%, less than 0.001-0.025 wt% and less than 0.01-0.07 wt%, respectively. The mean values for the analysis of each gas were as follows:

	Oxygen, %	Hydrogen, %	Nitrogen, %
Carbonyl	0.26	0.004	0.03
Electrolytic	0.55	0.006	0.014
Oxide-reduced	1.16	0.0038	0.03
Abrasion	1.47	0.005	0.02

Part III – The Properties of Sintered Compacts

Standard test-compacts, 6.5 × 1 × 1 cm, were prepared from each powder after compacting at 30 tons/sq in. and sintering for

1 hr at 1050°C in a hydrogen atmosphere. This pressure and this sintering temperature were chosen as being representative of those used in industry for the preparation of iron compacts for engineering applications. The properties quoted below are for compacts prepared under these standardized conditions.

Mechanical Properties. The values for the ultimate tensile stress were in the range 4.09-20.07 tons/sq in., or 4.09-11.5 tons/sq in. if powder No. 25 is excluded, since this powder was unique in the present series in possessing exceptionally high strength; the values for the yield point ranged from 2.0 to 13.1 tons/sq in. for 12 powders; in the remainder the yield point was close to the ultimate stress. Elongation values ranged over 1.1-8.3%. The strength and ductility of these sintered compacts are thus low, and only one powder, No. 25, possesses notable strength for pure iron. The elongation values of the electrolytic-iron powders, although low, are superior to those of the oxide-reduced irons (see Tables 11 and 12).

Hardness. The diamond pyramid indentation values (Tables 11 and 12) ranged from 22.5 to 61.2, excepting powder No. 25 which possessed a value of 113. The correlation between hardness and tensile strength for sintered iron compacts was less constant than it is for wrought materials, and ranged from 0.10 to 0.29.

Density. The densities ranged from 4.60-7.05 g/cc, corresponding to porosity values of 41.55-10.42%.

Volume Change. During sintering the compact undergoes a change in dimensions, and this can take the form of either an expansion or a contraction. As most iron powders contract on sintering, the dimensional changes are recorded as percentage shrinkages ranged from 0.0 to 10.10%, and the breadths (dimension between die-faces) from 0.0 to 10.3%, while the height (dimension between punches) changed from an expansion of 2.13% to a contraction of 9.28%. For these various conditions the volume changes ranged from an expansion of 1.56% (No. 5) to a contraction of 27.0% (No. 25). Twenty-one specimens (nine of them oxide-reduced) developed their maximum shrinkage along the height (see above), five along the breadth, and one along the length.

Gas Content. Since microsection of the compacts sintered in hydrogen showed appreciable oxide inclusions, it was considered desirable to determine the oxygen content, and samples adjacent to the microsections were examined.

TABLE 12. Summary of Experimental Results

Property	Type of Powder						
	Electrolytic		Oxide-Reduced		Carbonyl	Abrasion	Chloride-Reduced
	Range	Average	Range	Average			
Shape factor	1·72-1·76	1·74	1·38-1·63	1·49	1·06	1·50	⋯
Particle density, g/c.c.	7·45-7·89	7·77	7·26-7·86	7·60	7·84	7·60	⋯
Particle porosity, %	0·78-3·81	2·13	0·19-3·31	1·97	0·03	1·57	⋯
Apparent density, g/c.c.	2·05-3·37	2·77	0·97-3·03	2·17	3·40	2·49	1·09
Density ratio, tap/apparent	2·22-4·29	3·35	1·77-4·00	3·12	4·02	3·52	⋯
Flow cone No.	1·21-1·55	1·35	1·25-1·82	1·40	1·18	1·41	⋯
Flow coefficient	1-8	2·31	1-8	3·08	5·0	6·0	⋯
Median size, μ	1·24->6·4	2·62	1·42->6·4	3·15	2·72	2·64	⋯
Specific surface, sq. cm/g	47-114	66	6-95	64	7	41	97
Specific surface, sq. cm/g	265-1149	534	448-5161	1036	3459	585	⋯
Density, g/c.c., pressed at 30 tons/sq. in.	4·54-6·32	5·30	5·15-5·84	5·49	5·69	5·79	⋯
Porosity, %, pressed at 30 tons/sq. in.	19·76-42·36	26·22	25·83-34·48	30·27	27·73	26·44	⋯
Compression ratio	1·34-2·94	2·18	1·93-5·31	2·71	1·67	2·33	⋯
Iron content, %	98·54-99·64	99·22	96·70-98·69	98·01	99·53	97·89	99·17
Oxygen, unsintered, %	0·10-1·23	0·55	0·71-2·09	1·16	0·26	1·47	⋯
Oxygen, sintered, %	0·06-0·19	0·11	0·29-1·18	0·56	0·05	0·48	⋯
Hydrogen, unsintered, %	0·001-0·025	0·006	0·001-0·007	0·004	0·004	0·005	⋯
Hydrogen, sintered, %	0·002-0·016	0·009	0·003-0·089	0·047	0·008	0·031	⋯
Nitrogen, unsintered, %	0·01-0·02	0·014	0·01-0·07	0·03	0·03	0·02	⋯
Nitrogen, sintered, %	0·01-0·02	0·014	0·01-0·05	0·018	0·01	0·02	⋯
Ultimate tensile strength, tons/sq. in.	4·09-10·31	7·11	4·19-20·07	7·65	8·08	5·85	11·50
Yield point, tons/sq. in.	4·40-7·30	6·12	2·0-13·1	6·57	7·68		⋯
Elongation, %	0-8·3	5·04	0·0-7·15	2·86	3·35	1·1	2·2
Hardness (Vickers diamond)	22·5-61·2	34·5	26·9-113·0	42·0	49·8	32·3	39·5
Density, g/c.c., sintered	4·60-6·42	5·81	5·24-7·05	5·80	5·98	5·79	6·38
Porosity, % sintered	18·43-41·55	26·20	10·42-33·42	26·31	24·02	26·43	18·94
Shrinkage, %: Length	0·00-5·99	1·08	0·55-10·10	1·88	1·87	0·57	0·98
Breadth	0·13-7·06	1·25	0·55-10·30	1·89	2·22	0·72	0·00
Height	-2·13-7·36	1·60	0·20-9·28	2·19	1·54	0·91	0·70
Volume	-1·56-18·37	3·80	1·66-27·00	5·65	5·52	2·22	1·20

The oxygen contents ranged from 0.05 to 1.18% and showed
a decrease as compared with the unsintered content, with the exception of No. 25 (1.18% of oxygen) which showed no change. In
contrast to this, the hydrogen contents increased for every powder
by at least 100% to the values 0.002-0.089%. The nitrogen contents
ranged from less than 0.01 to 0.05%. The increase in nitrogen content for some powders was probably due to changing the furnace
atmosphere from hydrogen to nitrogen when cooling below 600°C.

Part IV – Discussion of Results

1. Summary and Comparison of the Data

Average values for all the properties have been summarized
in Table 12, together with the range of values obtained for each
property. For the purposes of comparison, it is unfortunate that
only single sets of data have been obtained for the carbonyl, abrasion, and chloride-reduced powders; the chief comparison in this
investigation is intended to be between electrolytic and oxide-reduced powders, but the data for the remaining types of powder have
been included for the sake of completeness.

2. Correlation of Various Pairs of Properties

A high degree of correlation between any two properties cannot be expected, in view of the numerous factors which enter into
the control of any general property of powders. Graphical correlation yielded points with appreciable scatter, so that the best approach was considered to be a statistical correlation.

The correlation coefficient is defined [21] as:

$$r = \frac{\Sigma(x - \bar{x})(y - \bar{y})}{\sqrt{\Sigma(x - \bar{x})^2(y - \bar{y})^2}}$$

where r = the correlation coefficient, x, y = any single value of the
two variables, $\bar{x}$, $\bar{y}$ = the average values of the data for the two
variables, Σ = the sum of any values, and $\Sigma(x-\bar{x})$, $\Sigma(y-\bar{y})$ = the sum
of the individual deviations from the mean.

Coefficients derived by the above method of simple correlation possess values ranging from +1 to -1. Generally, a positive
r denotes that one factor increases simultaneously with an increase
in the value of the second factor; this, however, must not be ac-

TABLE 13. Correlated Properties in Order of Decreasing Coefficients

No.	Combination of Properties for which r is Positive	Value of r	Combination of Properties for which r is Negative	Value of r
1	Volume shrinkage and increase in density in sintering	0·98	Apparent density and compression ratio	0·90
2	Apparent and tap densities	0·96	Tap density and compression ratio	0·87
3	Density increase on sintering and tensile strength	0·90	Median size and specific surface	0·81
4	Ratio of tap/apparent densities and flow coefficient	0·36	Apparent density and flow coefficient	0·60
5	Specific surface and volume shrinkage in sintering	0·36	Apparent density and tensile strength	0·59
6	Compression ratio and tensile strength	0·35	Median size and density increase in sintering	0·56
7	Tensile strength and Vickers diamond hardness	0·82	Apparent density and volume shrinkage in sintering	0·56
8	Density after sintering and tensile strength	0·81	Median size and flow coefficient	0·52
9	Compression ratio and Vickers diamond hardness	0·80	Median size and volume shrinkage in sintering	0·48
10	Compression ratio and volume shrinkage in sintering	0·79	Median size and compression ratio	0·47
11	Specific surface and tensile strength	0·77	Median size and Vickers diamond hardness	0·47
12	Density after sintering and Vickers diamond hardness	0·69		
13	Compression ratio and density after sintering	0·61	Tap density and volume shrinkage in sintering	0·40
14	Densities after pressing and sintering, respectively	0·60		
15	Flow coefficient and tensile strength	0·52		
16	Volume shrinkage in sintering and tensile strength	0·50		
17	Specific surface and flow coefficient	0·50		

TABLE 14. Effect of Various Treatments of Powders on the Mechanical Properties of Sintered Compacts

Powder No.	Type	Conditions	Tensile Strength, tons/sq. in.	Yield Point, tons/sq. in.	Elongation, %	Vickers Diamond Hardness	Porosity, %	Volume Shrinkage, %
13	Electrolytic	As-received	Did not compact satisfactorily					
		Annealed	4·28	...	...	37·9	25·00	1·55
16	Oxide-reduced	As-received	6·8	6·5	3	32·9	30·00	2·07
		Annealed	10·5	7·4	4	46·4	23·70	1·51
17	,,	As-received	8·9	6·8	3	33·4	24·53	5·93
		Annealed	10·9	6·6	4·5	43·2	22·30	2·92
20	,,	As-received	6·48	6·00	2·2	35·0	30·50	3·10
		Annealed	8·81	...	2·2	53·6	23·00	1·18
24	,,	As-received	Did not compact satisfactorily					
		Annealed	9·12	...	2·2	69·0	15·6	2·28
25	,,	As-received	20·07	13·10	7·15	113	10·42	27·00
		Ground in mortar	19·05	13·74	8·89	92·6	...	27·80
		Annealed	18·45	9·86	17·80	85·8	7·50	26·18
		Annealed and oxidized	4·55	...	...	24·2	29·20	30·45
		Annealed, oxidized, hydrogen-reduced	13·40	9·50	6·0	64·8	14·10	21·67
26	Carbonyl...	As-received	8·08	7·68	3·35	49·8	24·02	5·52
		Annealed	10·07	8·0	6·0	46·7	21·60	3·34

cepted as a rigorous relationship, since a possible third and fourth variables may be changed by a change in either the first or second variables included in r. Again, in the general example, a negative r denotes that one factor decreases while the other increases; this is subject to the same qualification as a positive r.

The correlation coefficient of any two properties from a large number can be placed therefore into one of two classes depending on whether one property increases or decreases, respectively, with regard to the other. The values of twenty-nine significant correlation coefficients out of a total of forty-six calculated combinations of properties are given in Table 14 in decreasing order for the two arbitrary classes. It is necessary to emphasize that no causal relationship obtains for any one combination of properties because, in the majority of instances, many factors contribute to produce any one final property. The significant positive and negative values of r ranged from 0.98 and 0.50 and from 0.90 to 0.40, respectively.

3. Some Factors Controlling the Properties of Sintered Compacts

In order to obtain a direct correlation between some of the factors controlling the general properties of powders and of pressed and sintered compacts, tests were carried out to determine to what extent the mechanical properties were influenced (a) by mechanical and thermal treatment of the powder before pressing and sintering, and (b) by the particle size of the powder.

Leafing Characteristics

During the examination of the individual particles of powders No. 25, the presence of groups consisting of 2-4 particles was noticed; the capacity to form groups is sometimes described as the "leafing characteristic." The groups were broken up in a mortar until an appreciable change had occurred in their appearance as viewed under the microscope. After compacting and sintering at the standard conditions, the tensile strength was of the same order as before (Table 14).

Annealing the Powder

Annealing some of the powders used in the present work at 600°C for 1 hr in an atmosphere of hydrogen produced significant effects on their properties either in rendering a noncompactible powder readily compacting or in increasing the ultimate tensile strength. As examples of the former, powders 13 and 24 were quite friable after pressing in the as-received condition but, when annealed before pressing, they yielded satisfactory compacts having tensile strengths of 4.28 and 9.12 tons/sq in. Similarly, powder No. 20 gave compacts having tensile strengths of 6.48 and 8.81 tons/sq in., respectively. In some instances the elongation was improved more than the ultimate tensile strength.

Surface Oxidation of the Powder

The deleterious effect of surface oxidation was demonstrated by slightly oxidizing a sample of powder No. 25 after annealing in hydrogen at 600°C by withdrawing it from the furnace at 300°C and exposing it to the air. The powder particles developed a blue tint and appeared harder to the touch than the original powder. After pressing and sintering (30 tons/sq in., 1050°C for 1 hr in hydrogen) the maximum strength was reduced from 18.45 to 4.55 tons/sq in. while the elongation was reduced from 17.8% to nil. A considerable improvement was effected by removing this oxide from the powder by reduction in hydrogen and preparing a compact in the usual manner; the ultimate strength and elongation were increased from 4.55 to 13.4 tons/sq in. and from 0 to 6%, respectively. X-ray examination showed that the iron oxide in the powder in the as-received condition was Fe_3O_4, whereas a mixture of Fe_3O_4 and Fe_2O_3 occurred in the oxidized sample of powder.

Size of the Individual Particle

Electrolytic (No. 9) and oxide-reduced (No. 18) powders were separated into seven and six fractions by air elutriation and mechanical sieving, since the size distributions occurred mostly in the subsieve (0-56 μ) and sieve (greater than 56 μ) ranges.

Details of the nominal size ranges and median sizes of the various fractions, together with the tests carried out and the results obtained, are given in Table 15.

A correlation has been made between the mean particle size of the fractions and the general properties, and the results may be summarized as follows.

Apparent Density. The apparent density is directly proportional to increasing particle size up to the subsieve range (No. 9)

TABLE 15. General Properties of the Size Fraction of Powders and Their Compacts

Powder No.	Fraction No.	Nominal Size, μ Range	Nominal Size, μ Median	Density, g./c.c. Apparent	Density, g./c.c. Tap	Flow Min. Cone	Flow k value	Shape Factor on Diameter Martin	Shape Factor on Diameter Max.	Pressed at 30 tons/sq. in. Density, g./c.c.	Pressed at 30 tons/sq. in. Porosity, %	Pressed at 30 tons/sq. in. Compression Ratio	Sintered Shrinkage, % Length	Breadth	Height	Volume	Density, g./c.c.	Porosity, %	Ultimate Tensile Strength, tons/sq. in.	Yield Point, tons/sq. in.	Elongation, %	V.D.H.
9 (as-received)	...	...	51	2·08	3·18	8	6·40	...	...	6·11	22·42	2·94	0·35	0·30	1·20	1·84	6·18	21·48	10·31	7·3	8·3	36·2
9 (separated into fractions)	1	+56	65	2·86	3·31	1	1·31	2·10	2·26	6·44	18·17	2·25	0·39	0·40	0·67	1·46	6·59	16·21	7·9	6·9	3	39·1
	2	42-56	49	2·41	2·95	4	2·57	1·76	3·25	6·32	19·69	2·62	0·36	0·43	0·93	1·70	6·47	17·78	10·2	7·1	7	41·5
	3	28-42	35	2·12	2·82	8	3·53	2·09	2·83	6·34	19·40	2·99	0·32	0·43	1·57	2·29	6·54	16·93	12·1	8·6	7	44·6
	4	20-28	24	1·93	2·78	8	>3·53	2·19	3·87	6·41	21·47	3·20	0·67	0·90	1·59	3·13	6·41	18·51	12·2	8·0	...	43·8
	5	14-20	17	1·78	2·63	8	>3·53	1·82	3·09	5·92	24·81	3·32	0·75	1·08	1·81	3·62	6·11	22·34	10·5	7·7	...	38·8
	6	10-14	12	1·68	2·63	8	>3·53	1·83	2·05	5·96	24·25	3·55	2·12	2·53	2·85	7·29	6·39	18·86	13·4	9·5	...	51·7
	7	0-10	5	1·36	2·65	8	>3·53	1·62	1·67	5·85	25·73	4·30	6·25	6·79	6·71	18·49	7·05	10·47	17·7	12·8	7	99·3
18 (as-received)	...	...	95	2·17	2·86	1	2·59	...	...	5·36	31·91	2·47	2·44	2·42	2·74	7·29	5·71	27·45	6·3	...	1·1	41·8
18 (separated into fractions)	1	+124	140	2·02	2·14	1	2·42	...	...	5·27	33·00	2·62	2·95	2·89	3·29	8·86	5·73	27·19	4·7	...	<1·0	43·2
	2	104-124	114	2·05	2·22	1	1·97	...	...	5·39	31·52	2·63	2·75	2·62	3·61	8·66	5·86	25·53	5·3	...	2·0	40·8
	3	89-104	96	2·04	2·38	1	1·81	...	...	5·60	28·82	2·75	2·24	2·23	2·70	7·01	6·02	23·51	6·1	...	2·0	45·6
	4	76-89	83	2·19	2·50	1	1·75	...	...	5·73	27·15	2·61	2·28	2·26	2·48	6·84	6·16	21·73	8·1	7·3	2·0	49·6
	5	66-76	71	2·12	2·58	1	1·74	...	...	5·70	27·58	2·70	2·18	2·38	2·51	6·87	6·13	22·10	7·8	7·2	2·0	48·4
	6	44-66	55	2·01	2·76	8	4·17	...	...	5·74	27·10	2·86	2·74	2·90	3·00	8·35	6·25	20·58	10·5	7·0	6·0	53·2

but remains reasonably constant in the sieve range for No. 18. This
tendency in powder No. 18 may be caused by the presence of a
greater proportion of porosity or of unreduced oxide in the coarse
particles.

Tap Density. A greater degree of packing was effected in
powder No. 9 than in No. 18, a difference which was maintained
throughout the size range. The values for No. 9 bear a general
parallel relationship to the apparent densities, that is, increase
with particle size, but those for No. 18 indicated a possible difference in packing mechanism, since the degree of packing decreased
with increasing particle size.

Flow Behavior. The flow coefficient k increased from 1.31 to
a value greater than 3.53 for No. 9, indicating a progressive increase with decreasing particle size, while a minimum value of
1.74 was obtained for No. 18 at a particle size of 71 μ.

Shrinkage. Significant changes were found in the dimensions
of compacts of No. 9, the total shrinkage varying from 1.46 to
18.49%. In contrast with this, however, the compacts of powder No.
18 developed only small differences in change of dimensions
throughout its size range.

Tensile Strength. Marked increases were obtained by decreasing the particle size; No. 9 increased in strength from 7.9 to
17.7 tons/sq in. for the range 65-5 μ and No. 18 from 4.7 to 10.5
tons/sq in. for the range 140-150 μ. It is of interest to note that,
although these two powders are of different types and size ranges,
they show a similar relationship between particle size and tensile
strength.

4. General Considerations

The phenomenon of shrinkage is of supreme importance during sintering, and it is significant from the standpoint of design and
manufacture that no law exists for defining its type or degree. The
pressing operation leads to (1) conjunction of the particles at the
existing pores, (2) a higher degree of contiguity between particles
so that the number of points of contact and consequently the surface area are greater, and (3) marked plastic deformation.

The net result is a state of high physical instability on account of the work expended in increasing the internal energy of the
system; the increase in the area of surfaces of common contact,

coupled with the smoothing of surface asperities, results in a penetration of surface films, so that increased cohesion is likely. The
degree to which the particles of any powder will become adjusted
to the conditions envisaged in the above mechanism is clearly dependent upon the external and internal properties of the particles,
the compacting pressure, and the sintering temperature. When a
highly deformed aggregate of metal particles in the form of a
pressed compact is heated, the strained adjacent particles undergo
recrystallization and the crystal boundaries will migrate, resulting in incipient cohesion at some of the points of contact between
the particles.

The cohesion of the compact is greatly affected by the degree
of surface oxidation of the powder; an annealing treatment of the
powder in hydogen increased the tensile strength, whereas oxidation of the powder before the usual processing produced a substantial decrease in tensile strength. It is of interest that powder No.
25 was exceptional in two respects: Its tensile strength, unlike that
of other powders, was not improved by an annealing treatment in
hydrogen before processing, and its oxygen content remained unaffected by sintering oxidation of the powder particles, so that
maximum cohesion would be attained. Mechanical strength of a
sintered compact is dependent upon a number of factors among
which may be included the particle size, the degree of cohesion between any two individual particles, and the degree, type, and loca-
tion of porosity in the material.

Thus, the tensile properties of sintered irons range from
4.09 to 20.07 tons/sq in., and the corresponding porosities from
42 to 10%. It has been shown that, in addition to porosity, the particle size of the powder has an important influence on properties
and that the median sizes in the iron powders examined range from
114 to 6 μ. The influence of crystal size on the properties of metals has been studied by a number of workers; Edwards and Pfeil,
[22] for example, have found that in iron of fairly high purity the
tensile strength ranged from 10.6 to 18.7 tons/sq in. when the number of crystals per square millimeter ranged from about 0.01 to
194.

Comment is sometimes made on the low values of the
strength of sintered compacts. The present work shows that, although the average strengths obtainable are less than those of
wrought materials, the differences are not great and are explicable

in terms of the form of the entrapped impurities and the porous nature of the sintered materials, as compared with the general purity and soundness of iron in the wrought condition. With the appreciable porosity resulting from the compacting and sintering treatment employed, the bulk strength will be affected by the concentrations of stress at the pores and filmlike inclusions and it is for this reason that the method of failure is different from that in wrought material.

Conclusions

In order to determine the properties of commercial iron powders, investigation and development of methods of testing were found to be necessary. The following powder properties have been studied: appearance, particle density, apparent density, tap density, flow behavior, compression ratio, particle size and size distribution, specific surface, shape factor, surface texture, microstructure, and chemical composition. It has been shown that suitable methods and apparatus for the examination of all varieties of iron powders required to be standardized and it is recommended that this aspect of the testing of metal powders be considered and undertaken by interested bodies. Attention has also been paid to the properties of sintered compacts prepared by a simple pressing and sintering technique and a wide range of properties was obtained.

It is evident that a number of factors govern the ultimate properties of sintered compacts for such widely divergent applications as machine parts possessing high strength and bearing materials of high porosity. The behavior of these components is dependent upon such powder properties as particle size, hardness, and surface purity, surface oxidation having a markedly deleterious effect on the compactibility and mechanical properties. A judicious choice of the powder properties and processing conditions is therefore necessary to develop a component for a specific purpose.

Acknowledgments

Thanks are due to the Superintendent, Metallurgy Division, National Physical Laboratory, for accommodation and facilities, and for arranging for the chemical and gas analyses; and to the Superintendent, Engineering Division, for the mechanical tests.

Acknowledgments are also made to Mr. R. Chandler for help in the experimental work and to the Chief Scientific Officer, Ministry of Supply, for permission to publish the paper.

References

1. C. Hardy, Symposium on Powder Metallurgy. American Society for Testing Materials, Philadelphia, Pa. (1943), p. 3.
2. H. W. Greenwood, Metallurgia, 30:178 (1944).
3. W. E. Deming and A. L. Mehring, Industrial and Engineering Chemistry, 21:661 (1929).
4. E. C. Bingham and R. W. Wikoff, J. of Rheol., 2:395 (1931).
5. Symposium on New Methods for Particle Size Determination in the Subsieve Range. American Society for Testing Materials, Philadelphia, Pa. (1941).
6. J. M. Dallavalle, Micromeritics. Pitman Publishing Corporation, New York (1943).
7. J. Wulff, Powder Metallurgy. American Society for Metals, Cleveland, Ohio (1942).
8. F. M. Lea and R. W. Nurse, J. of the Soc. of Chem. Ind., 58:277 (1939).
9. Symposium on New Methods for Particle Size Determination in the Subsieve Range. American Society for Testing Materials, Philadelphia, Pa. (1941).
10. "Determination of Particle Size in the Subsieve Range." British Colliery Owners Research Association and British Coal Utilization Research Association, London (1944).
11. G. L. Fairs, J. of the Soc. of Chem. Ind., 62(40):374 (1943).
12. P. G. W. Hawksley, British Coal Utilization Research Association, 8(9):245 (1944).
13. E. M. Chamot and C. W. Mason, Handbook of Chemical Microscopy, Vol. 1. John Wiley and Co., Ltd., London (1938).
14. P. R. Kalischer, Trans. of the Electrochem. Soc., 85:153 (1944).
15. G. Martin, C. E. Blythe, and H. Tongue, Trans. of the Ceram. Soc., 23:61 (1923-24).
16. H. Heywood, Proc. of the Inst. of Mech. Eng., 125:383 (1933).
17. R. H. Robertson and B. Emodi, Nature, 152:539 (1943).
18. H. A. Sloman, J. of the Iron and Steel Inst., I:298 P (1941).
19. H. A. Sloman, J. of the Iron and Steel Inst., II:235 P (1943).
20. H. A. Sloman, J. of the Inst. of Metals, 71:391 (1945).

21.   R. A. Fisher, Statistical Methods for Research Workers. Oliver and Boyd, Ltd., London (1944).

22.   C. A. Edwards and L. B. Pfeil, J. of the Iron and Steel Inst., II:79 (1925).

The Pressing and Sintering Properties of Iron Powders

G. Zapf

Sintermetallwerk Krebsöge GmbH
Krebsöge/Rheinland, Germany

I. Introduction

Raw materials for the manufacture of sintered bearings and structural parts are available in the form of a great variety of iron powders. In Europe alone there is a choice of some twenty different iron powders for powder-metallurgical purposes, and it is known that a number of new production processes are under development which will eventually lead to the introduction of further types of powders on the market.

The individual powders show considerable differences in chemical composition, particle size, particle-size distribution, and particle shape. According to the production process used, the powders may be grouped under four headings: (1) Reduced, (2) atomized, (3) comminuted, and (4) electrolytic.

Each of these groups includes powders of differing chemical composition, particle size, and particle shape.

Prices and methods for the production of iron powders are closely related, powders of high chemical purity being considerably more expensive than those containing permissible impurities. In each group fine powders are dearer than coarse ones.

Powders used in the manufacture of sintered bearings and structural parts of low and medium density belong mainly to group (1), (2), or (3). Powders in group (4) are used in general for the production of compacts demanding high density, high tensile strength, or special electrical and magnetic properties.

Today powder made by the reduction process is the raw material most widely used in the powder-metallurgical industry. Atomized powders are gaining ground rapidly, but comminuted powder has lost a considerable part of its importance. The growing use of electrolytic powder will, in all probability, be bound up with the expansion of markets for high-density parts having exceptional physical characteristics.

By reason of the different types of iron powders from which they are made, the physical properties of sintered materials tend to vary greatly, and the conditions of compacting and sintering have to be adapted accordingly. Satisfactory quality of the sintered product and economy in operation are dependent to a large extent on the choice of the most suitable raw material for a given purpose.

Since its earliest days, the metal-powder industry has spared neither time nor effort in finding a solution to the problems involved in testing iron powders, and a great number of papers and articles on the subject are to be found in the technical literature and in the sales publications of the industry.

In evaluating the possible use of a given metal powder for powder-metallurgical purposes, care must be taken to distinguish between those characteristics which tend to influence compacting and sintering conditions, e.g., the pressing operation or tool design, and those which affect sintering properties and thus the physical values of the finished product. Some characteristics may have a beneficial effect in both these directions, whereas others may have a favorable influence on compaction, but a detrimental effect on sintering, or vice versa. The selection of a given powder and of suitable compacting and sintering conditions must, therefore, be based on the most careful testing and assessment of the factors involved.

II. Determination of Standard Powder
Characteristics

Table 1 lists the powder characteristics normally subject to test. Columns (3) and (4) show whether the individual characteristic tends to influence the pressing process or the sintering process, or both.

In some countries definite standards for testing these characteristics have either been issued or are in the course of preparation, e.g., the Metal Powder Association (M.P.A) standards in the U.S.A., British Standard No. 3029, and the Prüfblätter of the Verein Deutscher Eisenhüttenleute in Germany.

In practice, particular importance is attached to the test for compressibility and sinterability. Methods designed to comply with the required physical values are governed by these characteristics to a far greater extent than by all the other relevant factors.

In the present paper it is intended to summarize the experience and data accumulated during the course of a series of test conducted in the powder metallurgical laboratories of Husqvarna Vapenfabriks AB, Sweden, and of Sintermetallwerk Krebsöge GmbH, Germany, on a number of different commercial iron powders, with a view to their possible use in the manufacture of sintered products. The test data quoted were collected between 1950 and 1960, and it must be pointed out that in the case of a number of the powders tested, technological developments have led to a certain changes in characteristics and to an adaptation of these characteristics to the requirements of practical operation. As a result, some of the powders, e.g., HVA Crown and Simetag RZ 100, are no longer commercially available in the particle shape and particle-size distribution tested.

The test procedure for determining powder characteristics 1, 2, 3, 5, 6, and 9 (Table 1) was in strict conformity with those laid down in M.P.A. standards.

To test the compressibility, a cylindrical die set with surface area of 3 cm^2 was employed. On plotting the measured values for pressed density against the specific pressure, a diagram of the type shown in Fig. 1 is obtained [9]. This figure shows clearly that the relationship between specific pressure and pressed density

TABLE 1. Types of Test Carried Out on Metal Powders

Test No.	Type of Test	Influence on:		Ref. No.
		Sintering Process	Pressing Process	
1	Chemical analysis	×	×	1
2	Hydrogen loss	×	×	2
3	Screen analysis	×	×	3
4	Particle shape	×	×	...
5	Flow characteristics	...	×	4
6	Apparent density	...	×	5
7	Compressibility	...	×	6
8	Sinterability	×	...	7
9	Green strength	...	×	8

is not linear. When plotted by means of a system of Cartesian co-ordinates, the curves show initially a steeply rising trend, leveling off progressively later in close proportion to the increasing specific pressure, and eventually approaching asymptotically a final value which varies for each individual powder and, in the case of nearly all the powders, is lower than the absolute density of iron.

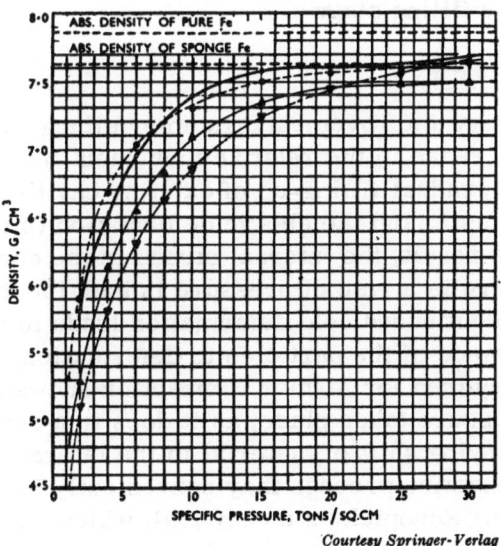

Courtesy Springer-Verlag

Fig. 1. Effect of specific pressure on the pressed density of various iron powders.

Powder	Powder
——O—— Hametag	—▲—Sponge iron
—●—— RZ	—·—▼—·—DPG shotted

This behavior is due to the fact that the compacting process for metal powder is divided into several consecutive stages:

1. Filling

When the die has been filled, the degree of compaction of the individual powder particles is low. The filling will be less complete in proportion as the powder used is of small particle size and of irregular particle shape. As a result of the pressure applied, bridges formed in the course of filling the die will break down, and greater compactness will be attained as a result of the ensuing shift of powder particles and the filling up of existing cavities.

The principal object of compacting pressure at this stage is to overcome the internal friction between the particles and the friction between the die wall and the compact. The result is both a lateral distribution of pressure and a lateral transfer of material. The pressure distribution may properly be described as hydrostatic. Relatively low specific pressure will be required before the onset of deformation. This part of the compacting process may be described as the filling stage.

2. Deformation

At this stage improvement in Raumerfüllung (R)* leads to a progressive decrease in the relative movement between the powder particles. The gradual filling-up of existing cavities is dependent upon both the plastic and the elastic deformation of the individual particles. As a result, the internal stress in the compact is increased, and there is a rise of pressure in both the die wall and the compact. The continuous cold deformation leads to a steadily increasing resistance of the particles to deformation, and the specific pressure required for further compaction shows a rapid rise. Transfer of pressure takes place, as it does in a rigid body, without an accompanying shift of material in the lateral direction. This part of the compaction process may conveniently be termed the deformation stage. Konopicky's theory [9a], which will be discussed later, on the interrelation between specific pressure and pressed density, would appear to justify the conclusion that, starting at the filling stage with a value of R in the die of between 25 and 43% (the precise percentage being governed by the powder used), a value may be reached of approximately 60%. In general the deformation

---

* R = (Pressed density/Absolute density) x 100.

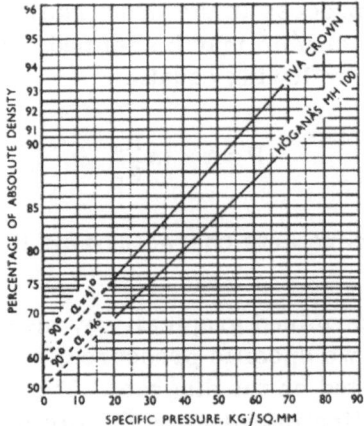

Fig. 2. Compressibility diagram
(specific pressure plotted against percentage of absolute density) for
HVA Crown and Höganäs MH 100
powders.

	R_0, %	A, kg/mm^2
HVA Crown	60	39.2
Höganäs MH 100	52	46.7

stage starts at a value of R = 50%, and it is, therefore, evident that
at a value of R of between 50 and 60% simultaneous filling and deformation may be taking place.

Comprehensive tests have been carried out and a number of
papers have been written on the relationship between specific pressure and pressed density. Ready access to most of this information
can be had in the literature. Excellent review articles have been
published by Kieffer and Hotop [10] and by Goetzel [11].

The determination of compressibility can be considerably
simplified by the use of logarithmic plots of the specific pressure
vs. pressed density relationship (see Fig. 2).

Values of R can be determined from the pressed density by
means of the equation already quoted.*

Figure 2 is a diagram of this kind and shows the specific
pressure vs pressed-density curves for two separate iron powders.
The diagram is based on the theory of Konopicky, who has shown

---

*See footnote on previous page.

that the compaction process in the industrially important pressure range of 2-8 metric tons/in^2 may be defined with sufficient accuracy by means of the following equation:

$$P = A \times \ln\frac{R_0}{R_p}$$

where A and R_0 are constants characterizing the pressing properties of the powder.

Values are determined grapically by means of the diagram. R_p is measured for two specific pressures, e.g., 2 and 5 metric tons/cm^2, and is plotted on the diagram. The two values are connected by a straight line intersecting the ordinate, the point of intersection giving the R_0 value. This value indicates the degree of compaction that can be achieved without applying appreciable pressure. The gradient of the straight line in relation to the ordinate shows the progressive increase in density with specific pressure. The steeper the gradient, the smaller will be tan $(90° - \alpha)$ and the better the compressibility.

The value of A can also be determined on the basis of the diagram the scale of which has been so chosen that a straight line tan $(90° - \alpha) = 1$ will be equivalent to an A value of 45 kg/mm^2. The A value applicable to any given powder with an angle of $(90° - \alpha) = x$ can be computed by the following equation:

$$A = 45 \times \tan x \text{ kg/mm}^2$$

For the example shown in Fig. 2, the values for the two powders concerned are as follows:

	R_0, %	tan $(90° - \alpha)$	A, kg/mm^2
HVA Crown	60	0.869	$45 \times 0.869 = 39.2$
Höganäs MH 100	52	0.039	$45 \times 1.036 = 46.7$

The compressibility of any given powder will be better the higher is R_0 and the lower is A.

The sinterability of the individual powders was determined by assessing the tensile strength of test bars which had been pressed under a given specific pressure and sintered under given sintering conditions. A linear relationship exists between pressed density and tensile strength for each powder and for each set of sintering conditions.

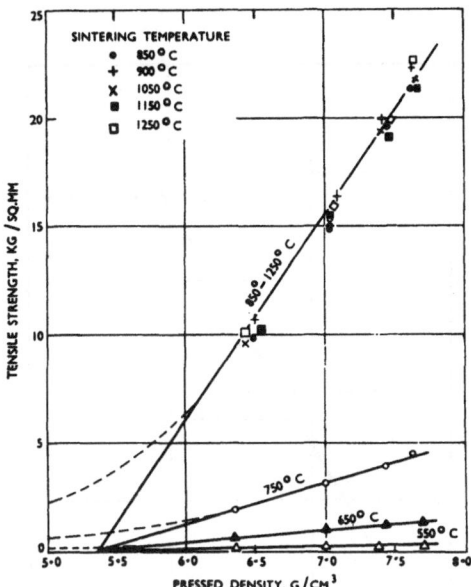

Fig. 3. Effect of pressed density on tensile strength for various sintering temperatures. M.P.A. tensile testbar.

In Fig. 3 the pressed density versus tensile strength is plotted for the sintering of an electrolytic iron powder at a number of different temperatures.

Bockstiegel [12] has pointed out that, irrespective of the sintering temperature employed, the curves for pressed density plotted against tensile strength will always cut the abscissa at a density of 5.25 g/cc, i.e., two-thirds of the theoretical density. He recommends that this fact should be used as a basis for the simplified determination of the sinterability coefficient b, which is a measure of the ratio between sintered density and tensile strength:

$$b = \frac{\text{Tensile strength}}{\text{Sintered density}} - 5.25 \ \frac{\text{kg}}{\text{mm}^2} \bigg/ \frac{\text{g}}{\text{cc}}$$

As shown in Fig. 4, only one test value, which can be determined by examination of between 5 and 10 tensile test bars, will be required.

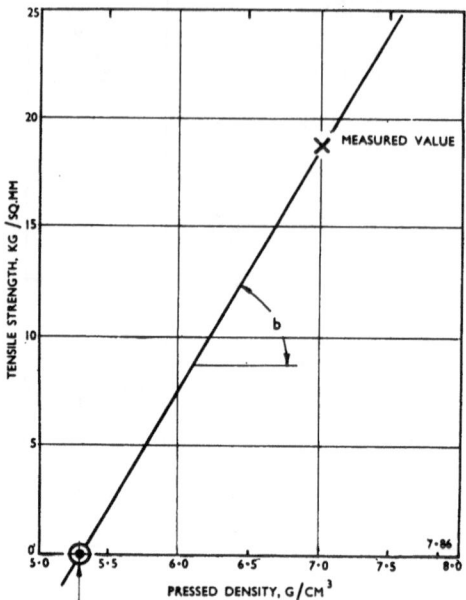

Fig. 4. Illustrating method of determining the
sinterability coefficient b.

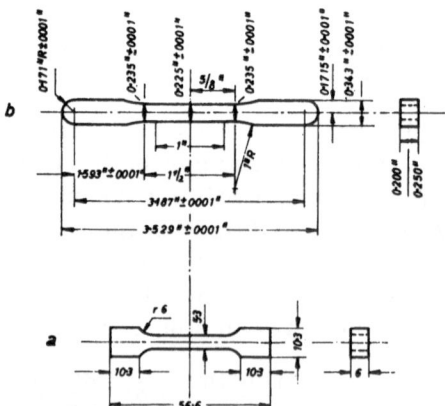

Fig. 5. Form and dimensions of test bars for
determining sinterability.

When the double pressing and sintering technique is used, there will not be a unique intersection of the curve with the abscissa, and it will therefore be necessary to establish two test values for the determination of the sintering behavior. This method is well suited to provide a simple test of the sinterability of iron powders.

To determine the physical characteristics of the powders use was made in general of the procedures described in the M.P.A. standards. However, in a number of individual test series, which are clearly indicated, test bars were used for determining tensile values which differed from the relevant M.P.A. standard bars. The two bars in question are illustrated in Fig. 5. Test bar a is a development of the round tensile test bar as standardized in Germany, and is of identical length and cross section. It is thus possible to use this bar for testing elongation for L = 5d, in conformity with the prescribed German test procedure for round bars. Test bar b conforms to the American M.P.A. standards.

III. Influence of Powder Characteristics on Pressing and Sintering Properties

The procedures described above were used to test a great number of different powders, and an attempt was made to derive from the data so compiled conclusions as to the extent of the influence of each of the powder characteristics on the following properties: (a) Filling behavior, (b) pressing behavior, (c) green strength, and (d) sintering behavior.

1. Chemical Composition

With the exception of the electrolytic powders, all commercial iron powders made either by reduction or by atomization contain, in addition to between 0.3 and 1% oxygen, other extraneous matter not exceeding 0.6%, which is present in the powder either in the form of alloying constituents or oxides. These substances excercise a considerable influence on the pressing and sintering properties of the powder.

(a) Filling Behavior

It is very unlikely that chemical composition will influence the filling behavior of commercial iron powders. In any event, however, such an effect would be so greatly overshadowed by the physical properties of the powder agglomerate (particle shape, particle size, and particle-size distribution) that a more detailed examination of this problem has not been considered necessary.

(b) Pressing Behavior

In the case of pressing behavior also there is a combined effect of chemical composition of the powder particles and physical characteristics of the powder agglomerate used for compaction. There is definite evidence, however, both practical and theoretical, of a relationship between pressing behavior and the chemical composition of the powder.

All alloying constituents tend to raise the compression strength of iron and to increase the resistance of the individual particles to plastic deformation. Their deleterious influence on compressibility is therefore evident. Even the small percentage of alloying contituents and nonmetallic admixtures present in commercial iron powders is sufficient to exert this influence, and the compressibility of powders made either by reduction or atomization is therefore less favorable than that of electrolytic powders, which are of higher chemical purity. Evidence of this is given in Fig. 6,

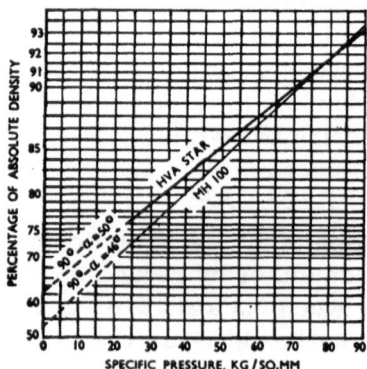

Fig. 6. Compressibility diagram for two sponge iron powders of different chemical composition.

	R_0, %	A, kg/mm²
HVA Star	62	53.55
Höganäs MH 100	52	46.7

TABLE 2. Physical Properties and Chemical Analyses
of Various Iron Powders

	HVA Crown	HVA Sponge	HVA Standard	Höganäs MH 100	Simetag RZ 100
Screen Analysis: (mesh)					
+ 48	...	...	...	...	...
+ 65	...	...	0·2	...	...
+100	1·3	0·1	12·4	1·1	0·1
+150	13·2	6·3	17·1	27·8	20·0
+200	22·8	13·3	19·2	26·0	31·8
+270	25·3	25·1	17·1	19·4	25·9
−270	37·4	55·2	34·0	25·7	22·2
Apparent Density, g/cm³	2·60	2·80	3·43	2·46	2·60
Flow Rate, sec/50 g	30·2	28·8	24·9	37·2	32·4
Rattle Test,* %	0·12	0·15	0·46	0·19	0·18
Loss on Heating in Hydrogen, %	0·22	0·38	0·23	0·65	0·46
Composition:					
Carbon, %	Max. 0·05			0·1	0·09
Silicon, %	,, 0·02			0·2 (SiO₂)	0·03
Manganese, %	,, 0·06			...	0·20
Copper, %	,, 0·01			...	...
Sulphur, %	,, 0·001			0·015	0·03
Phosphorus, %	,, 0·01			0·015	0·03
Chloride, %	,, 0·1			0·015	0·03
Other oxides and silicates				0·25	0·08

*See M.P.A. Standard No. 15-51.

where a comparison is made between two sponge-iron powders
(Höganäs MH 100 and HVA Star) of as similar a particle shape,
particle-size distribution, and apparent density as can be attained
in any two powders made by different production processes. Their
physical properties and chemical analyses are given in Table 2.
Figure 6 shows the compressibility curves for the two powders.

The R_0 value for the electrolytic iron powder is some 10%
higher than that of its counterpart. Even though the A value of the
MH 100 powder is somewhat more favorable, it will be noted that
the point of intersection of the two compression curves lies in the
vicinity of a specific pressure of 80 kg/mm², i.e., in a range no
longer of practical interest for industrial purposes.

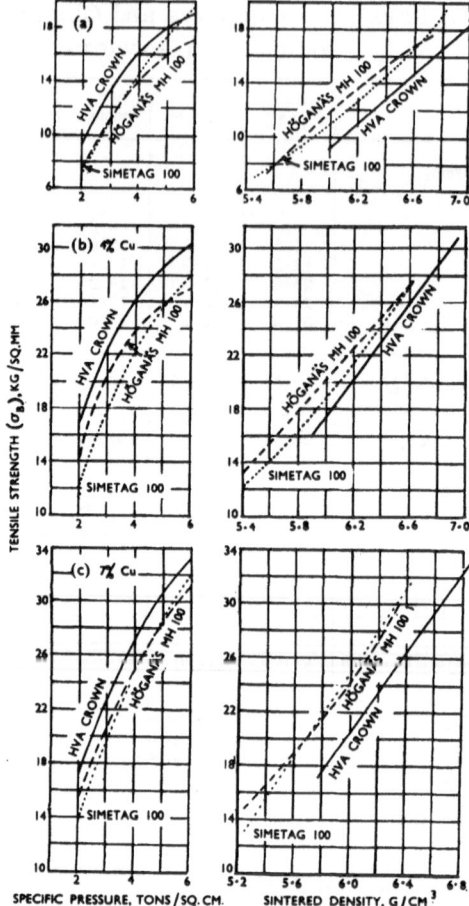

Fig. 7. Tensile strength of samples of three
different iron powders, with and without additions of copper, in relation to specific
pressure and sintered density. (a) No copper;
(b) 4% Cu; (c) 7% Cu.

(c) Green Strength

This is governed by considerations similar to those elabora-
ted under Section III. 1 (a).

(d) Sintering Behavior

The presence of alloying constituents in iron powders tends
to delay diffusion and, therefore, constitutes an obstacle to sinter-

ing. In particular, constituents which oxidize may be the cause of oxide-film formation and so impair the physical qualities of the sintered compacts.

Compacts made from sintered iron alloy, on the other hand, possess, at the same density, better physical characteristics than those made from unalloyed iron, and may thus, in many cases, be a better choice. When similar sintering conditions are employed, even the small percentage of alloying constituents contained in reduced and atomized powders will lead, at an identical density, to higher physical characteristics than those obtained with electrolytic powders.

The practical effects of a low content of alloying constituents are illustrated in Fig. 7 with respect to HVA Crown (electrolytic), Höganäs MH 100 (reduced), and Simetag RZ 100 (atomized) powder. The chemical analyses of the three powders are given in Table 2.

In Fig. 7(a) the tensile strength of test bars made from the three basic powders, without the addition of copper, is shown as a function of specific pressure and of sintered density. The test bars were sintered for 1 h at 1120°C.

Because of its better pressing properties, the electrolytic-iron powder, under identical pressing conditions, shows a higher strength than that of the other two powders. However, when strength is plotted as a function of density, the two powders not made by the electrolytic process show a higher tensile strength at an identical density, on account of the admixtures which they contain.

TABLE 3. Influence of Annealing Temperature on the Flow Rate of a Number of Different Iron Powders

Condition	Flow Rate, sec/100 g		
	Hametag Powder	Sponge-Iron Powder	DPG Shotted Powder
As delivered	25·8	34·0	18·0
Annealed at:			
600°C	25·8	31·1	17·4
700°C	30·0	29·0	17·0
800°C	31·2	29·0	17·0
900°C	33·0	28·0	19·2
1000°C	34·0	27·5	19·2

Figure 7(b) relates to powders containing 4% copper and
Fig. 7(c) to powders with 7% copper. These figures also serve
to show that under the same specific pressure, compacts of higher
tensile strength can be obtained from electrolytic iron powder.
When considering strength in relation to density, however, better
physical characteristics can be obtained with the two powders made
by the nonelectrolytic processes.

2. Loss on Heating in Hydrogen

In commercial iron powders the content of constituents
prone to volatilization by hydrogen reduction fluctuates, in general,
between 0.3 and 0.8%, and it is standard practice to process powders in this condition. By subjecting the powder to further annealing in hydrogen the losses may be reduced still further and the
processing properties of the powder consequently improved. This
treatment is expensive, however, and therefore not commonly applied. It is important to remember the considerable influence of
losses in hydrogen on the pressing and sintering properties of the
powders.

(a) Filling Behavior

Within normal production tolerances, the apparent density of
iron powder is not affected by losses in hydrogen; the flow rate,
on the other hand, is detrimentally influenced by an increased
oxygen content, and more time will be required for filling the dies.
A good example of this effect is cited by Kieffer and Hotop [13] and
the pertinent data are given in Table 3.

(b) Pressing Behavior

Formation of oxide skin and oxide inclusions tends to lower
the plasticity of the powder particles and thus to have an adverse
effect on the pressing properties. The influence of annealing
losses on compressibility is illustrated in Fig. 8 with reference
to powder RZ 150, both as delivered and after a further annealing
treatment in cracked ammonia, as a result of which the R_0 value
is raised from 48 to 52%, and there is also a slight increase in the
A value. However, within the whole industrial pressure range the
compression values for the annealed powder do not reach the level
for powder that has not been annealed.

Another example of the effect of hydrogen loss on the
pressing properties of iron powder is shown in Table 4, which

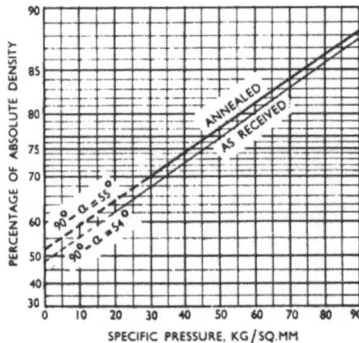

Fig. 8. Compressibility diagram for
Simetag RZ 150 powder as received
and after annealing at 850°C.

	R_0, %	A, kg/mm^2
As received	48	61.65
Annealed	52	64.3

gives the pressed-density values for compacts made from three
different HVA sponge-iron powders with differing hydrogen losses,
i.e., 0.05, 0.4, and 0.88%, respectively. The progressive drop in
the pressed density with rising hydrogen loss is clearly apparent.

(c) Green Strength

No tests were made.

(d) Sintering Behavior

The sintering behavior of iron powders is strongly influenced
by their oxygen content. Depending upon the form the oxide takes,
the result may be either to promote or to impede sintering.

TABLE 4. Pressed Density of Test Bars
Made from Three HVA Electrolytic
Sponge-Iron Powders of Differing Hydro-
gen Loss

Specific Pressure, metric tons/cm²	Density, g/cm³		
	Annealing Loss in Hydrogen, 0·05%	Annealing Loss in Hydrogen, 0·4%	Annealing Loss in Hydrogen, 0·88%
2	5·96	5·91	5·79
4	6·80	6·70	6·56
6	7·17	7·15	6·95
8	7·39	7·37	...

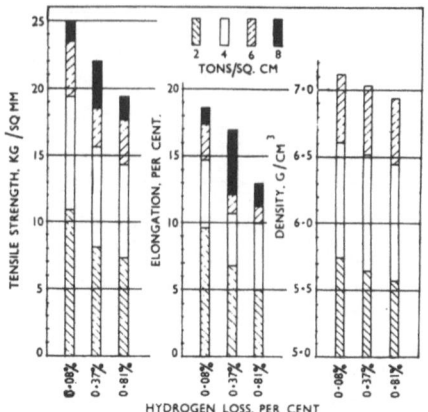

Fig. 9. Tensile strength (σ_B), elongation (δ_5), and sintered density (γ) of test bars made from electrolytic iron powder with differing hydrogen loss.

Wiemer and Hanebuth [14] have shown, for example, that during sintering pre-oxidized compacts attain a greater tensile strength than those not subjected to such treatment. Zapf [15], on the other hand, has reported that oxidized iron powders exhibited less favorable sintering properties than powders with no oxygen-bearing constituents. These conflicting observations may well be interpreted to mean that formation of oxide skin on powder particles tends to prevent metal-to-metal contact during pressing and thereby to reduce the number of points of contact at which sintering can start; but that oxidation and reduction of powder particles after compaction cannot affect existing metallic bridges and can only tend to activate the remainder of the particle surface.

An example of the detrimental effect of increased hydrogen loss is given in Fig. 9, which shows the tensile strength, elongation, and sintered density of test bars made from three HVA electrolytic iron powders with respective hydrogen losses of 0.08, 0.37, and 0.81%. The specific pressure applied can be seen from the diagram. The bars were sintered for 1 h at 1200°C. It is clear that the improvement in physical properties is directly related to the decrease in hydrogen loss.

3. Sieve Analysis

The majority of commercial iron powders are marketed in a number of different degrees of fineness. There is, for example, no difference in chemical composition or particle shape between the well-known powders MH 300, MH 100, and MH 60, but there are considerable differences in both sieve analysis and average particle size of the three powders. Similarly differences exist between RZ 150 and RZ 400. Average sieve analyses for these two makes of powder are shown in Table 5. The influence of particle-size distribution on the filling, pressing, and sintering properties of the powders is considerable.

(a) Filling Behavior

The flow rate of a fine powder is, in general, less favorable than that of a coarse one. Kieffer and Hotop [16] cite an example (Table 6). The filling of the die is therefore frequently very difficult, particularly in the case of tools with narrow, deep cavities. Fine powders in general are of lower apparent density and therefore require die assemblies with larger cavities. For these reasons, it is evident that all the disadvantages will be found to apply in the case of fine powders which are discussed in detail in Section III, 6 with reference to powders of low apparent density.

TABLE 5. Sieve Analyses of Various
Iron Powders

Sieve Fraction, mm	Höganäs MH 300	Höganäs MH 100	RZ 150	RZ 400
	%	%	%	%
+0·3	...	...	...	6·82
+0·2	...	0·1	0·08	15·54
+0·15	...	0·8	0·2	19·98
+0·10	...	26·52	12·64	25·08
+0·075	0·14	27·02	26·14	10·04
+0·06	9·02	13·02	12·5	9·74
−0·06	90·24	31·94	48·2	12·26

TABLE 6. Effect of Particle Size on the Flow
Rate of Various Iron Powders

Particle-Size Range, mm	Flow Rate, sec/100 g			
	Hametag Powder		Sponge Iron Powder	DPG Shotted Powder
	Unannealed	Annealed		
0·30–0·15	24·5	27·5	27·0	21·0
0·15–0·06	37·0	45·5	25·5	17·7
0·06	No flow	No flow	34·5	19·2

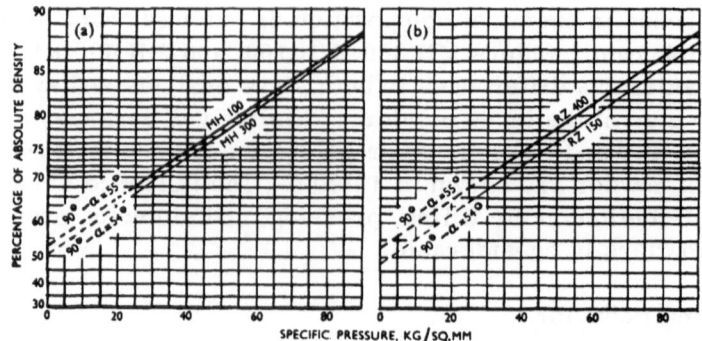

Fig. 10. Compressibility diagrams for: (a) two Höganäs MH powders;
(b) two Mannesmann RZ powders as received.

	R_0	A, kg/mm^2
MH 300	51	61.65
MH 100	53	64.35
RZ 150	48	61.65
RZ 400	53	64.35

There is, in addition, in the case of powders with a high flow-ability a tendency for powder particles to find their way into the spaces between the die and the punch; this may give rise to sticking between the individual members of the die assembly and impede working.

(b) Pressing Behavior

The pressing properties of fine powders tend to be less favourable than those of powder with large particles. This is true of all types of iron powder. Figure 10 shows compressibility diagrams for (a) MH 300 and MH 100, and (b) for RZ 150 and RZ 400 powders. In both cases the powder with the larger particle size gives the better results.

Figure 11 shows the compressibility of the individual screen fractions of (a) MH 100, and (b) HVA Standard powders. Here, too, there is clear evidence of the superior pressing properties of the larger fractions.

Conditions encountered with powder RZ 150 are not quite so predictable. Figure 12 shows the compressibility diagram for the powder as received and for three different fractions of it. The original powder has the lowest and the fine fraction the highest compressibility.

(c) Green Strength

No quantitative determinations were made of the relationship between green strength and particle size.

(d) Sintering Behavior

The sintering properties of fine powders are superior to those of powders of large particle size. The physical properties of test bars made from fine powders under identical specific pressure and the same processing conditions are, therefore, in spite of a lower density, better than those attainable with powders of large particle size. Examples are shown in Table 7, which gives the physical properties of test bars made in a single pressing and sintering operation from electrolytic iron powders of various particle sizes. The bars were sintered for 1 h at 1120°C. In spite of a lower sintered density, the finer powder shows in each case the higher physical properties.

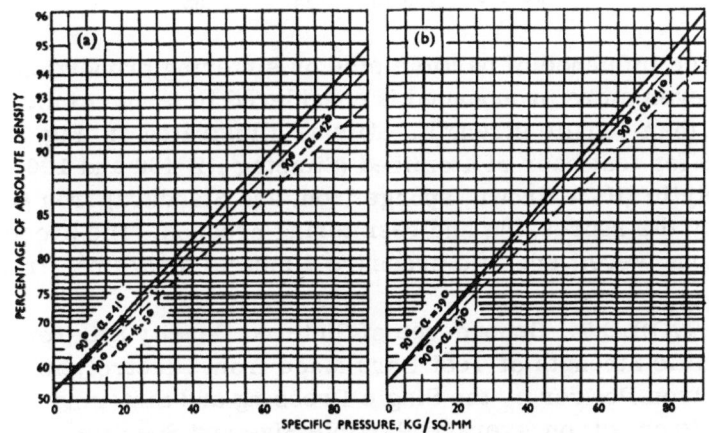

Fig. 11. Compressibility diagrams for screen fractions of: (a) Höganäs MH 100; (b) HVA Standard powder.

	Mesh	Particle Size, μ	R_0, %	A, kg/mm
		Höganäs MH 100		
———	-100 + 150	-147 + 104	52	39.15
—·—	-150 + 270	-104 + 53	52	40.5
———	270	-53	52	45.8
		HVA Standard		
———	-100 + 150	-147 + 104	55	36
—·—	-150 + 270	-104 + 53	55	38.7
———	-270	-53	55	41.85

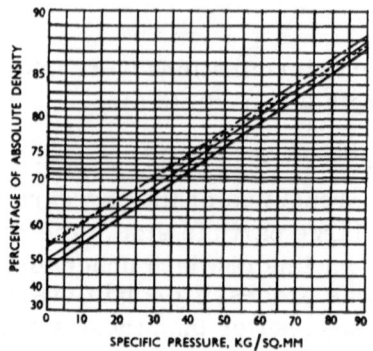

Fig. 12. Compressibility diagram for
screen fractions of powder RZ 150.

Particle Size, mm	R_0	A, kg/mm²
——— As received	47	61.65
——— < 0.06	50	64.35
—·— 0.06-0.1	53	64.35
———— 0.1-0.15	54	64.35

The influence of particle size on sintering properties becomes even more pronounced if considered in terms of elongation
rather than tensile strength. This is illustrated in Fig. 13 and
Table 8 for test basr of atomized (Simetag RZ 400 and 100) and
electrolytic (HVA Crown, sponge iron, and Standard) iron powder
made by the repressing and resintering process. The processing
conditions were: specific pressure 4 metric tons/cm², first sin-

TABLE 7. Dependence of Sintered Density and Ten-
sile Strength of Test Bars Made from Electrolytic
Iron Powder by the Single Pressing and Sintering
Process, on Average Particle Size and Specific
Pressure

Powder	Average Particle Size, μ	Specific Pressure, metric tons/cm²			Specific Pressure, metric tons/cm²		
		2	4	6	2	4	6
		Density, g/cm²			Tensile Strength, kg/mm²		
HVA Sponge	59	5·76	6·59	6·96	8·4	16·7	19·7
HVA Standard	90	5·89	6·64	6·98	5·2	10·7	14·0
Cohen Sintrex 200	61	5·93	6·67	7·04	8·2	15·0	18·9
Cohen Sintrex 100	85	5·99	6·72	7·06	7·3	13·0	15·1

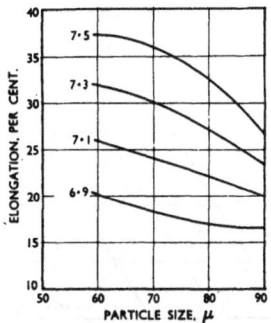

Fig. 13. Relationship between elongation and particle size in test bars made from HVA sponge-iron powder by the double pressing and sintering process. Density (g/cc) indicated on curves.

tering temperature 800°C, first sintering time 1 h, resintering temperature 1200°C, resintering time 1 h.

In Fig. 13 elongation is shown graphically as a function of sintered density and mean particle size for four density values: 7.5, 7.3, 7.1 and 6.9 g/cc. Table 8 contains data illustrating the relationship between elongation and density at a number of different densities and mean particle sizes. The values quoted show an unmistakable connection between particle size and elongation. In assessing the absolute values, however, it must be remembered that the particle shape of the three electrolytic iron powders was

TABLE 8. Relationship between Elongation, Resintered Density, and Particle Size in Test Bars Made from Various Iron Powders by the Repressing and Resintering Process

owder	Average Particle Size, μ	Density, g/cm³			
		6·9	7·1	7·3	7·5
		Elongation, %			
HVA Sponge	59	20·6	26·0	31·8	37·4
HVA Crown	75	17·4	23·2	28·9	34·6
HVA Standard	90	16·3	19·8	23·2	26·6
		Density, g/cm³			
		6·6	6·8	7·0	7·2
		Elongation, %			
Simetag RZ 100	90	10·0	12·5	15·1	17·7
Simetag RZ 400	220	9·0	9·8	11·7	13·5

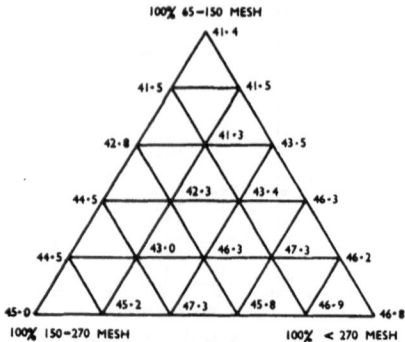

Fig. 14. Relationship between elongation and particle size for type a test
bars made from electrolytic iron powder by the double pressing and sintering
process.

not identical, and that to some extent the test data, therefore, reflect the influence of this factor. It would be necessary to carry
out further tests to be able to appraise the two factors separately.

The influence of particle size on elongation is also shown in
Fig. 14, which is based on tests made on samples of 21 mixtures of
electrolytic iron powder of different particle sizes. The powder
was separated into three sieve fractions: 65-150, 150-270, and
270- mesh. In addition to the pure fractions, 12 binary and 6
ternary mixtures of fractions were also made into tensile test bars.
The processing conditions were: pressed first at 6 metric tons/cm^2;
sintered first at 1050°C for 1 h; repressed at 8 tons/cm^2; resintered at 1250°C for $2^1/_2$ h; atmosphere dried hydrogen. The
figure shows clearly how the progressive rise in elongation values
is related to the increasing content of fines.

The discussion of the problem of the particle-size effect has
shown the necessity in practice of arriving at a compromise between the larger particle size which is desirable on account of the
good pressing properties it confers and the smaller particle size
which gives the advantage of better sintering properties. It must
be remembered, however, that fine powder is more expensive on
account of higher production and handling costs.

It is more than doubtful whether the commercial iron powders
available today offer a real solution to the problem of the best

particle-size distribution. Further tests in this particular field, closely related to the experience gained in practical operation, seem to be highly desirable.

4. Particle Shape

There are considerable variations in the particle shape of commercial iron powders. Reduced powders are microporous and of highly irregular shape. In the case of atomized powders, both the surface tension of the liquid metal and the atomizing conditions used are decisive factors in determining whether the powder obtained will be of irregular shape with a large surface area, or will comprise more compact particles of spherical form. For electrolytic powders particle shape is governed both by the conditions of electrolysis and by the milling and annealing procedure employed for converting the cathode deposit into powder. It is possible to produce by this method dendritic, compact, flake, or sponge powder.

A large body of qualitative test data tends to show that the influence of particle shape on the pressing and sintering properties of iron powders is a factor of considerable importance. However, in the test data available there is frequent superposition of the in-

TABLE 9. Particle Shape and Green Strength
of Various Commercial Iron Powders

Powder	Average Particle Size, μ	Apparent Density, g/cm³	Rattle Test,* %	Particle Shape
Höganäs MH 100	90	2·46	0·19	Microporous, spongy
Höganäs MH 300	60	2·45	0·26	,, ,,
Simetag RZ 100	90	2·60	0·18	Irregular
Simetag RZ 150	70	2·46	0·16	,,
Simetag RZ 400	180	2·46	0·23	,,
HVA Crown	75	2·60	0·12	Flaky
HVA Star Sponge	64	2·80	0·15	Spongy
HVA Star < 0·06 mm	60	2·80	0·14	,,
HVA Standard	100	3·42	0·46	Irregular, compact
Merisinter Standard	95	3·41	0·38	Irregular, compact
Cohen Sintrex	75	3·16	0·65	Dendritic

*See M.P.A. Standard No. 15-51.

fluence of particle size and shape, and it is therefore not possible
to draw definite quantitative conclusions.

(a) Filling Behavior

A microporous and irregular particle shape results, in general, in a powder of low apparent density. A good example of this
can be found in the low apparent density of the Höganäs powders
and of the HVA sponge-iron powder. Compact powder particles,
on the other hand, form the basis of powders of high apparent
density, e.g., Hametag Eddy-Milled and HVA Standard powder.

The apparent densities of a number of different powders are
given in Table 9, which also indicates the respective particle
shape.

(b) Pressing Behavior

The compressibility of microporous and irregular powders
is less favorable than that of compact and flaky types.

(c) Green Strength

The green strength of compacts made from bulky and irregular powder is considerably higher than in the case of compact powders. Data illustrating this point are included in Table 9.

(d) Sintering Behavior

The test data available do not permit definite quantitative
conclusions to be drawn as to the influence of particle shape on the
sinterability of powders. This is because of the difficulty in
separating the effects of chemical composition, particle size, and
particle shape. It is certain, however, that better sinterability can
be obtained with bulky and irregular powder particles than with
powders of smooth and globular shape. The excellent sintering behavior of HVA Crown powder is, however, in contradiction to general practical experience. It would therefore appear advisable to
investigate further the influence of particle shape.

5. Flowability

The influence of the flowability of any powder is limited to its
filling behavior. The flowability is decisive as regards the time

required for filling a tool, and in particular as regards the degree of packing that can be obtained in narrow cavities.

It is, therefore, important to select powders with the best possible flow characteristics. A satisfactory flow rate reduces the time required to fill the die and thus tends to increase the output of any individual press. The higher the flow of a given powder, the less may be the depth of the die cavity. The apparent density of the powder poured into the die is increased and approaches the values determined by standard tests.

An exceptionally high flow rate may, however, lead to certain disadvantages in pouring powders with a large content of fines. This is true, in particular, if the powder contains any particles of dimensions smaller than the space between the individual components of the die assembly, e.g., the punch and die. Should this be the case, the powder may find its way into the crevices and interfere with the operation of the tool by sticking either to the punch or to the die. This defect can be remedied by mixing with the powder a lubricant designed to reduce the flow rate. The lubrication method described by Ljungberg and Arbstedt[17] has proved very suitable for this purpose.

6. Apparent Density

The apparent density of any given powder is a vital factor in determining the dimensions of the cavities in tools designed for pressing metal powders. Iron powders exhibit considerable variations in this respect. Figure 15 shows three containers holding 100 g each of different iron powders: (a) Höganäs MH 100, apparent density 2.5 g/cc; (b) HVA sponge-iron electrolytic powder,

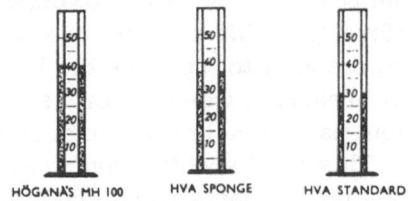

HÖGANÄS MH 100   HVA SPONGE   HVA STANDARD

Fig. 15. Volume occupied by 100 g of each of the following powders: (a) Höganäs MH 100; (b) HVA Sponge; (c) HVA Standard.

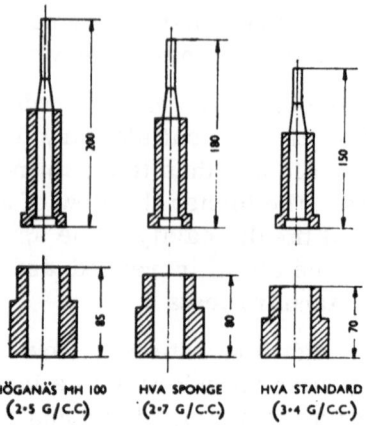

HÖGANÄS MH 100   HVA SPONGE   HVA STANDARD
(2·5 G/C.C)   (2·7 G/C.C.)   (3·4 G/C.C.)

Fig. 16. Length of tool segments (mm) required for processing three iron powders of different apparent densities.

apparent density 2.7 g/cc; (c) HVA Standard electrolytic powder, apparent density 3.4 g/cc. The differences in apparent density of the three powders are clearly visible.

It is only natural that these differences should constitute a factor of considerable importance both in the design of the tools and in the determination of the length of the stroke required for compaction. Figure 16 shows the movable punch and the fixed die, as well as the die insert, for pressing a structural part. Indicated on the figure are the respective dimensions of the individual components required to compact each of the three powders shown in Fig. 15, i.e., with apparent densities of 2.5, 2.7, and 3.4 g/cc, respectively. It will be noted that the length of these tool components is reduced progressively in proportion to the rise in the apparent density of the powder. By using powders of higher apparent density it is, therefore, possible not only to save material and time in tool construction, but also to increase the life of the tools and to diminish the risk of breakage, which otherwise arises as a result of axial compression, particularly when high specific pressures have to be applied. This is of special advantage when the production of high-density sintered parts requiring high specific pressure is involved. It must be emphasized, however, that in general powders of high apparent density are of coarser particle size and thus have less favorable sintering properties. As a result, the tensile strength and elongation values of structural parts made from these powders will be lower, particularly when only a single sintering process is employed.

The relationship between density and the length of stroke required for compaction is illustrated in Table 10, which shows that a considerably shorter stroke will be adequate when a powder of high apparent density has to be processed. This is of particular advantage in press design when parts of considerable height are to be compacted.

Powders of high apparent density, however, possess a certain disadvantage in the compaction process, since considerably more care is needed in setting the tools and presses than when powders of lower apparent density are to be used. With the latter type there is a considerably longer compacting action before the onset of plastic deformation, and the lateral transfer of material is continued during a greater part of the stroke. This, again, tends to balance any fluctuations in the density caused by inaccurate setting of the tools. When the design of the tools and presses available permits, this represents a very great advantage in practice, particularly when parts of low individual height and of low pressed density are to be produced.

To summarize, it may be said that (1) powders of high apparent density are to be preferred in all cases where the manufacture of parts of high density and considerable individual height is involved, (2) powders of high apparent density require greater accuracy in setting the tools, (3) powders of low apparent density

TABLE 10. Relationship Between Length of Stroke and Apparent Density at a Pressed Density of 6.8 g/cc

Height of Compact, mm	Apparent Density, g/cm^2		
	2·5	2·7	3·4
	Length of Stroke Required, cm		
10	17	15	10
20	34	30	20
30	51	45	30
40	68	60	40
50	85	75	50

make possible a better balancing of density in the compaction process, and thus a lesser degree of accuracy can be tolerated in the
setting of the tools, and that (4) powders of low apparent density
require tools of greater height and a greater length of stroke.
They are thus less suited for making parts of high density and
considerable individual height.

7. Compressibility

It has been pointed out in discussing chemical composition
(Section III, 1) and annealing loss (Section III, 2) that compressibility is a powder characteristic of particular importance. Its influence is confined to the compaction process.

For each individual iron powder there is a direct relationship between density and tensile strength and elongation for each
sintering temperature and time employed. A demand for improved
physical properties thus leads inevitably to a need for higher density. Higher density, however, is equivalent to increased specific
pressure, and increasing specific pressure means heavier wear
and more breakage of tools. It is therefore natural to look for
powders which, with a minimum specific pressure, give a compact
of the highest possible density. That this requirement can be met
in a particularly satisfactory manner by the use of electrolytic
iron powders has been shown repeatedly in Sections III, 1 and
III, 2.

The practical results of the pressing behavior of a number
of different iron powders are illustrated in Table 11, which gives
the specific pressure needed to produce a cylindrical compact with
a surface area of 3 cm^2. Specific pressures which, in the interest
of tool life, should be avoided are printed in italics.

The influence of compressibility on the capabilities of the
press is illustrated in Table 12. This shows the maximum surface
(diameter of compact) that can be pressed with the individual
presses for the five different powders at a pressed density of 6.8,
7.0, and 7.2 g/cc. Blanks have been left in the table where the
capacity ranges involve specific pressures which, in accordance
with the practice laid down above, are not permissible in mass
production.

The potential savings both in press capacity and in cost of
tools are so great that use of the expensive electrolytic iron pow-

TABLE 11. Pressure Requirements in Pressing a
Cylinder of 3 cm^2 Surface Area from a Number of
Iron Powders of Differing Compressibility

Density Required, g/cm³	Pressure Required, metric tons/cm²				
	HVA Crown	HVA Standard	HVA Sponge	Höganäs MH 100	Allevard*
6·2	2·5	2·8	3·1	3·8	3·4
6·4	3·0	3·3	3·5	4·5	4·0
6·6	3·5	3·9	4·1	5·3	4·6
6·8	4·2	4·5	4·8	6·1	5·4
7·0	5·0	5·4	5·6	7·4	6·3
7·2	6·2	6·5	6·8	...	7·7

der may be considered desirable, even in some cases where this
would not be strictly necessary for metallurgical reasons.

8. Green Strength

The strength of compacts in the unsintered condition greatly
affects the care required in handling them before sintering; the
greater the green strength, the fewer precautions will be needed.
This is applicable in particular to operations which involve presses

TABLE 12. Maximum Diameter of Compacts That
Can Be Pressed from Five Different Iron Powders
on Presses of Varying Capacities

Density Required, g/cm³	Capacity of Press, metric tons	Maximum Diameter of Compacts, mm				
		HVA Crown	HVA Standard	HVA Sponge	Höganäs MH 100	Allevard*
6·8	20	25	23	21·7	20	21·5
	40	36	33	30·5	28	30·5
	80	50·5	46·5	43·5	40	43
	120	62	57	53	49	53
	200	80	73·5	68·5	63	68
	400	113	104	97	89	96
	600	138	137	119	110	118
7·0	20	22·5	22	21·5	...	20
	40	32	30·5	30	...	28·5
	80	45	43·5	42·5	...	40
	120	55	53	52	...	49
	200	71	68·5	67·5	...	63·5
	400	101	97	95·5	...	90
	600	123	118	117	...	110
7·2	20	20	20	19·5	...	...
	40	28·5	28	27	...	...
	80	40·5	39·5	39	...	...
	120	49·5	48·5	47·5	...	...
	200	64	55·5	61·5	...	...
	400	91	88·5	87	...	...
	600	111	108	106	...	...

*Atomized powder produced in France by the Mannesmann-Simetag
process.

equipped with automatic feeding and stripping devices. Here, the
feeding device serves also to remove the compacts from the press
table, and provision is often made for them to roll down a ramp
into a container. It is essential that at this stage compacts shall
be damaged neither visibly nor in the form of invisible cracks
which cannot be closed up in the subsequent sintering process,
but that the compacted structure should be preserved unimpaired.

Green strength is governed to a large extent both by the particle size and by the particle shape of the powder used. Powders
of irregular shape are the more suitable. The best behavior in this
respect is found with microporous reduced powders which, not
least for this reason, are so widely used in the mass production of
sintered parts of low density.

Acknowledgments

The test data which form the basis of this paper were obtained in the laboratories of Vapenfabriks AB and Sintermetallwerk
Krebsöge GmbH. The author is particularly grateful to his collea-
gues and assistants in both companies for their untiring assistance.

References

1.   M.P.A. Standard No. 6-S4T; V.D.E. Prüfblatt No. 0091-52.
2.   M.P.A. Standard No. 2-48.
3.   M.P.A. Standard No. 5-46; V.D.E. Prüfblatt No. 0081-58.
4.   M.P.A. Standard No. 3-45.
5.   M.P.A. Standard No. 4-45; V.D.E. Prüfblatt No. 0083-58.
6.   British Standard No. 3029: 1958; V.D.E. Prüfblatt No. 0088-
     53.
7.   Brish Standard No. 3029; 1958; V.D.E. Prüfblatt No. 0086-53.
8.   M.P.A. Standard No. 15-51; V.D.E. Prüfblatt Nos. 0088-52;
     0087-52.
9.   R. Kieffer and W. Hotop, "Sintereisen und Sinterstahl,"
     p. 124. Springer-Verlag, Vienna (1948).
9a. R. Kieffer and W. Hotop, ibid., p. 126.
10. R. Kieffer and W. Hotop, ibid., p. 118.
11. C. G. Goetzel, "Treatise on Powder Metallurgy," Vol. I,
     pp. 259-312. Interscience Publishers, New York and London
     (1950).

12. G. Bockstiegel, unpublished HVA Report.
13. R. Kieffer and W. Hotop, ref. (9), p. 114.
14. H. Wiemer and R. Hanebuth, ref. (9), pp. 533-535.
15. G. Zapf, Métaux, Corrosion-Ind., 26:10 (1951).
16. R. Kieffer and W. Hotop, ref. (9), p. 115.
17. I. Ljungberg and P. G. Arbstedt, Proc. Metal Powder Assoc., 1956, p. 78.

High Velocity Compaction of Iron Powder

Eugene M. Stein and John R. Van Orsdel
Battelle Memorial Institute
Columbus, Ohio

and

Peter V. Schneider
International Business Machines Corp.
Endicott, New York

Large differences in mechanical properties usually exist between a wrought component and one of the same composition made from metal powders by conventional pressing and sintering. The lower properties of the latter stem primarily from their lower density. Some metal components, such as bearings and filters, require low density (high porosity) to retain oil, but in most applications the highest practical density is desirable for strength, ductility, corrosion resistance, and surface quality.

As the strength and ductility requirements for a powder metal part increase, the density must also be raised. For several years iron powder parts with densities of about 7.65 g/cc have been produced in quantity by double pressing and double sintering. Metal with this high density, 97.2% of theoretical for pure iron, exhibits excellent properties. However, for some applications, including those where a uniform carburized case must be obtained, a density of 7.65 g/cc is not high enough.

The main stimulus to reach even higher densities is an economic one. Substantial savings (as much as 200% or more) can be realized by reducing or eliminating machining operations for complex parts which can be made to high density by powder methods. With potential manufacturing economies and quality as the ultimate goals, International Business Machines Corp. sponsored an investigation at Battelle Memorial Institute to determine the feasibility of high velocity compacting for achieving densities over 99% of theoretical in iron base powders.

High energy rate (HER) equipment available at Battelle can produce velocities up to 167 ft/sec. However, the minimum energy that can be released with this equipment is far too high for the small test specimens. Consequently, we developed an apparatus which utilizes explosives to propel a ram at high velocity against

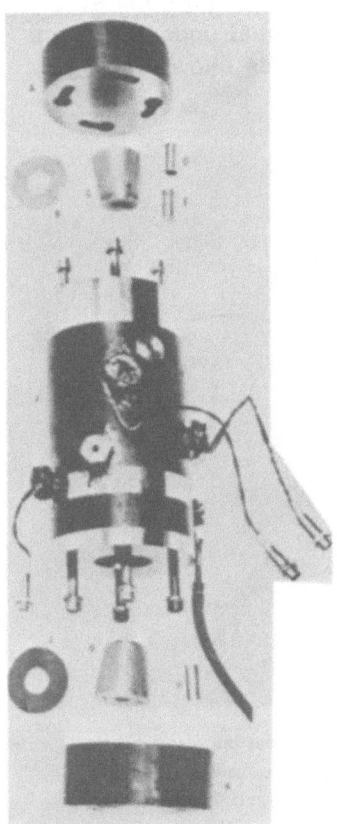

Fig. 1. Apparatus developed at Battelle Memorial Institute for explosive compacting of metal powders is shown about one third actual size. (A) Top cap is 5 in. in diameter and 2 in. thick. Bottom end of firing pin can be seen in hole at center of cap. (B) Rubber gasket, used for vacuum runs, is placed between top cap and top cone frustrum. (C) Top cone frustrum, 2 in. high, fits into mating cavity in die body and has 0.070-in.- diameter hole for polystyrene pin that holds ram before firing. (D) Cartridge case (45 caliber Colt regular) fits into top cone frustrum so that top of cartridge is flush with top of cone. "Bull's-eye" smokeless powder is held near primer end by three copper gas checks pressed into the case. (E) Ram is held in top cone by polystyrene pin. Bottom end of ram protrudes into main die barrel. (F) Die body is 5 in. outside diameter by 0.500 in. inside diameter by 10 in. long. It contains two rupture ports (closed by copper discs), three electromagnetic pickups for pulsing an electronic counter for time interval measurements and a vacuum port. (G) Bottom cone frustrum. (H) Base holds metal powder or precompacted specimens. (J) Rubber gasket. (K) Bottom cap. Perhaps a forerunner of commercial equipment, this apparatus will compact powders of iron base alloys to densities which are close to those of wrought metals.

a powder compact. The apparatus is shown in Fig. 1, along with a brief description of each part.

Operation of the device is quite simple. A smokeless powder charge can be varied to produce a wide range of ram velocities These speeds are calculated from electronically measured time-of-flight intervals between accurately spaced electromagnetic pickups (see Fig. 1). For the most part, the middle and bottom pickups were used, as these were closest to the impact and should indicate the terminal velocity of the ram. Where the specimens consisted of loose powder, the die chamber was evacuated to about 0.2 mm of mercury before being fired.

Characteristics of Metal Powders

We studied the compaction characteristics of five metal powders, two of which were iron (electrolytic and sponge). The other three were AISI 4620 steel, type 304 and type 410 stainless steel. These, when pressed at 40,000 psi by conventional opposing rams in a floating die, had thickness-density curves as shown in Fig. 2.

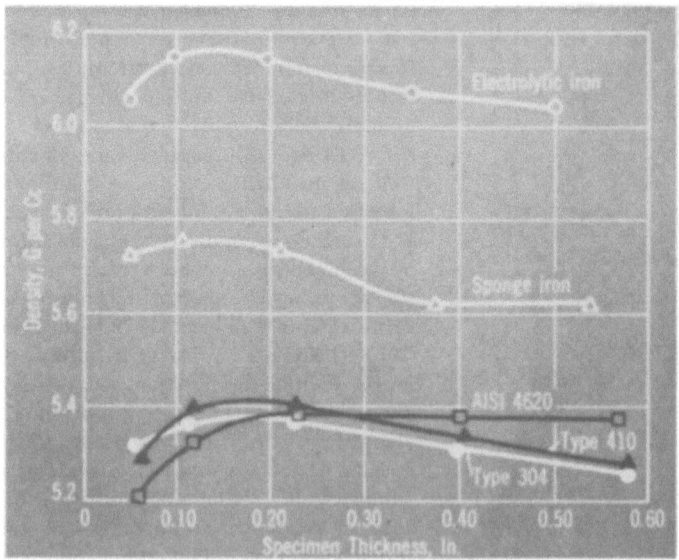

Fig. 2. Curves compare effect of specimen thickness on green density of five powders pressed conventionally. Specimens weighing 1, 2, 4, 7, and 10 g (from left to right), each 0.505 in. in diameter, were pressed at 40,000 psi.

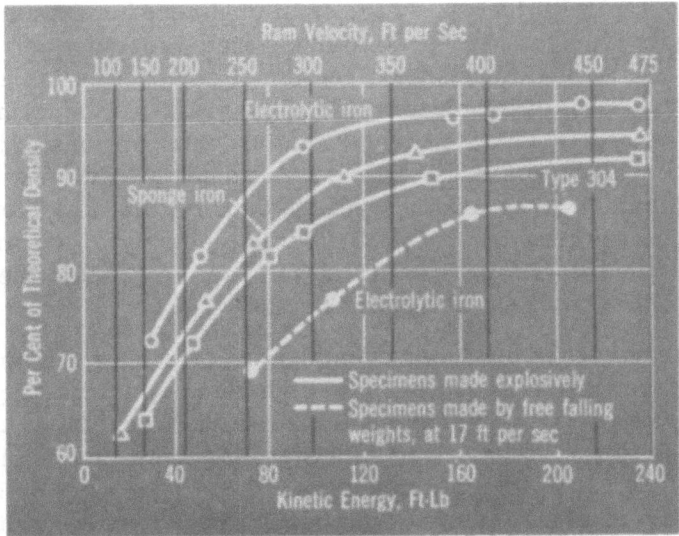

Fig. 3. Curves compare effect of impact energy on density of
4-g, $\frac{1}{2}$-in.-diameter samples made by explosive compacting and
by free falling weights. Kinetic energy values obtained from formula $E = mv^2/2$, where E is kinetic energy in ft-lb, m is mass of
ram (weight in pounds divided by 32.2 ft/sec^2), and v is ram velocity in ft/sec.

The electrolytic iron had by far the highest peak density
(78.1% of theoretical) of any of the powders. Next in density, at
73.0%, was the sponge iron. The other three materials were in a
relatively narrow grouping at a density of about 69%.

The effect of kinetic energy on the density of compacts made
by the explosive technique is shown in Fig. 3. Included in the graph
is a curve for specimens of electrolytic iron powder that were
compacted by falling weights at a velocity of 17 ft/sec. The densities reached by high velocity compaction ranged from 92% for
type 304 stainless to 98% for electrolytic iron. The curves for the
4620 steel and type 410 stainless coincided closely with that shown
for the sponge iron, especially above 150 ft-lb of energy.

The effect of velocity on the density of electrolytic iron compacts is pronounced. For 205 ft-lb. of energy delivered at 17 ft/sec,
a density of 87% is obtained. Only 65 ft-lb is necessary to attain
the same density when the velocity is 250 ft/sec. Energy inputs
of 205 ft-lb delivered at velocities of 440 ft/sec. and 17 ft/sec.
gave respective densities of 98 and 87%.

The typical microstructure of an unsintered specimen of electrolytic iron compacted explosively from loose powder is shown in Fig. 4. Although the direction of compaction is unmistakable, the ratio of length to width of the grains is lower than would be expected. This indicates that under conditions of high velocity impact the stress pattern approaches the uniformity of that in a liquid. Since oxides (average density, 5.2) are present as gray inclusions, the density (7.68) of this electrolytic iron powder is lower than the 7.87 g/cc accepted for pure iron; thus, the true density of this compact is higher than 98%.

Further work produced even higher densities when specimens were pressed and sintered prior to high velocity compacting. Several electrolytic iron specimens – pressed at from 20,000 to 60,000 psi, sintered 1 h at 2000°F in hydrogen, and compacted explosively at 400 ft/sec 120 ft-lb) – had densities of 7.81 to 7.86 g/cc or 99.3 to 99.9% of theoretical density. Specimens of 4620 steel, treated similarly, reached about 97% of the density of wrought ingots of this steel.

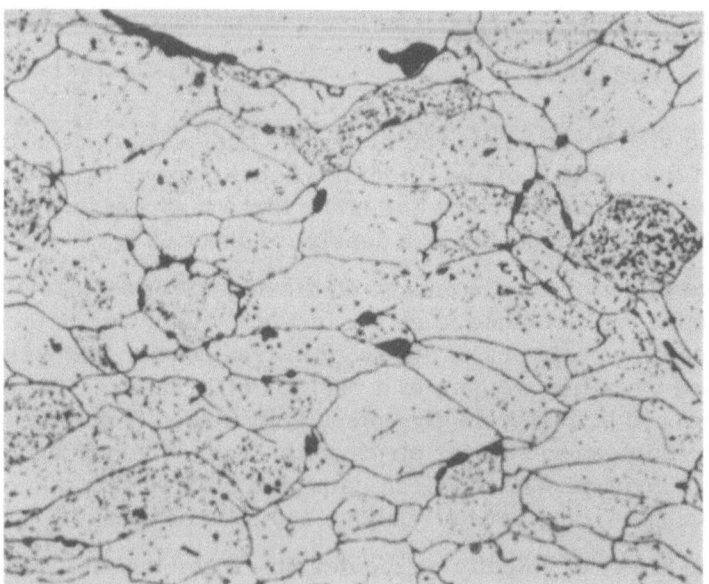

Fig. 4. Microstructure of unsintered electrolytic iron made by high velocity compacting of loose powders contains some oxide (gray) inclusions. Density: 7.68 g/cc. Etchant, Picral; 500 X.

Fig. 5. Printing character made by high velocity compaction illustrates sharp definition of
detail that can be obtained.

Evacuating the Die Chamber

Evacuation of the die chamber is essential for compacting
loose powders to high densities. Without evacuation, horizontal
separations are obtained in the microstructures, indicating entrap-
ment of air. However, there is no noticeable beneficial effect of
evacuation for material that is pressed and sintered prior to high
velocity compaction.

Although no quantitative data on mechanical properties of the
explosively made compacts were obtained, many specimens were
notched and bent to obtain indications of general behavior. Those
in the unsintered condition were relatively brittle, while the sintered pieces had relatively high strength and ductility.

Some specimens were carburized, and those with densities over 99% had excellent cases with uniformity of case depth comparable to that obtainable with wrought steel.

The Future

Electrolytic iron compacts can be produced with densities near theoretical for pure iron by high velocity compaction—without subsequent sintering. (However, parts are generally sintered to increase their strength and ductility.) The advantages accruing to the powder metallurgist through the use of high velocity compacting are many. Sharp definition of detail, as shown in the printing character part in Fig. 5, and close adherence to dimensional tolerances are rewards of the process. More work needs to be done to determine the mechanical properties of parts made by the process, but we believe they will be within a few percent of those obtained in wrought ingot parts.

Die materials are the limiting factors as far as energy input is concerned. However, our work indicates that die materials now available will perform satisfactorily if a certain energy level, which depends on the shape of the part, is not exceeded. For example, a simple $1/2$-in.-diameter die with flat-ended punches made from an AISI-SAE A6 tool steel will absorb about 200 ft-lb of energy without damage in repeated use. However, the complex die and punches used to make the type character mentioned previously could only absorb about 100 ft-lb per stroke. It can be expected that die and punch materials will be improved to the point where they are of no more concern in high velocity compaction than in regular pressing operations.

Study on Continuous Rolling Fabrication of Iron Strip from Metal Powders

Takashi Kimura
Toyota Central Research and
Development Laboratories, Inc.
Nagoya, Japan
and
Heihachiro Hirabayashi and Mutsuo Tokuyoshi
Electrical Communication Laboratory
Nippon Tel. and Tel. Public Corp.
Tokyo, Japan

Introduction

The continuous production method of manufacturing metal strips by rolling strips directly from metal powders is not a new technique historically. The patent for this method was announced about sixty years ago [1]. However, for a long time it has been buried under the conventional melting method. Since G. Naezer and F. Zirm published a paper on powder rolling of RZ iron powder, many experiments were carried out with various kinds of powders, such as iron [2, 3], copper [4, 5], nickel [6], and stainless steel [7] powders, in the United States and England. It has been reported that an attempt to produce copper [8] and nickel[9] powder rolled strips on a pilot plant scale proved successful, but no detailed reports have been published yet. The advantage of this process is

the lower cost of equipment and operation, as compared with that
in the conventional melting method. If the metal can be refined
from the ore in a form of powder in the near future, the powder
rolling process will be advisable from an economic viewpoint.

     In this paper, application to the manufacturing of materials
for electrical communication use is described. First, the fundamental of the powder rolling process was studied using Höganäs
iron powders. The effects of various factors, such as roll gap
dimensions, powder feed rate, rolling speed on the powder rolling
load, and the properties of green, sintered, and cold-rolled strips
were investigated. Second, the physical and mechanical properties
of cold-rolled strips made by the powder rolling process from the
electrolytic, nitride process, mill-scale reduced, and sponge iron
powders were compared with each other. Finally, silicon-iron
strips containing a high silicon content, which are hard to cold
roll from the conventional molten ingots, were test manufactured
by the powder rolling of a mixture of electrolytic iron and silicon
powders.

     Properly speaking, in the case of the powder rolling process,
the metal powder is directly converted to a green strip by roll
compacting, and it is continuously sintered, hot or cold rolled,
cooled, and coiled up. However, in this report the sintering and
cold rolling were carried out in separate steps.

1. Equipment, Powders, and Measurements
of Properties

1.1 Powder Rolling Mill

     The type of experimental powder rolling mills used in the
past depended on the type of powder used and on the design considerations. From these data, a horizontal rolling mill was test
made, of which the specification and design are shown in Table 1
and Fig. 1 respectively.

1.2 Hopper

     As a result of preliminary experiments [10], the precise
control of the powder feed rate, which would most effectively determine the thickness and properties of powder-rolled green
strips, was determined. The hopper as shown in Fig. 2 was

TABLE 1. Specifications of Powder
Rolling Mill

Roll arrangement	Horizontal
Roll diameter	150 mmϕ (200 mmϕ changeable)
Roll width	200 mm
Roll material	Chromium steel
Roll speed	5.3—47.7 r.p.m. (stepless)
Horsepower	15 HP
Manufacturer	Ohno Roll Manufacturing Co.

specially designed. The width of the hopper is 50 mm. The powder feed is controlled by two-step adjusters, that is, an upper powder feed rate adjuster controls the powder feed to the lower hopper and a lower hopper powder-feed rate adjuster controls the powder feed to the horizontal rolling mill. Both adjusters can be regulated with the accuracy of 5/100 mm gap. The angle of the plastic plate shown in Fig. 2 is changeable, but in this experiment the angle was fixed at 90° to the tangent line of the roll surface.

1.3 Metal Powders

The metal powders used in this experiment are summarized in Tables 2 and 3. In the case of silicon-iron alloy strip, the

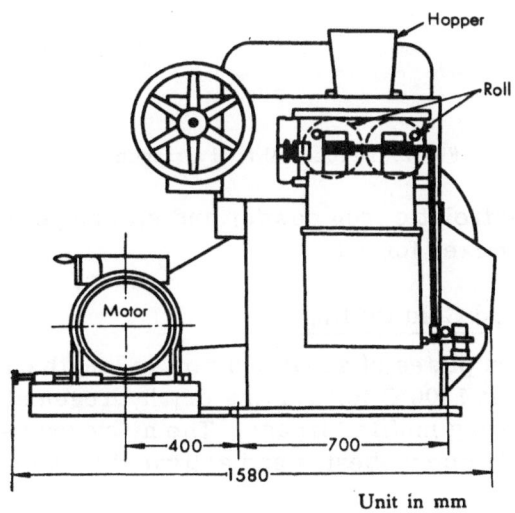

Unit in mm

Fig. 1. Diagram of powder rolling mill.

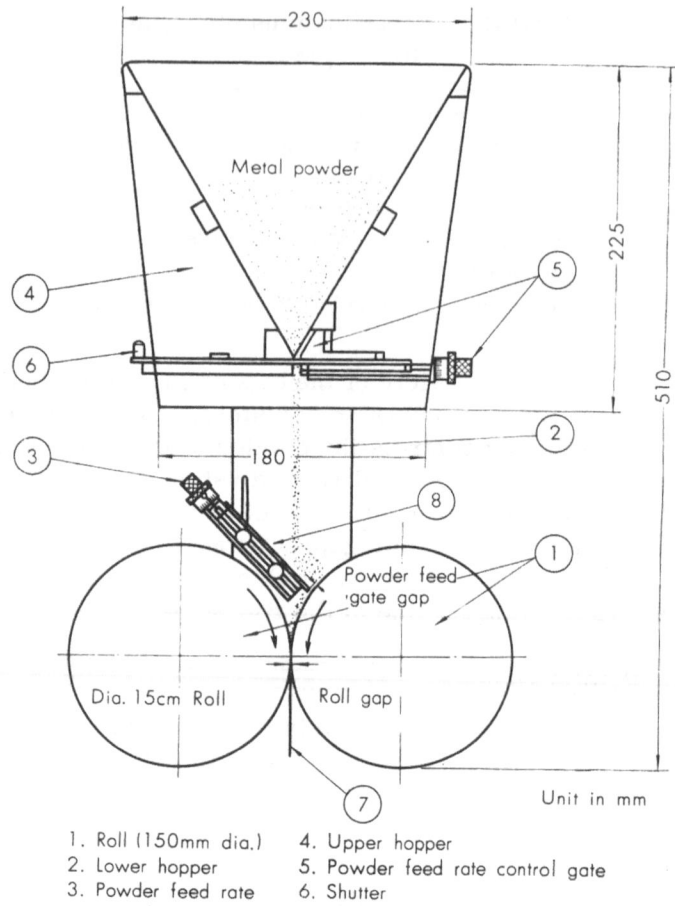

Fig. 2. Powder roll feed arrangement.

1. Roll (150mm dia.)   4. Upper hopper
2. Lower hopper   5. Powder feed rate control gate
3. Powder feed rate   6. Shutter
   control gate   7. Green strip   8. Plastic plate

"Shoden"* electrolytic iron powder and silicon powder were mixed
with a V-type mixer for 1 h.

1.4 Sintering and Cold Rolling

The green plates of about 300 mm in length cut from strips
were sintered at 1100°C for 1 h in a dry hydrogen atmosphere with
an electric heating muffle furnace. The hydrogen used was purified by the palladium asbestos and activated alumina method. The

* Showa Denko Ltd., Tokyo, Japan.

TABLE 2. Properties and Particle Size Distribution of Various Iron Powders

Property	Electrolytic iron powder		Nitride* process iron powder	Mill-scale reduced iron powder	Höganäs sponge iron powder			Sponge** iron powder
	Shoden†	HVA-star††			MH-100-24	MH-100-28	MH-300P	
Apparent density (g/cm³)	2.41	2.61	2.33	2.39	2.45	2.45	2.74	3.12
Tap density (g/cm³)	2.98	3.30	2.90	2.96	2.96	2.96	3.26	4.02
Flow (sec/50 g, 3 mmφ)	23.0	24.0	35.2	28.1	28.1	23.0	※	26.6
Powder viscosity factor (k)	7.3	6.8	7.9	6.7	7.1	6.2	※	6.19
Particle size distribution +80 mesh	—	—	—	—	—	—	—	—
80~100	—	3.0	—	—	0.7	0.3	—	—
100~150	14.0	12.7	—	33.0	22.1	22.3	—	0.1
150~200	76.3	28.8	8.7	34.8	27.9	29.4	—	15.6
200~250	8.2	13.3	42.4	12.4	9.5	10.4	0.1	26.2
250~325	2.3	19.9	44.3	12.8	17.8	17.8	1.1	52.3
−325	0.2	22.3	4.3	7.0	22.1	19.8	98.8	5.8

* Iron nitride prepared by nitriding of iron strip scraps is pulverized and reduced by hydrogen.
** Iron ore is reduced to iron powder with a rotary kiln.
† Showa Denko Ltd. (Japan). †† Husquvarna (Sweden). ※ Impossible to measure.

sintered specimens were cold rolled by a four-high 7.5 HP rolling-mill.

1.5 Measurement of Various Properties

(a) Apparent Density. The specimens for measuring apparent density were prepared by immersing a part of the green strip into molten paraffin, and the excess paraffin on the surface of spec-

TABLE 3. Chemical Compositions of Various Powders

Element (%)	Electrolytic iron powder		Nitride process iron powder	Mill scale reduced iron powder	Höganäs sponge iron powder			Sponge iron powder	Silicon powder
	Shoden	HVA-star			MH-100-24	MH-100-28	MH-300P		
Fe	98.99	99 up	99.05	98.39	98 up SiO₂	98 up SiO₂	99 up SiO₂ ※	98.23	0.91
Si	0.005	0.02	0.03	0.16	※ 0.30	※ 0.30	0.10~0.15	0.49	98.30
C	0.03	0.05	0.010	0.01	<0.1	<0.1	0.01	0.04	0.054
P	0.002	0.01	0.010	0.007	0.015	0.015	0.005 ~0.01	0.03	0.009
S	0.009	0.005 Ca	0.010	0.004	0.015 O	0.015 O	0.005 ~0.01	0.008	※ Ca
Cu	0.047	※ 0.01 Ni	0.10	—	※<0.8	<0.8	—	0.01	※ 0.37 Al
Mn	0.008	※ 0.01	0.32	0.36	※	※	※	0.07	※ 0.19

Analyzed by manufactures. ※ No. data.

imen was shaved off by a razor blade after freezing. In reality, these specimens include the weight of paraffin. However, its density is about one eighth that of iron, and the porosity of the green strip is 10-30%, and it is apt to remain on the surface of specimen, so that an error of apparent density measurement seems to be 1-2%. This error, however, is low enough to permit comparison of the apparent density of the specimens. The volume of specimens was determined by the difference between their weight in air and that in water, and the apparent density was calculated from their weight per volume. The apparent densities of sintered and cold-rolled strips were directly measured in the same way without paraffin treatment.

(b) Mean Thickness. The mean thickness of the green strips was determined by measuring the thickness at three points parallel and perpendicular to the rolling direction.

(c) Tensile Properties. The dimensions of tensile test specimens are as shown in Fig. 3. The test specimens prepared from the sintered and cold-rolled strips were annealed at 700°C for 1 h in vacuo. The tensile tests were carried out with an INSTRON universal tester at the rate of 10 mm/min. The elongation was calculated from $\Delta l/30$ mm, where Δl is the difference between marked lines on the specimens before and after tests.

(d) Electrical Resistivity. The specimens for electrical resistivity measurements, 7 mm in width and 100 mm in length, were prepared from the sintered and cold-rolled strips and annealed at 700°C for 1 h in vacuo. The electrical resistivity was measured and calculated by the Kelvin bridge method, using 0.1 Å current, and the potentiometer.

(e) Magnetic Properties. Thin strips of below 0.2 mm in thickness were prepared by repeated cold rolling and annealing.

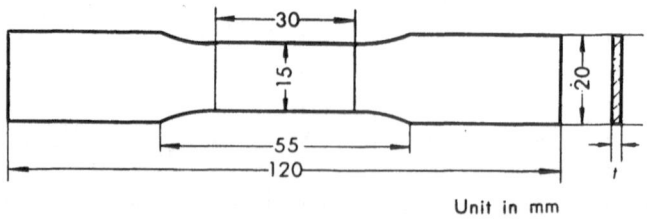

Unit in mm

Fig. 3. Dimensions of tensile test specimen (t = 0.1-0.8 mm).

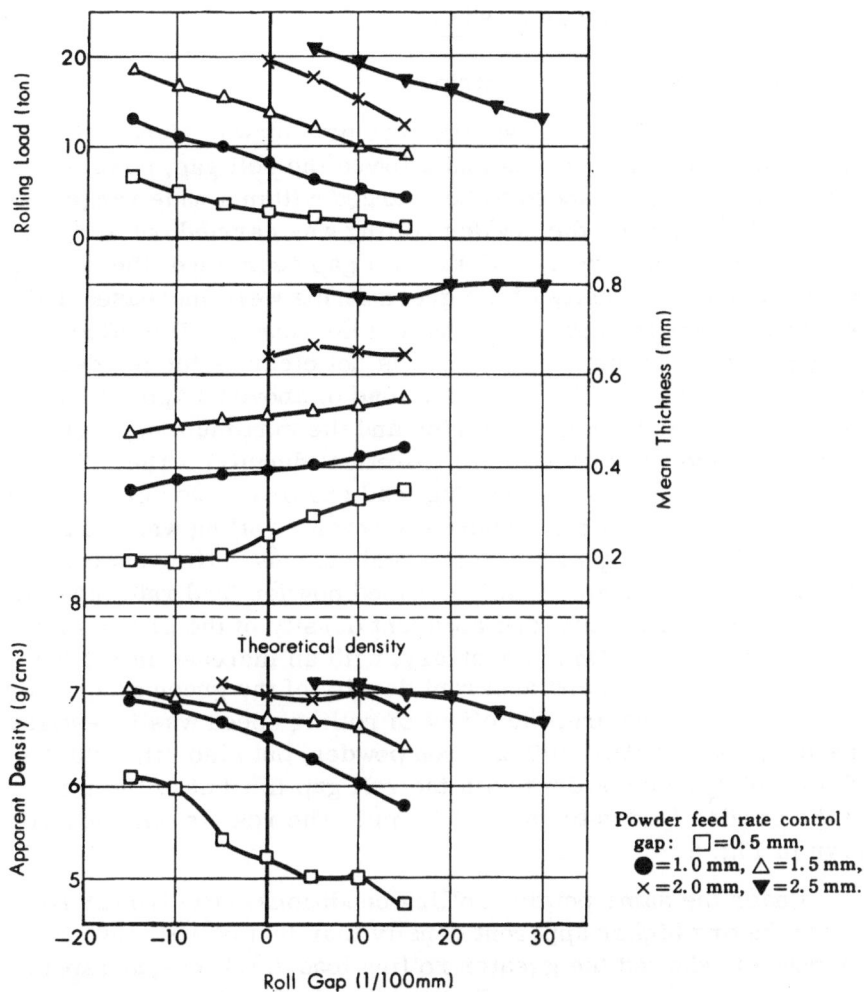

Fig. 4. Effects of initial roll gap and powder feed rate on rolling load and properties of green strip made from Höganäs MH·100·24 iron powder (150-mm-diameter roll, 12.5 rpm).

They were then cut to 10 mm in width and wound three or four layers deep so that the inside diameter was 26.5 mm. Such specimens were used for magnetic property measurement. The specimens were annealed at 880°C to 1200°C for 1 h in a dry hydrogen atmosphere. B-H curves were recorded automatically by an auto fluxmeter, and the values of B_5, B_r, μ_m, μ_0, and H_c were determined from the B-H curves.

2. Results and Discussion

2.1 Powder Rolling of Green Strips

Höganäs MH·100·24 sponge iron powder was found to be quite suitable in this process and allowed the roll gap, powder feed rate, and rolling speed to be changed within a wide range. As shown in Fig. 4, when the powder rolling was carried out at a rolling speed of 12.5 rpm, with the roll gap decreased, the rolling load and apparent density of the green strips were increased and the mean thickness of green strips was decreased. The effect of roll gap on the properties of strip was not clear as the powder feed rate increased. With a rolling load of above 22 t per 50 mm, cracks occurred in the green strips and the maximum apparent density was not over 7.3 g/cc (93% theoretical density). The effect of the powder feed rate on the rolling load and properties of the green strips are shown in Fig. 5, where the powder rolling was carried out with a fixed roll gap, 5/100 mm and various rolling speeds 5.7, 12.5, and 17.6 rpm. With increased powder feed rate the rolling load, mean thickness, and apparent density of the green strips were all increased. On the contrary, with an increase in rolling load, mean thickness, and apparent density of the green strips were decreased. Futhermore, the effect of rolling speed was investigated regarding not only MH·100·24 iron powder, but also MH·100·28 and MH·300P iron powders with the roll gap fixed at 5/100 mm and the powder feed gate gap at 1.5 mm. The results obtained are shown in Fig. 6.

Under the same powder rolling conditions, MH·100·28 iron powder, having higher apparent density than that of MH·100·24 iron powder, showed the greater rolling load, thickness, and apparent density of green strips. On the other hand, MH·300P iron powder which flows less than the others could be rolled at a relatively low rolling speed and larger powder feed rate. The thickness and apparent density of these green strips were similar to those of MH·100·24 iron powder green strips.

(a) Effect of Roll Gap. According to the papers reported before, the maximum apparent density approached the theoretical density[2], or, in another case, it was impossible to reach the theoretical density even at a minus roll gap, that is, by adding mutual roll pressure before powder rolling began [5]. Although these results may depend upon the construction of the rolling mill, it seems

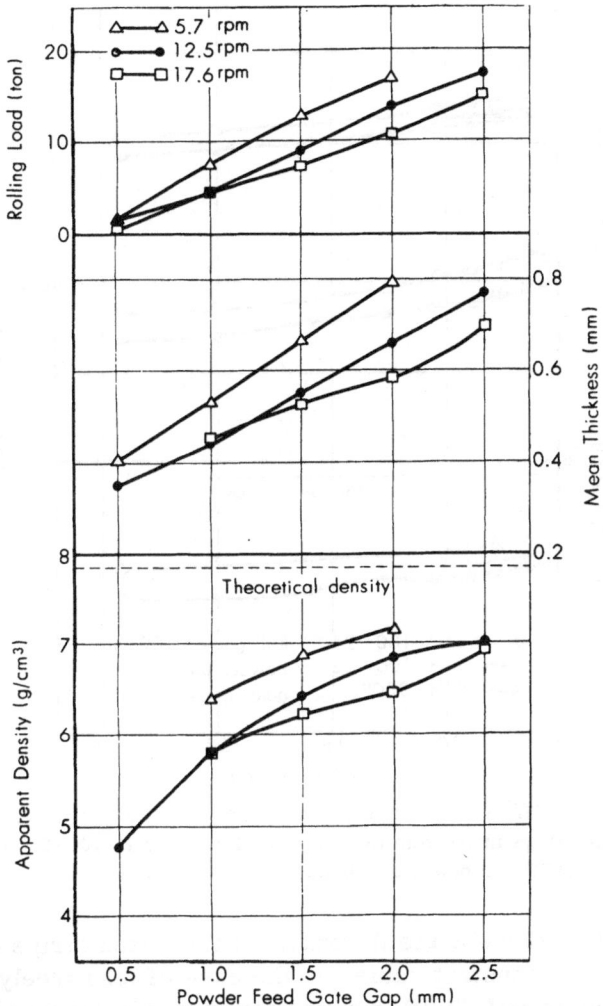

Fig. 5. Effects of powder feed rate and rolling speed on rolling load and properties of green strip made from Höganäs MH·100·24 iron powder (150-mm-diameter roll, roll gap = 15/100 mm).

to be difficult to obtain a strip of theoretical density by rolling alone, since the strips break by overloading. It is impossible to change the roll gap in the course of powder rolling, so that the control of thickness of green strips by the roll gap is not practicable.

(b) Effect of Powder Feed Rate. Unlike the case of the roll gap, the powder feed gate gap in the course of powder rolling is

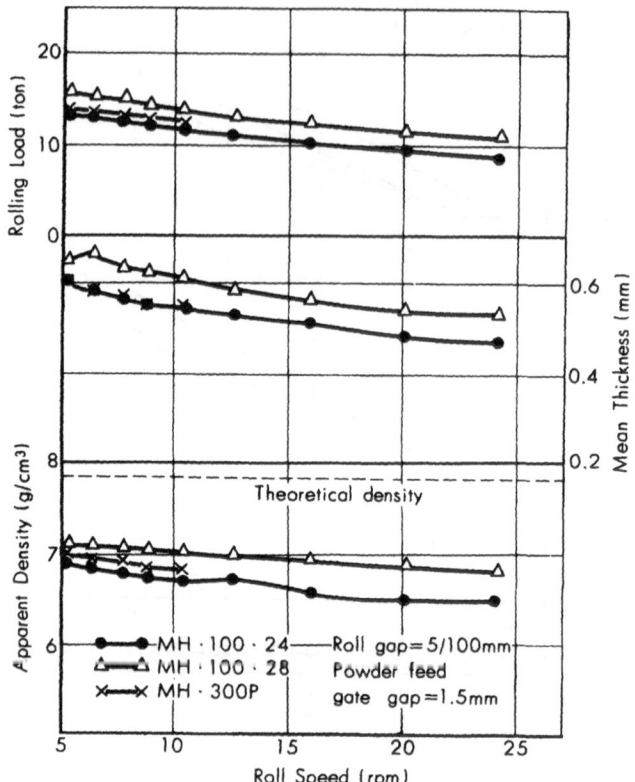

Fig. 6. Effect of rolling speed on rolling load and properties of green strips made from Höganäs MH·100·24, MH·100·28 and MH·300P iron powders (150-mm-diameter roll).

easy to control thereby the thickness of the green strips can be maintained at a constant value. In the case of less freely flowing powder, such as MH·300P iron powder, not only the powder feed rate should be increased, but also the height of the powder in the lower hopper is important. In any case, it is desirable to keep the height of powder in the lower hopper constant as far as possible for stabilizing the rolling load and the thickness of the green strips.

(c) Effect of Rolling Speed. The rolling speed has a close relation to the powder feed rate. In the case of powders which flow more freely the loss of powder through roll gap in the course of rolling was increased at a low rolling speed and the green strips became smaller in width. On the contrary, in the case of less freely flowing powders, such as MH·300P iron powder, rolling to strips was impracticable at a high rolling speed, because of the insufficient powder feed rate for rolling. The balance between the powder feed rate and the rolling speed, which determines the product rate of green strips, is a key to the effective production of desired green strips.

(d) Relationship between the Rolling Load and the Thickness and Apparent Density of Green Strips. As shown in Fig. 7, if the thickness and apparent density of green strips produced with a constant roll gap are plotted against the rolling load which was obtained from the data on MH $\cdot$ 100 $\cdot$ 24 powder rolled at 12.5 rpm, the thickness increases linearly in proportion to the rolling load as the

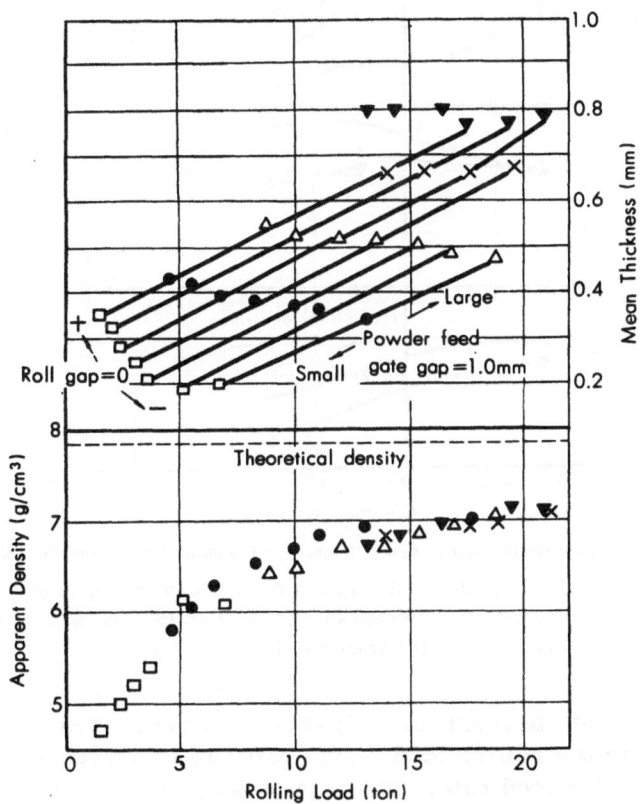

Powder feed rate control gap:
□=0.5 mm, ●=1.0 mm, △=1.5 mm, ×=2.0 mm, ▼=2.5 mm.

Fig. 7. Relationship between powder rolling load, initial roll gap, powder feed rate and properties of green strip made from Höganäs MH $\cdot$ 100 $\cdot$ 24 iron powder (150-mm-diameter roll, 12.5 rpm).

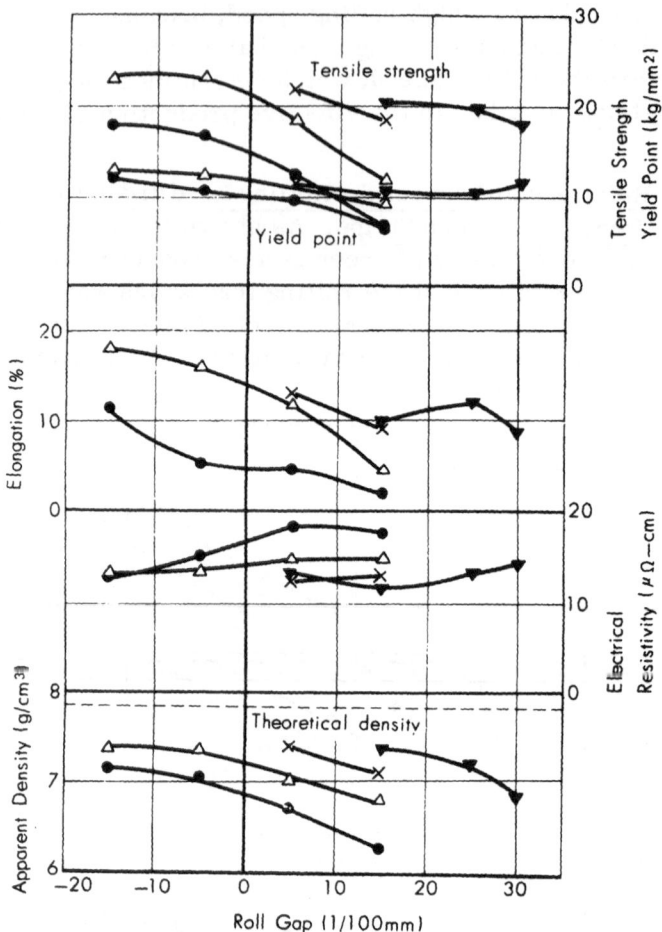

Powder feed rate control gap: ● = 1.0 mm, △ = 1.5 mm, × = 2.0 mm, ▼ = 2.5 mm

Fig. 8. Effects of initial roll gap and powder feed rate in powder rolling on properties of sintered strip made from Höganäs MH · 100 · 24 iron powder (150-mm-diameter roll, 12.5 rpm).

powder feed rate is increased. However, it decreases reversely in proportion to the rolling load as the roll gap is decreased with a constant powder feed rate. The relationship between the rolling load and the apparent density is also shown in Fig. 7. The following experimental equation was obtained from the experimental function curve of Fig. 7 with 50 mm wide strip.

$$d = 4.5 + 0.9 \ln P$$

where d is the apparent density of the green strip of MH · 100 · 24 iron powder and P is the rolling load at 12.5 rpm.

With a rolling load of above 15 t per 50 mm, the apparent density reached a saturated value and with that of above 22 t per 50 mm the green strips broke. Therefore, the optimum rolling load for rolling MH · 100 · 24 iron powder is deduced to be 15–20 t per 50 mm in width, that is 3–4 t/cm. This value is much smaller as

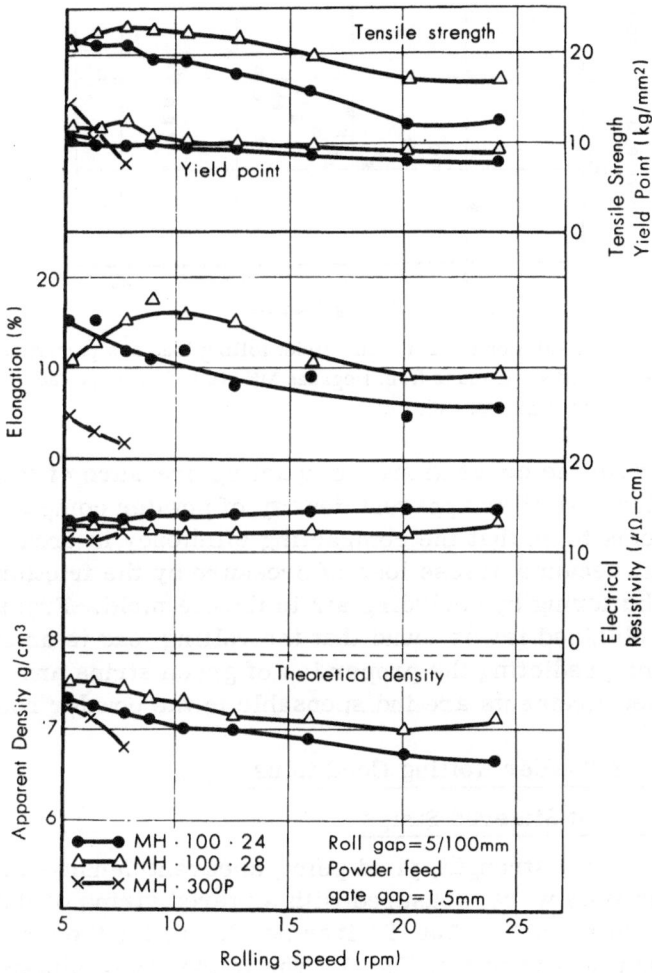

Fig. 9. Effect of powder rolling speed on properties of sintered strips made from Höganäs MH · 100 · 24, MH · 100 · 28, and MH · 300P iron powders (150-mm-diameter roll).

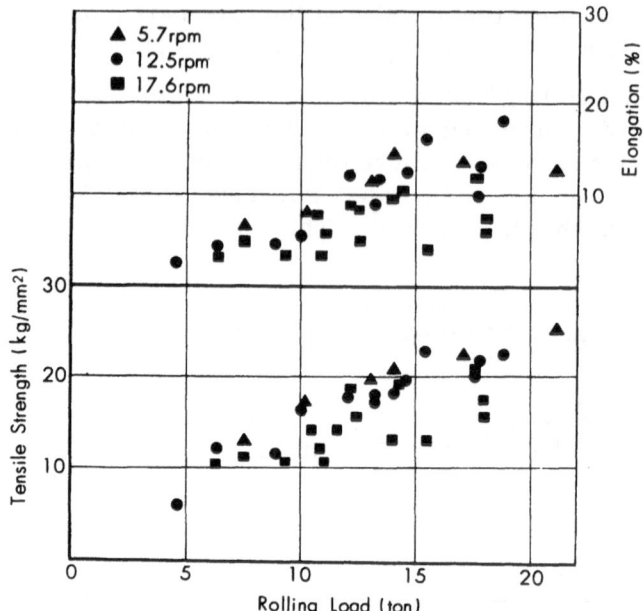

Fig. 10. Relationship between powder rolling load and properties of sintered strips made from Höganäs MH · 100 · 24 iron powder (150-mm-diameter mm roll).

compared with the conventional compacting pressure of the die for obtaining a similar apparent density of powder compacts. The reason seems to be that the compacting efficiency is great in powder rolling, because of less loss of pressure by the friction of die walls and buffering by occluding air in the die mold. From the results thus obtained it was found that the rolling load is an effective standard for predicting the properties of green strips and the rolling load measurements are indispensable to the powder rolling.

2.2. Effect of Powder Rolling Conditions

on Properties of Sintered Strips

The tensile strength, yield point, apparent density, and electrical resistivity were measured with sintered strips of 0.2–0.8 mm prepared from MH · 100 · 24 iron powder under the powder rolling conditions shown in Fig. 5. The thickness of sintered strips was almost the same as that of green strips, although some of them which were rolled under heavy rolling load showed a slight expansion. In the case of iron powder sintered strips no blister, as was

seen in the case of copper ones [11], occurred on the surface of strips
after sintering. The best properties of MH · 100 · 24 iron powder
sintered strips were as follows: Apparent density was 7.5 g/cc
(95% theoretical density), tensile strength was 25 kg/mm^2, yield
point was 15 kg/mm^2, elongation was 19%, and electrical resistivity
was 12.5 $\mu\Omega$-cm.

A similar tendency, as in the effect of powder rolling conditions on the green strips, was observed in various properties of
sintered strips. That is to say, as shown in Figs. 8 and 9, the apparent density, tensile strength, and elongation were increased,
while the electrical resistivity was decreased as the roll gap and
rolling speed decreased and the powder feed rate increased in the
course of powder rolling. Generally speaking, the properties of
MH · 100 · 28 iron powder sintered strips were better than those of
MH · 100 · 24 ones, but the tensile strength and elongation of the
former showed their maxima under a rolling load of 10 rpm and
were inferior to those of the latter under a rolling load of below 7
rpm, as shown in Fig. 9. The properties of MH · 300P iron powder
sintered strips were markedly deteriorated as the powder rolling
speed was increased.

If the relationship between the tensile strength and elongation
of sintered strips of MH · 100 · 24 iron powder and the powder rolling load is plotted against various powders rolling speeds Fig. 10
can be obtained. The greater tensile strength and elongation of
sintered strips are shown as the powder rolling load is increased,
that is the apparent density of the green strips is larger.

2.3. Effect of Powder Rolling Conditions

on Properties of Cold-Rolled Strips

Cold-rolled strips were prepared through the following procedure: Green strips prepared under the powder rolling conditions
mentioned above were sintered at 1100°C for 1 h in dry hydrogen,
cold rolled with 10% reduction, annealed at 1100°C for 1 h in dry
hydrogen and again cold rolled with 30% reduction. In the case of
Höganäs MH · 100 · 24 powder cold-rolled strips, the maximum yield
point, tensile strength, and elongation and the minimum electrical
resistivity obtained were 16 kg/mm^2, 26 kg/mm^2, 38%, and 11 $\mu\Omega$-
cm, respectively, although the apparent density was limited to 7.7
g/cc (98% theoretical density). The results obtained are shown in
Figs. 11 and 12. The effect of powder rolling conditions on various

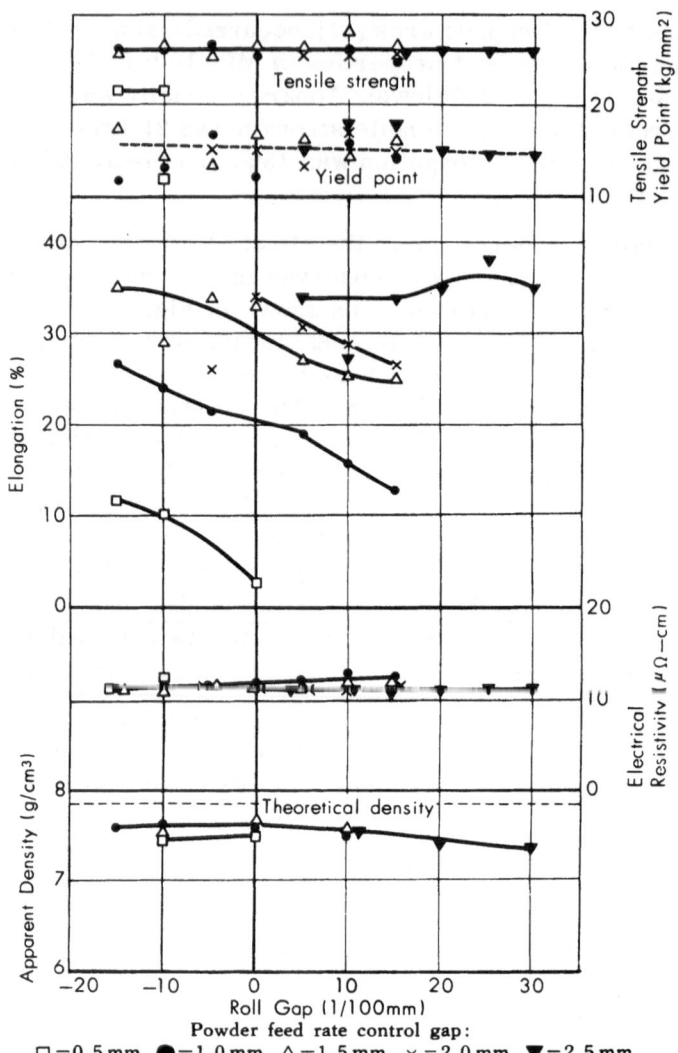

Powder feed rate control gap:
□=0.5 mm, ●=1.0 mm, △=1.5 mm, ×=2.0 mm, ▼=2.5 mm.

Fig. 11. Effects of initial roll gap and powder feed rate in powder rolling on properties of cold-rolled strip made from Höganäs MH·100·24 iron powder (150-mm-diameter roll, 12.5 rpm).

properties except elongation of the cold-rolled strips was less than in the case of sintered strips. The elongation was increased as the roll gap or roll speed decreased and the powder feed rate increased.

In the case of MH·100·28 powder cold-rolled strips, a maximum value of elongation was observed at the roll speed of 10 rpm, as shown in Fig. 12. The difference between the properties of both cold-rolled powder strips was small; MH · 300P cold-rolled powder strips could not be prepared from the sintered strips as shown in Fig. 9, which were powder rolled and sintered under the same conditions as the other two powders. When MH · 300P powder was powder rolled under the optimum condtions such as a powder feed rate gap of 2.5 mm, a roll gap of 25/100 mm, and a rolling speed of 5.3 rpm, the properties of MH · 300P cold-rolled strip were similar to those of other cold-rolled powder strips, as shown in Table 3.

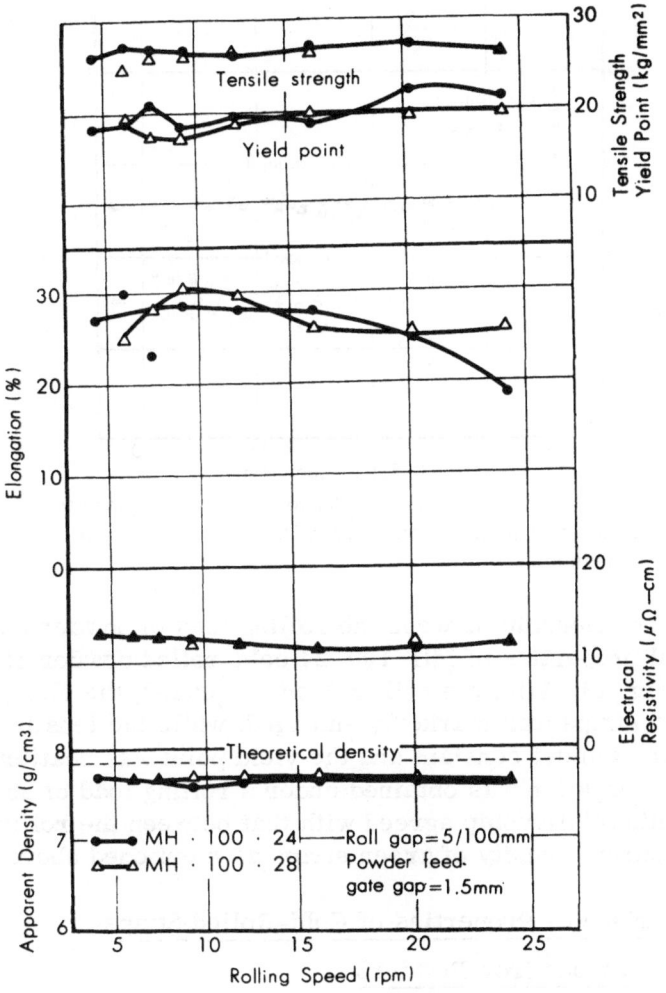

Fig. 12. Effect of powder rolling speed on properties of cold-rolled strips made from Höganäs MH · 100 · 24, MH · 100 · 28, and MH · 300P iron powders (150-mm-diameter roll).

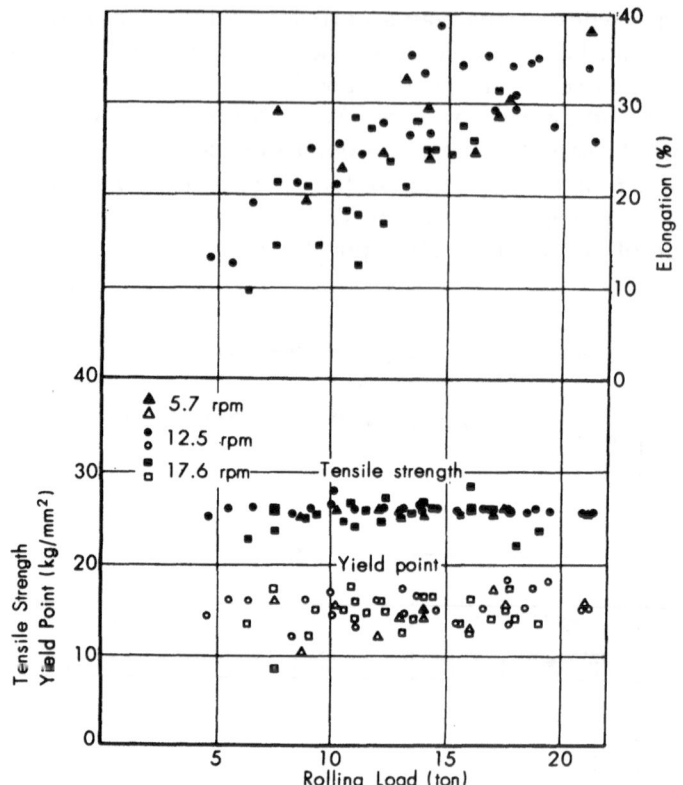

Fig. 13. Relationship between powder rolling load and properties of cold-rolled strip made from Höganäs MH · 100 · 24 iron powder. (150-mm-diameter roll).

The relationship between the rolling load of powder and the mechanical properties of MH · 100 · 24 cold-rolled powder strips is shown in Fig. 13. With the rolling load increased, the elongation of cold-rolled strips was markedly enlarged, while the tensile strength was almost constant and the yield point was scattered. Maximum elongation was obtained under a rolling load of above 3 t/cm. This relationship agreed with that between the rolling load and the apparent density of green strips as mentioned above.

2.4. Comparison of Properties of Cold-Rolled Strips

Made from Various Iron Powders

The mechanical and physical properties of cold-rolled strips made from various iron powders are summarized in Tables 4 and

5. In general, the mechanical and physical properties of cold-rolled strips were dependent upon the purity of powder. Especially the elongation is markedly affected by such purity. The electrical resistivity, however, is influenced by the specific impurities. For example, in the case of cold-rolled strip nitride process iron powder the apparent density is equal to the theoretical density and the tensile strength is superior to those of other powders, but the electrical resistivity is greater than that of electrolytic cold-rolled iron powder strips. The reason seems to lie in the effect of impurities, that is, nitride-process iron powder contains a larger amount of manganese with 0.02% of nitrogen (although the latter is not shown in Table 3), as compared with electrolytic iron powder. The nitrogen content must not be overlooked in view of the manufacturing method of nitride-process iron powder.

The effect of impurities on the magnetic properties of cold-rolled strips is much more marked than on the mechanical properties. In general, with increased purity of iron powder the magnetic properties of cold-rolled strips will deteriorate. In spite, however, of the purity of Höganäs MH·300P iron powder being higher than that of MH·100·24 iron powder, the magnetic properties of both cold-rolled iron powder strips are similar. The factors

TABLE 4. Mechanical and Physical Properties of Various Iron Strips Produced by Powder Rolling Method

Powder		Apparent density (g/cm³)	Tensile strength (kg/mm²)	Yield point (kg/mm²)	Elongation (%)	Electrical resistivity (μΩ-cm)	Heat treatment (in vacuo)
Electrolytic iron powder	Shoden	7.86	25.3	18.6	38.9	10.5	700°C, 1 hr
	HVA-Star	7.86	25.2	21.4	42.3	10.6	⁄⁄
Nitride process iron powder		7.82	30.5	26.2	44.0	11.6	⁄⁄
Mill-scale reduced iron powder		7.84	26.5	21.2	31.0	11.2	⁄⁄
Höganäs MH·100·24 iron powder		7.66	25.6	18.1	39.3	11.3	⁄⁄
Höganäs MH·100·28 iron powder		7.60	25.0	14.8	40.0	11.3	⁄⁄
Höganäs MH·300P		7.63	26.4	18.9	42.5	10.7	⁄⁄
Sponge iron powder		7.54	29.0	20.8	16.0	12.2	⁄⁄

TABLE 5. Magnetic Properties of Various Iron Strips Produced by Powder Rolling Method

Powder		B_s (gauss)	B_r (gauss)	μ_m	μ_0	H_c (Oe)	Heat treatment in dry hydrogen
Electrolytic iron powder	Shoden	14000	12800	9800	250	0.80	880°C, 1 hr
		13000	10600	11600	350	0.44	1200°C, 1 hr
	HVA-Star	13600	13000	8900	300	0.85	880°C, 1 hr
		14000	12400	10000	350	0.60	1200°C, 1 hr
Nitride process iron powder		13000	12000	6100	400	1.09	880°C, 1 hr
		14800	10600	9100	450	0.68	1200°C, 1 hr
Mill-scale reduced iron powder		12000	11500	3800	400	1.45	880°C, 1 hr
		12800	10900	4800	250	1.10	1200°C, 1 hr
Höganäs MH·100·24 iron powder		12400	11200	3600	200	1.68	880°C, 1 hr
		13800	12000	4000	300	1.62	1200°C, 1 hr
Höganäs MH·300P iron powder		13000	11400	3600	200	1.79	880°C, 1 hr
Sponge iron powder		5500	4000	1000	—	3.50	880°C, 1 hr

other than the purity, such as particle size, shape, and microdefects in the powder rolling process, seem to influence the magnetic properties of the cold-rolled strips. Not only the impurities but also the manufacturing method are considered to be important factors in the properties of cold-rolled strip.

2.5. Trial-Manufacture of Silicon-Iron Alloy Strips from Powder

The flow of mixed powder was lessened by increasing the silicon powder content. The mixed powder containing 4% and 6% silicon powder was powder rolled with a roll gap of 5/100 mm, a powder feed rate control gap of 2.0 mm and a roll speed of 12.5 rpm. The green strips were sintered at 1050 to 1130°C for 1 h in dry hydrogen, and cold rolled and annealed at 100°C in dry hydrogen. The cold rolling of silicon-iron alloy strips was relatively difficult compared with that of iron powder sintered strips, and special care was required for the first cold-rolling pass. Whereas silicon-iron alloys containing above 4% silicon prepared by the conventional melting method was hard to cold roll, the sintered strips made by the powder rolling process were capable of being cold rolled to a thickness below 0.1 mm.

TABLE 6. Mechanical Properties and Electrical Resistivity of Sintered and Cold-Rolled Silicon-Iron Alloy Strips Produced by Powder Rolling Method

Specimen		Sintering temperature (°C)	Thickness (mm)	Tensile strength (kg/mm²)	Elongation (%)	Electrical resistivity (μΩ-cm)	No. of specimen
Powder		1050	0.15	47.8	6.8	52.2	3
	4% Si-Fe	1100	0.13	54.1	11.2	59.0	2
rolling		1100	0.10	56.5	12.5	57.8	2
		1130	0.14	40.6	3.6	51.9	3
method	6% Si-Fe	1050	0.13	40.1	0.9	77.6	2
Conventional*	4% Si-Fe	—	—	45.1	4.0	59	—

* ASM Metals Handbook (1961)

The mechanical and physical properties of silicon-iron alloy strips made by the powder rolling process are shown in Tables 6 and 7. The mechanical properties and electrical resistivity of powder-rolled strips were similar to those of the strips made from molten ingot. From this point of view, it may be considered that the homogenization between iron powder and silicon powder was completely accomplished by solid phase reaction occurring in the process of sintering. Regarding the magnetic properties, however, the coercive force of the former was inferior to that of the latter, and that of 6% Si-Fe alloy strip made by powder rolling process was not superior to that 4% Si-Fe alloy strip, despite the expectation that smaller coercive force and higher permeability should be obtained from the larger silicon content. The reason for this phenomenon is thought to be due to the coercive force of 6% Si-Fe alloy strip being increased because of the porosity due to its less workability and the remaining oxides. The oxides on the surface of silicon cannot be reduced by heat treatment in hydrogen. It is also reported that the coercive force depends upon the grain size, i.e., the smaller the grain size, the larger the coercive force. Further improvement may be expected by improving various fac-

TABLE 7. Magnetic Properties of Silicon-Iron Alloy Strips Produced by Powder Rolling Method

Specimen		B_s (G)	Br (G)	μ_m	μ_0	Hc(Oe)	Heat treatment
Powder	4% Si-Fe	14300 15200	12150 12500	6000 5800	800 600	0.93 0.81	900°C, 1 hr in vac. 1100°C, 1 hr in vac.
rolling method	6% Si-Fe	13800	11800	5700	650	0.85	1100°C, 1 hr in vac.
Conventional 4% Si-Fe*		(Bs) 10000	12000	6000	600	0.5	—

* ASM Metals Handbook (1961)

tors such as heat-treatment conditions, process of adding silicon powder, e.g., adding silicon-iron in the form of mother alloy powder, process of hot rolling of sintered strips, etc.

3. Conclusions

(1) The effect of powder rolling conditions on the rolling load and the thickness and apparent density of green strips made from Höganäs MH · 100 · 24 sponge iron powder has been found to be as follows:

	Powder feed rate Larger	Roll gap Larger	Rolling speed Larger
Rolling load	Larger	Larger	Larger
Thickness of green strip	Larger	Larger	Larger
Apparent density of green strip	Larger	Larger	Larger

The control of powder feed seems to be the most favorable manner in which to control the thickness and apparent density of green strips.

(2) Uniformly thick and crack-free green strips may be obtained with powder rolling by adapting rolling conditions to the characteristics of the iron powders. The rolling load is an optimum measure of estimating favorable rolling conditions.

(3) The mechanical and physical properties of sintered strips are affected by the powder rolling conditions. Better properties are obtained by sintering the green strip having higher density.

(4) The effect of powder rolling conditions on the mechanical and physical properties cold-rolled strips is not remarkable, except that the elongation is increased as the rolling load increases in the process of powder rolling.

(5) As a result of comparison between various cold-rolled strips made from various iron powders, it was found that the mechanical and physical properties of the strips cold-rolled by the powder rolling method were equal to those of conventionally produced materials. The magnetic properties were markedly influenced by the purity and characteristics of the iron powder used.

(6) The powder rolling method proved successful in cold rolling the silicon–iron alloy strips containing 4% and 6% silicon to a thickness of below 0.1 mm even though it is difficult to cold roll this material by the conventional method.

Acknowledgment

The authors take this opportunity to express their gratitude to Dr. M. Sugihara, Chief of the Magnetic Materials Section, and to the staff engineers of this section for their continued cooperation and valuable suggestions. The authors also wish to thank Professors I. Obinata and Y. Masuda, Tohoku University, for their helpful cooperation and advice on this work.

References

1. Siemens Halske, German Patent 154,998.
2. G. Naeser and F. Zirm, Stahl u. Eisen 70:995 (1950).
3. S. R. Crooks, Iron and Steel Eng. p. 72, Feb., 1962.
4. P. E. Evans and G. C. Smith, Symposium on Powder Metallurgy, 1954, J. Iron and Steel Inst, Special Rept. 58:131(1956).
5. P. E. Evans and G. C. Smith, Powder Met. 26(3) 1 (1959).
6. D. K. Worn and R. P. Perks, Powder Met., No. 3. p. 45, (1959).
7. S. Storchheim, J. Nylin, and B. Sprissler, Sylvania Technologist 8:42 (1955).
8. R. A. Smucker, Iron and Steel Eng., p. 118 July, 1959.
9. Canadian Chem. Process, 46:52 (1962).
10. T. Kimura, H. Hirabayashi, and M. Tokuyoshi, J. Jap. Soc. Tech. Plasticity, 4:259 (1963).
11. I. Obinata and Y. Masuda, Rept. contracted with Elect. Comm. Lab., 1961.

The Effect of Die-Wall Lubrication and Admixed Lubricant on the Compaction of Sponge-Iron Powder

Phillip M. Leopold

Babcock and Wilcox Research Center
Alliance, Ohio

and

Russell C. Nelson

Associate Professor
Mechanical Engineering Dept.
University of Nebraska

Effect of Lubrication on Powder Compaction

Introduction

During the process of compacting metal powders in a die, several types of friction act to decrease the effectiveness of the applied pressure. The frictional forces are: (a) between the moving punch and die wall; (b) between powder particles; (c) between powder particles and the die wall; and (d) between the compact and the die wall during ejection.

The first of these frictional forces depends on the die materials, on the surface conditions of the die parts, and the clearance between the punch and die wall. It is almost negligible in comparison to the other frictional forces involved. The interparticle friction is complicated in nature and is influenced by the powder material, particle size, shape, and surface conditions, and by the pressure between particles. Limited and contradictory investigations have been performed in this area but they point out that is is difficult to discern when particle sliding (friction) stops and plastic deformation begins (or, for that matter, if it occurs at all). The pressure losses which mainly occur during compaction at a given pressure are caused by the friction between powder particles and the die wall (a function of the particular materials involved); the number of powder particles in the mold, which is a function of the shape and the amount of powder (volume) being compacted; as well as the surface condition of both the powder particles and the die material. The value of the maximum die reaction which occurs as a result of die-wall friction is the result of the extent of cold welding between the compact and the die walls. It has been shown experimentally that the die reaction is a measure of the product of the shear strength of the metal under compression and the true area of contact between the metal and the die wall.

In order to decrease these frictional forces, lubricants are either applied to the die wall or admixed with the metal powder. Although lubrication plays an important role in the production of parts by powder metallurgy, very little is known about the best way to apply lubrication. Recent work by M. Burr [1, 2, 3] indicated that the admixture of the lubricant to the powder offers hardly any advantages compared to die-wall lubrication as regards pressure-density relationship, whereas Hausner and Sheinhartz [4] found that the addition of lubricant to the powder affects density and green strength of the pressed compact, depending on the amount of lubricant added. Other investigations by Ljungberg and Arbstedt [5] and by Yarnton and Davies [6] also covered the same subject from various angles.

No references were found where the amount of lubricant sprayed on the die wall was investigated as a variable. Techniques for measuring the coefficient of friction between the die wall and powdered metal were described by Train and Hersey [7], Long

[8], and Hausner and Sheinhartz [4]. The first two investigations measured the friction by moving the die past the powdered metal, while Hausner and Sheinhartz used the force transmitted to the bottom punch as an indication of the average frictional force. However, none of these investigations attempted to change this coefficient except by varying the percentage of admixed lubricant.

Because of the conflicting viewpoints regarding the effect of die-wall friction on the density of the powder metal compacts, it seemed desirable to study the effect of different values of frictional forces on the green compact density. Also, very little is known about how the ejection force varies with changes in die-wall lubricant and admixed lubricant.

Experimental Equipment

All pressing operations were carried out on an Instron Universal Testing Machine* with a 10,000 lb capacity in compression. The compaction was performed in a 0.400 in. (I.D.) by 2.0 in. long hardened (Rc 60) steel die. The inside of the die had a hard, chrome-plated finish. The punches were machined from 410 stainless steel which had a nominal yield strength of 125,000 psi. The clearance between the die and punches was 0.0006 ± 0.0001, which is consistent with the clearances recommended by Jones [10] and others. The maximum compaction pressure attainable was 80,000 psi.

In order to measure the load transmitted from the top punch to the die, a small load cell was designed and built (Fig. 1) which contained three strain gages installed at symmetrically similar points (designated by "A" in Fig. 1). These points were then connected in a series to eliminate eccentricity. A similar installation of three gages was made at points designated "B." Thus, a two-arm bridge was used with two series of gages, one of which was in tension and the other in compression, thereby eliminating any need for temperature compensation. The load cell was calibrated by means of a static strain indicator. The response was almost linear and the calibration curve obtained indicated that the cell had good sensitivity and stability, up to the maximum load of 6,000 lb.

A Brush amplifier and oscillograph was used to measure the strain and thus the load transmitted to the cell. This equipment

*Instron Engineering Corporation, Canton, Mass.

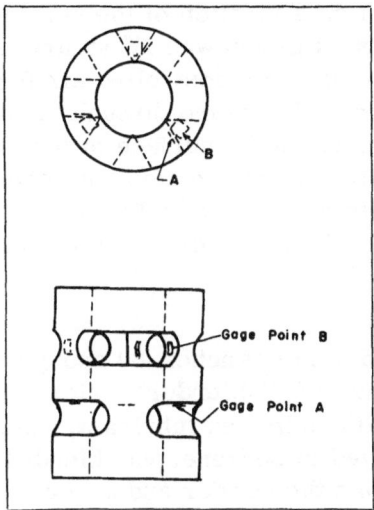

Fig. 1. Load cell (full size).

also offered a means of continuously recording strain as a function of time to complement the Instron chart, which recorded the applied load as a function of time.

In order to vary the amount of die-wall lubricant and to be able to reproduce the amount applied, an apparatus was designed and built which permitted a constant amount of lubricant to be ap-

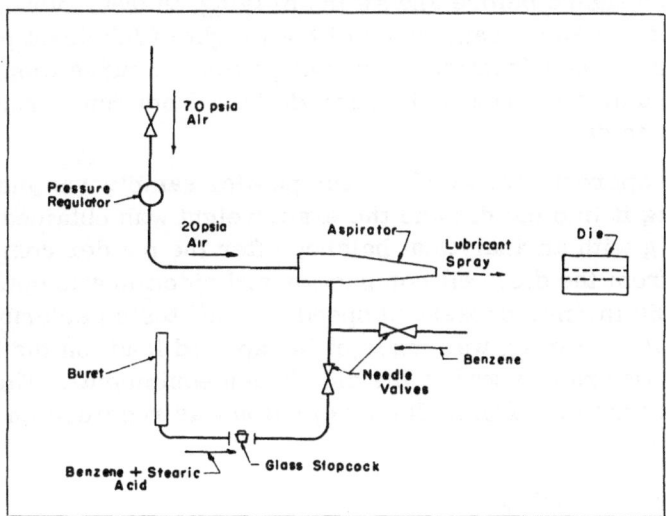

Fig. 2. Schematic of lubricant and air flow.

plied to the die wall as a function of the spraying time. The lubricant used in this investigation was 1% stearic acid dissolved in benzene. An aspirator and a controlled air flow were used to spray the benzene–stearic acid mixture through the die. The air stream evaporated the benzene on the surface of the die and thus left a fairly uniform coating of stearic acid. A schematic flow diagram of the spraying system is shown in Fig. 2. By means of a buret and a stop watch, the flow rate of the lubricant could be determined.

Test Procedure

The powdered metal (Anchor-80 sponge-iron powder*) was blended for one hour in a V-blender† prior to any compaction studies. However, the admixed lubricant, which was also 1% stearic acid dissolved in benzene, was blended into the powder by hand-mixing, because the powder and a liquid could not be mixed suitably in the blending equipment available. The percentage of admixed lubricant was varied simply by regulating the amount of stearic acid added.

By utilizing the spraying apparatus and a standardized flow rate of 11 ml/min, the amount of lubricant deposited on the die wall could be controlled by the spraying time. The evaluation of this method was carried out on a piece of glass tubing of similar dimensions to that of the die (i.e., 0.3975 in. I.D. × 2 in. long). The tube was weighed before and after spraying, thereby permitting calculation of an average value of the weight of lubricant per unit surface area as a function of spraying time. A curve was plotted from the data and found to be reproducible from run to run and from day to day.

An approximate weight of the powder sample was made prior to pouring it into the die and the exact weight was obtained to ± 0.0001 g with an analytical balance after the powder compact was ejected from the die. All compaction and ejection was done at the rate of 0.50 in./min crosshead speed. On all tests performed, a continuous recording was made of the applied load (on the Instron) and the load transmitted to the die (Brush equipment). The applied load as a function of time during ejection was recorded on the Instron chart.

*Höganäs Sponge Iron Corp.
†Patterson-Kelley.

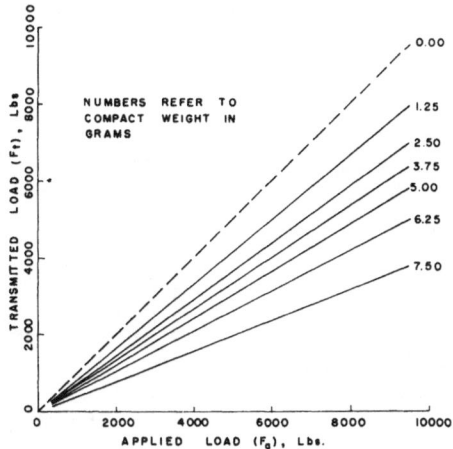

Fig. 3. F_a vs F_t for no lubrication.

Using a procedure developed by Heckel [9] and the continuous chart recordings of the applied load versus time, it was possible to compute the density of the compact as a function of several applied pressures from a single test run. All compact densities were converted to a true compact density by subtracting the weight of the admixed lubricant.

Test Results and Discussion

A total of 93 powdered iron compacts were pressed. The weight of powder, amount of admixed lubricant, and amount of die-wall lubricant were varied to find the relationship of these variables with green density and ejection force. There were six different weights of powder used, ranging from 1.25 to 7.5 g in 1.25-g increments. The amounts of stearic acid blended into the powder were: 2.00, 1.50, 1.00, 0.50, and 0.25 weight percent. The amount of die-wall lubricant was varied from 0.162 to 0.006 mg/cm^2, with five values investigated between these limits.

Analysis Using K-Factor

The transmitted load refers to the load transmitted to the bottom punch and is designated as F_t, and the applied load is designed as F_a. By plotting F_t versus F_a it is found that, for all conditions investigated, this is approximately a linear relationship (Fig. 3). Since this relation approximates a straight line it can be

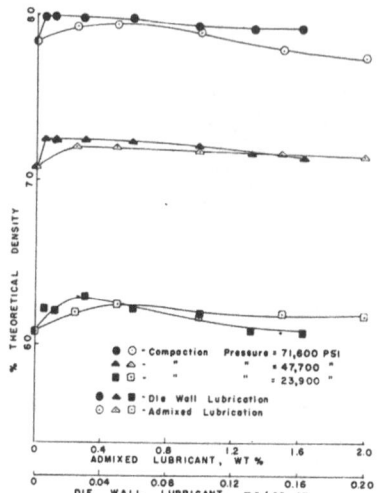

Fig. 4. Density vs amount of lubricant for 1.25-g compact.

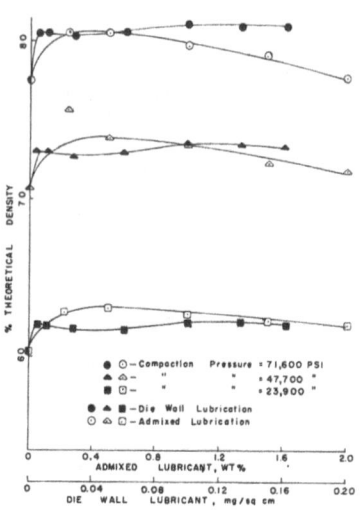

Fig. 5. Density vs amount of lubricant for 3.75-g compact.

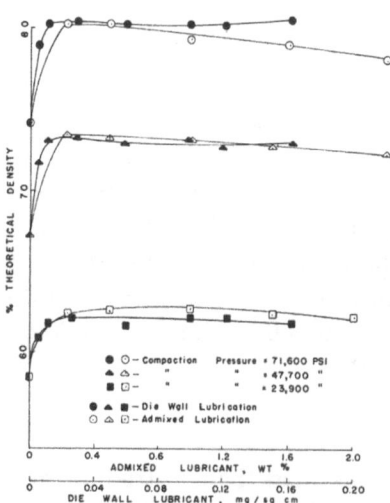

Fig. 6. Density vs amount of lubricant for 6.25-g compact.

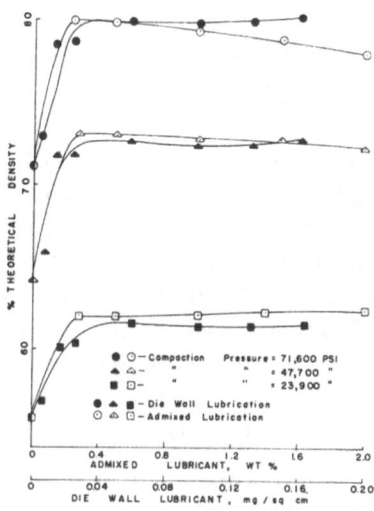

Fig. 7. Density vs amount of lubricant for 7.50-g compact.

represented by the equation

$$K = F_t/F_a \qquad (1)$$

where K is the slope of the line and is independent of applied pressure. This same relationship is also expressed by Jones [10] and Hausner and Sheinhartz [4]. An evaluation of the lubrication methods revealed that die-wall lubrication is more effective than admixed lubricant in increasing K (i.e., the transmitted load). This is reasonable because the admixed lubricant must be forced from the powder to the die wall during pressing in order to effectively increase the transmitted pressure. Within the range of admixed lubricant investigated there was little difference in the K-factor for a given compact weight. The amount of die-wall lubricant has a large effect on the K-factor which is more pronounced with increasing compact weight. Actually the K-factor increases rapidly with increasing lubricant at first, and then levels off, indicating that only a small amount of die-wall lubrication is necessary to give a high K-factor.

Green Compact Density

The measured compact densities were converted to a true compact density by subtracting the weight of the admixed lubricant. The compact volume was determined from measurements of compact diameter and height. The true compact density is reported as a percentage of the theoretical density of iron (7.874 ± 0.001 g/cc at 20°C).

Admixed Lubricant

A series of curves was obtained relating the true percent theoretical density and the weight percent of admixed lubricant for three different compaction pressures (Figs. 4-7). Each family of curves represents a different compact weight for the variables enumerated. The graphs show that each curve has a point of maximum density, where the lubricant has its maximum effectiveness. Also, it can be seen that this point of maximum lubrication effectiveness shifts with decreasing pressure towards a greater amount of lubricant, if only one given weight of powder is considered. These results confirm an earlier investigation by Jones [10].

By plotting the amount of admixed lubricant to give maximum density for different compact weights, Fig. 8 was obtained. By using this curve, one can determine the optimum amount of lubricant to give the maximum green density for a given combination of powder weight and compaction pressure. The low pressure curve diverges rapidly from the other two curves because at low pressure the lubricant is not forced to the die walls to decrease the die-wall friction. Thus a greater amount of lubricant would be necessary to effectively reduce the die-wall friction.

A second important point can be seen from Figs. 4-7. For small compacts (i.e., low weights), admixed lubrication has little effect on the true density, whereas for large compacts a very small amount of lubricant will markedly increase the true density.

Die-Wall Lubrication

The effect of die-wall lubrication on the density is also shown in Figs. 4-7, whereby the density continues to increase with increasing amounts of lubricant. There is apparently no optimum lubrication point which gives a higher density similar to the effect discussed for admixed lubricant. At high pressures (i.e., 71,600 psi), the die-wall lubrication method gives higher green density than admixed lubrication for all weights of powder investigated. At medium pressures there is little difference in the effectiveness of the different lubrication methods, since each method gives approximately the same curve. At low pressures, however, the ad-

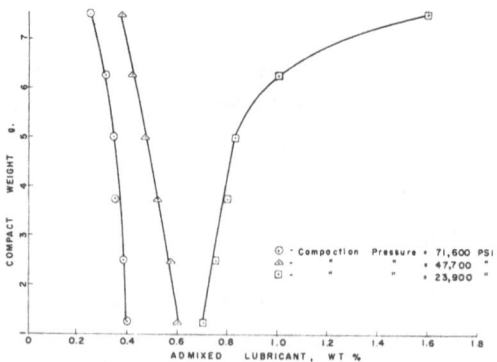

Fig. 8. Percent admixed lubricant to give maximum density for a given compact weight.

mixed lubrication gives much higher densities for all weights considered. These results are in agreement with those reported by Ljungberg and Arbstedt [5], but the data just presented show the effect found by these investigators as well as the opposite effect at higher pressures.

For higher weight compacts, the density can be increased quite significantly with either higher admixed lubricant or die-wall lubrication. These results are in agreement with Jones [10] and Hausner and Sheinhartz [4], but in disagreement with Burr [1] and Burr and Donachie [2, 3]. An interesting effect is noted regarding the compact weight and percent theoretical density as a function of the amount of die-wall and admixed lubricant. There is a weight of compact for which a maximum density occurs, which is not the smallest compact investigated. It is assumed that this is due to restricted particle movement with decreasing compact size, which does not permit satisfactory particle repacking and deformational bonding.

Conclusions

1. A linear relationship exists between the applied force and the transmitted force for all conditions investigated, as expressed by the K-factor.

2. Only a very small amount of either die-wall or admixed lubricant is necessary to give a high K-factor, increasing amounts of lubricants having little effect.

3. There is an optimum amount of stearic acid which gives a maximum true density for a given combination of powder weight and compaction pressure. The optimum amount of admixed lubricant increases with decreasing compaction pressure, decreases with increasing weight at the higher compacting pressures, and increases with increasing weight at low compaction pressures.

4. For small compacts, admixed lubrication has little effect on the true density, whereas for large compacts a very small amount of admixed lubricant will markedly increase the true density.

5. There appears to be no optimum amount of die-wall lubrication which gives a higher density than other amounts. Within the range investigated, the density increased with increasing amount of die-wall lubrication for all compacting pressures and compact weights.

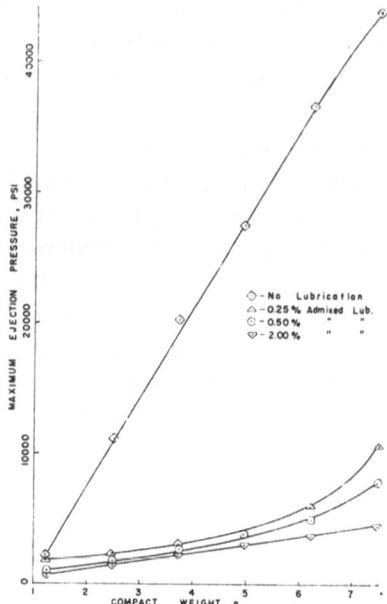

Fig. 9. Effect of admixed lubricant and
compact weight on the ejection pressure.

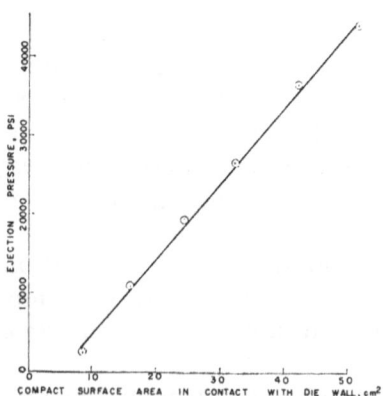

Fig. 10. Ejection pressure vs compact
surface area for no lubrication.

6. Die-wall lubrication gives higher densities at high compact-
    ing pressures, while at low compacting pressures, the den-
    sity is higher for admixed lubricants.
7. For the larger weight compacts, the density can be increased
    significantly by either die-wall or admixed lubrication,
    whereas for lubrication of the lesser weight compacts there
    is only a slight change in density for any compaction pres-
    sure.

Effect of Lubrication on Ejection Pressure

As pointed out previously, frictional forces are present during
ejection between the powder compact and the die wall. Furthermore, no comparison of lubrication methods and ejection pressure
had been made. The same set-up was used as for compaction and
the compact was ejected from the die in the same direction as it
was compacted. A small cylinder of the same dimensions as the
load cell, but without the machined slots, was used during ejection
to eliminate any possible damage to the load cell. This substitu-
tion was permissible since the Instron chart could be used to
measure the ejection force.

Ejection Pressures

The pressure required to eject a completed compact from
the die is affected inversely by the amount of lubricant. By plotting
the maximum pressure applied during the ejection cycle versus
the compact weight for various amounts of admixed stearic acid,
the relation shown in Fig. 9 was obtained.

No Lubrication

For no lubrication, the ejection pressure versus compact
weight is nearly a linear relationship. Since the surface area is a
function of the weight of the compact, by plotting the surface area
of the compact in contact with the die wall during ejection against
the ejection pressure (Fig. 10), essentially a straight line is obtained.
This graph indicates that for unlubricated conditions the ejection
pressure is directly related to the area over which friction is acting and that the coefficient of friction along the length of the compact is constant.

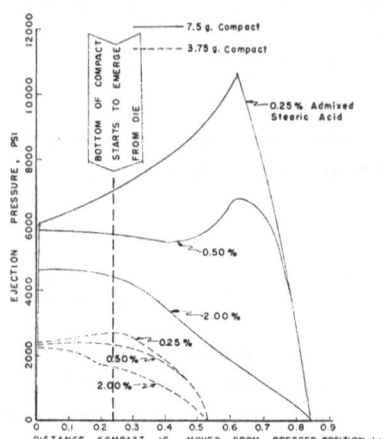

Fig. 11. Compact position vs ejection pressure as a function of admixed lubricant.

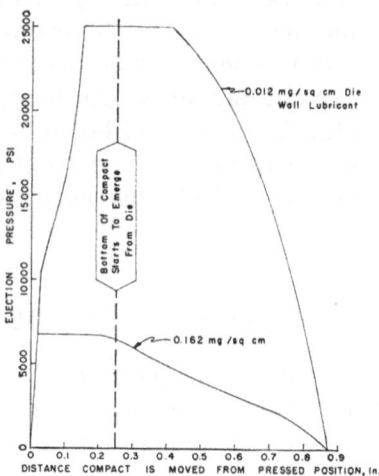

Fig. 12. Compact position vs ejection pressure as a function of die-wall lubrication (7.5-g compact).

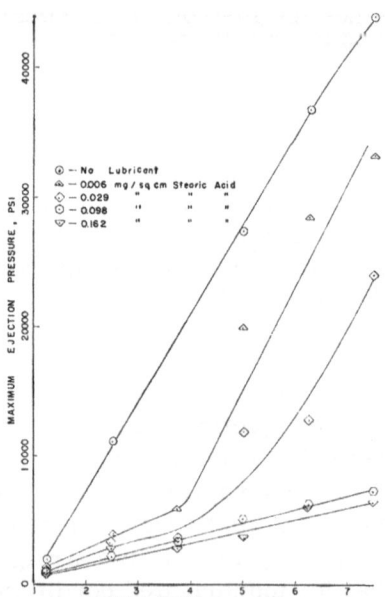

Fig. 13. Effect of die-wall lubrication and compact weight on ejection pressure.

Admixed Lubrication

When the lubricant is admixed with the powder, the coefficient of friction does not appear to be linear with either the weight or the surface area of the compact in contact with the die. In Fig. 11 the ejection pressure is plotted as a function of the distance the compact is moved in the die from its pressed position for both 3.75- and 7.5-g compacts.

One conclusion that can be drawn from Fig. 11 is that, for a compact with a small amount of admixed lubricant, the amount of lubricant squeezed onto the die walls during pressing is insufficient to maintain a low ejection force as the compact is moved. As the compact moves along the die, the lubricant is left behind and the new die-wall surface, where the bottom punch had previously been, is unlubricated and thus the ejection pressure increases with distance. In a compact with a large amount of admixed lubricant, there is enough lubricant squeezed out during the compaction operation to effectively reduce the ejection pressure during the entire ejection operation. The curves also show that for the smaller (3.75-g) compact 0.5% admixed lubricant is sufficient to reduce the ejection pressure during the ejection cycle, since the peak pressure occurs at the start of the ejection operation. This shows why the curves for 0.25% and 0.50% admixed stearic acid are not linear in Fig. 9, for the larger weight compacts since the pressure is a function of more than the surface area on which friction is acting.

Die-Wall Lubrication

This line of reasoning is substantiated by Fig. 12, which shows that for die-wall lubrication the maximum pressure incurred during ejection is at the beginning of the ejection cycle, even for a very small amount of die-wall lubrication. In the case of die-wall lubrication, the lubricant is on the die wall, where the bottom punch is placed during pressing and remains there to reduce the ejection pressure during the entire ejection cycle. In contrast to this, for the case of admixed lubricant, there is no lubricant in this region to decrease the ejection pressure.

The effect of die-wall lubrication on the ejection pressure is illustrated in Fig. 13. The points are scattered for the very small amounts of lubricant, because of the problem of controlling spraying time. A comparison of Figs. 9 and 13 shows that a small amount of die-wall lubricant is much more effective in decreasing the ejection force than a large amount of admixed lubricant.

The general trend of decreasing ejection pressure with increasing amounts of admixed lubricants is substantiated by Yarnton and Davies [6] and Ljungberg and Arbstedt [5]. However, neither of these investigations took into account the effect of compact weight, varying amounts of die-wall lubrication, or the position of the compact in the die.

Conclusions

1. Increasing the amount of lubricant (by either method) decreases the ejection pressure.
2. A small amount of die-wall lubrication is much more effective in decreasing the ejection pressure than a large amount of admixed lubricant (i.e., 0.012 mg/cm^2 die-wall lubrication is much more effective than 2.0% admixed lubricant).
3. For the higher weight compacts and admixed lubrication, the ejection pressure increases as the compact moves toward the end of the die. For all cases of die-wall lubrication investigated, the maximum ejection pressure occurred for the initial movement of the compact in the die.

References

1. Burr, M. F., "Compressibility of Copper Powders," Master's Thesis, Rensselaer Polytechnic Institute, Troy, N.Y. (June, 1961).
2. Burr, M. F., and Donachie, M. J., Jr., "Effects of Pressing on Metal Powders," J. Metals 15 (11) 849-54 (Nov. 1963).
3. Burr, M. F., and Donachie, M. J., Jr., "Effects of Pressing on Copper Powders," Trans. ASM Quarterly 56 (4) 863-6 (Dec., 1963).
4. Hausner, H. H., and Sheinhartz, I., "Friction and Lubrication in Powder Metallurgy," in Proc. of Metal Powder Assoc., 1954, p. 6-27.
5. Ljungberg, I., and Arbstedt, P. G., "Influence of Lubrication and Die Surface on the Pressing Characteristics of Metal Powders," Proc. of Metal Powder Assoc., 1956, p. 78-92.
6   Yarnton, D., and Davies, T. J., "The Effect of Lubrication on the Pressing of Metal Powder Compacts," Powder Met. No. 11, 1-22 (1963).

7. Train, D., and Hersey, J. A., "Some Fundamental Studies in the Cold Compaction of Plastically Deforming Solids," Powder Met. No. 6, 20-35 (1960).

8. Long, W. M., "Radial Pressures in Powder Compaction," Powder Met. No. 6, 73-86 (1960).

9. Heckel, R. W., "Density-Pressure Relationships in Powder Compaction," Trans. AIME 221 (4) 671-5 (Aug., 1961).

10. Jones, W. D., "Fundamental Principles of Powder Metallurgy," Edward Arnold Publishers, London (1960) p. 242-319.

Hot-Pressing of Iron Powders

Otto H. Henry

Associate Professor of Metallurgical Engineering
Polytechnic Institute of Brooklyn
Brooklyn, New York

and

J.J. Cordiano

Instructor
Polytechnic Institute of Brooklyn
Brooklyn, New York

Excluding the refractory metal field, most commercial powder metallurgical products made by a cold compressing and sintering operation contain voids. These voids can be substantially eliminated by subsequent severe working and heat treatment, but such practice has not proved economical in most applications.

By applying heat during the compressing operation, plastic deformation of the powder particles is substantially increased, and the resultant compact or briquette is practically free of voids. However, hot-pressing has many attendant problems, which at present preclude its use in the manufacture of precision parts. The more important of these are to devise:

1. A die steel capable of withstanding high pressures at elevated temperatures.

2. A high-temperature lubricant capable of preventing excessive die wear and welding of the compact to the die walls.

3. A method of retaining a neutral or reducing atmosphere in the compacting zone of the die assembly.

4. A means of ejecting the hot-pressed specimen into a neutral atmosphere until reasonably cold.

5. A means of confining to the die cavity the heat developed.

6. A design of a soundly engineered machine for mass production.

Apparatus

Three methods are generally used for supplying heat to the specimens being hot-pressed: (1) induction; (2) passing a current through the highly resistant powder mass; (3) enclosing the entire die assembly in a furnace. (Because of its relative simplicity, the latter method was use in this investigation.)

The details of the apparatus are shown in Figs.1 through 10. The die, upper ram, and lower punch, and sliding block were made of heat-resistant steel. The die block was made of cast iron and the chute of sheet steel. The chemical analysis and elevated-temperature properties of this steel are shown in Table 1.

Material and Procedure

The metal powder used in these tests was an electrolytic iron having the characteristics listed in Table 2.

1. To facilitate handling during hot-pressing, the iron powder was precompacted cold at 20 tons/sq in. into 3 by $^3/_8$ by $^5/_8$ in.

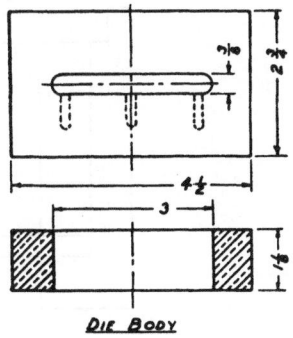

DIE BODY

Fig. 1. Die body.

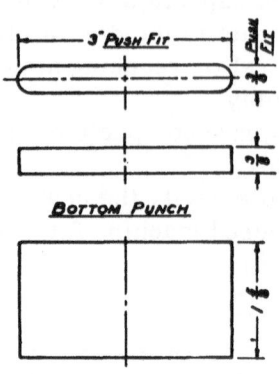

Fig. 2. Punches.

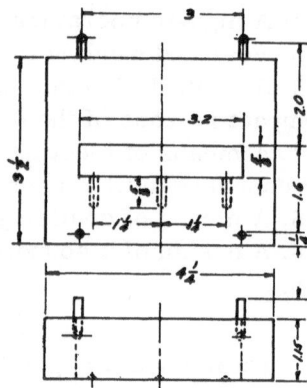

Fig. 3. Sliding block.

specimens. The die lubricant was composed of one gram of stearic acid dissolved in 20 cc of carbon tetrachloride.

2. The hot-pressing punches and die parts were rubbed with flake graphite to provide high-temperature lubrication.

3. With the die assembly as indicated in Fig. 7, the system was purged with dry nitrogen before the furnace was brought to temperature. Gas flowed through the system by passing up through

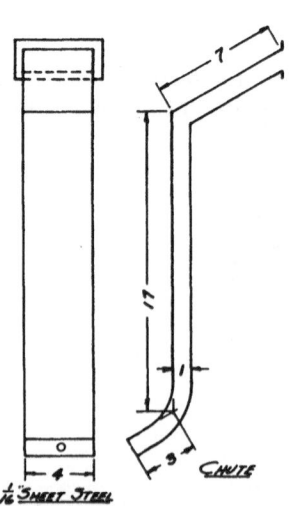

Fig. 4. Chute.

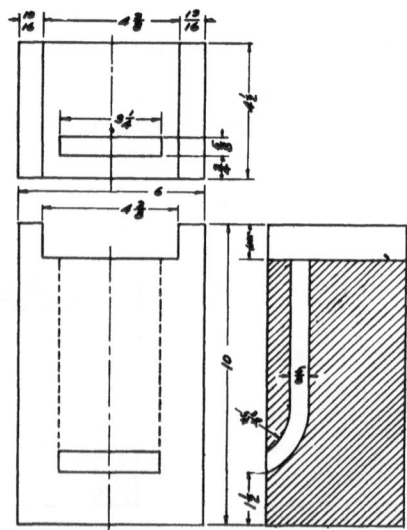

Fig. 5. Die block.

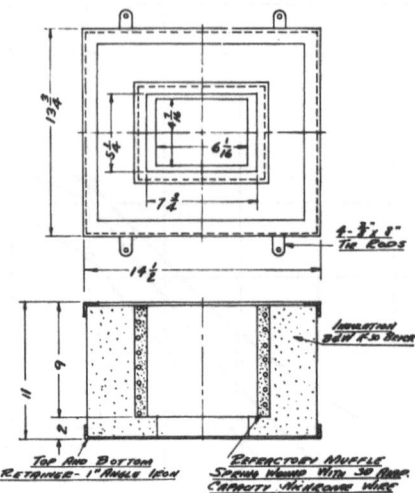

Fig. 6. Electric resistivity furnace.

the chute E and die block D into the sliding block B and through the
die cavity. Grooves cut into the bottom of the die permitted the
flow of gas from the sliding block into the die cavity. The nitrogen
served two purposes: to prevent oxidation of the die parts and to
provide a means for purging the hot die cavity with hydrogen without forming an explosive mixture.

    4. When the furnace had reached temperature, hydrogen was
passed through the system instead of nitrogen. Before it was led
into the chute, the hydrogen was purified by bubbling through sulfuric acid traps and passing through phosphorus pentoxide drying
towers. The hydrogen escaping from the upper portion of the die
assembly was ignited and allowed to burn.

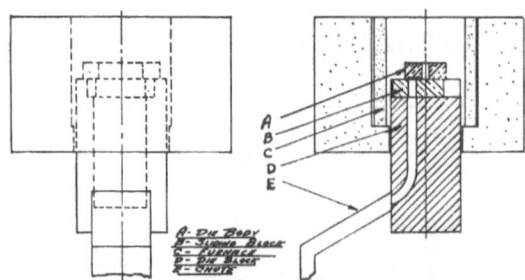

Fig. 7. Die assembly.

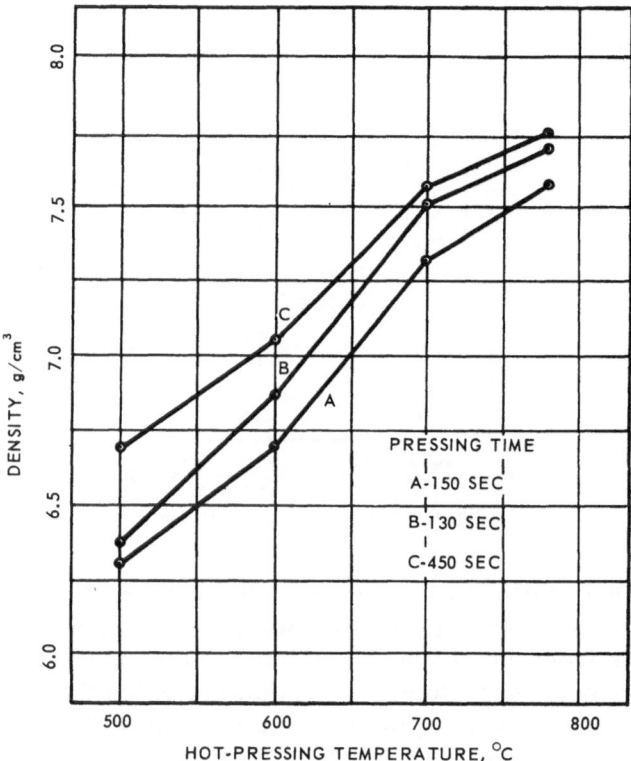

Fig. 8. Relation between density and pressing temperature for indicated time.

TABLE 1. Analysis and Properties of Punch and Die Steel Used in Apparatus

Chemical Analysis, Per Cent

C................... 0.33   Ni.............. 19.70
Mn................ 0.49   Cr.............. 8.31
Si................ 1.16   Fe.............. Bal.

Properties at Elevated Temperatures

Temperature, Deg. C	Tensile Strength, Lb. per Sq. In.	Proportional Limit, Lb. per Sq. In.	Elongation, Per Cent in 2 In.
20	112,500	45,200	21.0
205	102,000	45,600	20.0
480	90,800	40,100	19.7
595	80,000	35,500	19.7
650	74,000	35,800	19.3
705	65,400	26,200	18.5

TABLE 2. Characteristics of Electrolytic Iron Powder Investigated

Apparent density in hall flowmeter, grams per c.c. 2.4
Flow (for 50 grams) in hall flowmeter, sec. ... 46
Hydrogen loss at 850°C. for one hour at temperature in dry hydrogen, per cent 0.34

Chemical Analysis, Per Cent		Screen Analysis, Per Cent	
C.	0.005	+100.	1.0
Mn.	0.002	−100+150.	10.5
Si.	0.003	−150+200.	21.5
P.	0.001	−200+250.	11.0
S.	0.004	−250+325.	13.0
Ni.	0.008	−325.	43.0

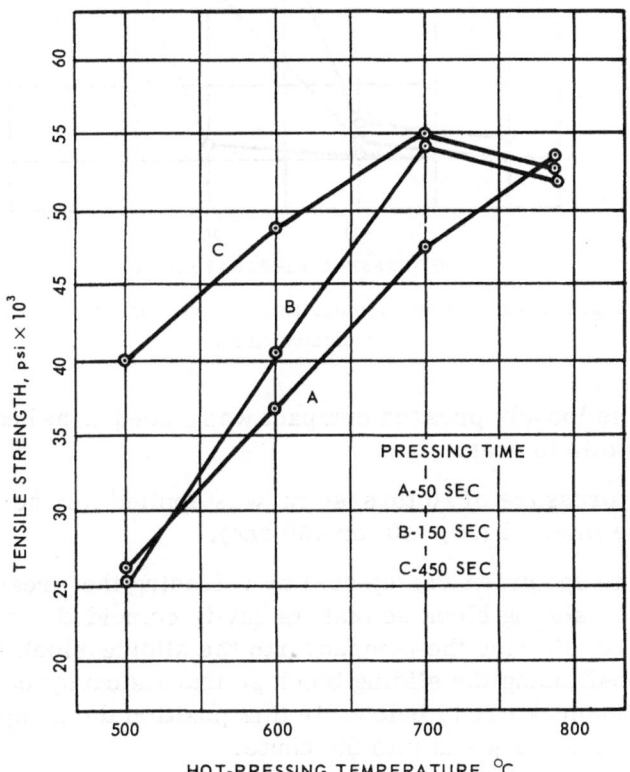

Fig. 9. Relation between tensile strength and pressing temperature for indicated time.

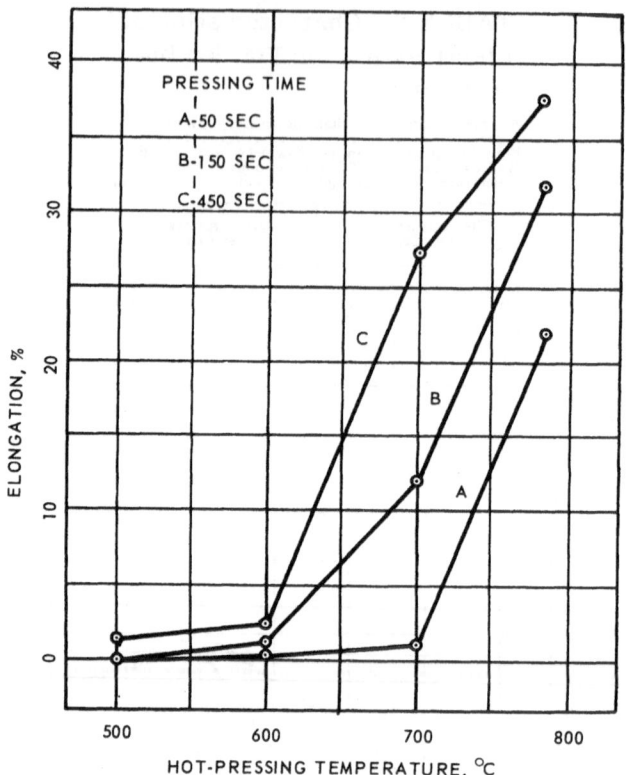

Fig. 10. Relation between ductility and pressing temperature
for indicated time.

5. The loosely pressed compact was placed in the heated die and left for 10 min to soak.

6. A pressure of 10 tons/sq in. was applied for the predetermined time interval (50, 150, or 450 sec).

7. The specimen was ejected by releasing the pressure, positioning the sliding block so that its cavity coincided with the die cavity above, ejecting the compact into the sliding-block cavity, and then positioning the sliding block so that its cavity coincided with the die-block cavity below. In this position the compact fell through the die block and into the chute.

8. When the specimen had cooled, the hydrogen was replaced by nitrogen, the specimen removed, and the system completely purged with nitrogen before hydrogen was again used.

9. The die was quenched in oil, rubbed with graphite again, and replaced in the assembly for the next run.

10. The compacts obtained by hot-pressing were machined to conform to standard A.S.T.M. specifications for tensile specimens of $1/4$-in. diameter.

11. Tension, density, and hardness tests were conducted on the specimens. The tensile tests were conducted on a 10,000-lb Riehle universal testing machine. The density determinations were made by the water-immersion method. To guard against absorption of water, the specimens were coated before immersion with a wax of known specific gravity. Hardness tests were conducted on a Brinell hardness tester using a 500-kg load and a 10-mm ball.

12. A microscopic examination was made of each specimen (Figs. 12 and 13).

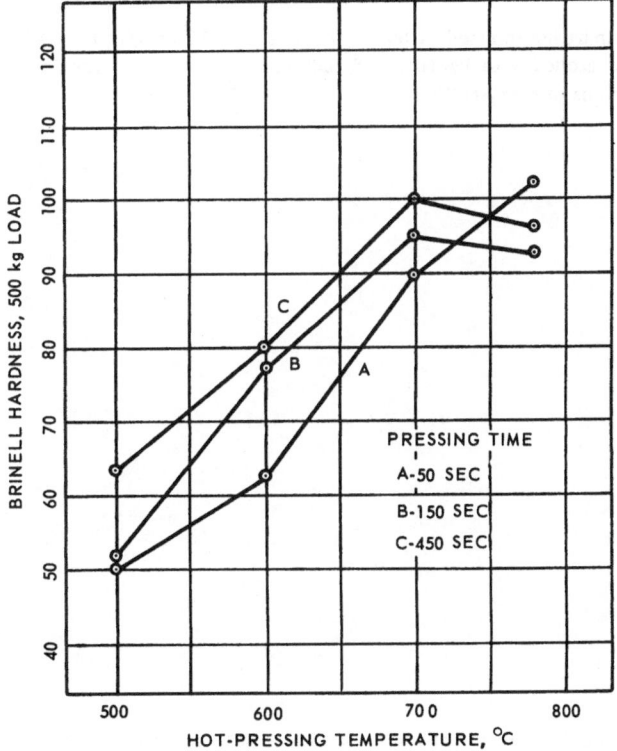

Fig. 11. Relation between hardness and pressing temperature
for indicated time.

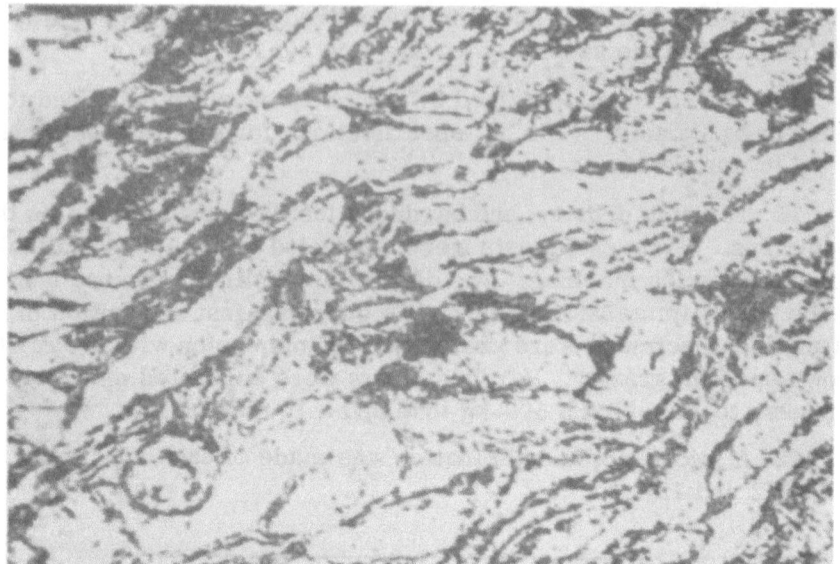

Fig. 12. Compart hot-pressed at 500°C for 50 sec at 10 tons/sq in. Original magnification 500 x. Etched with 1% Nital. Structure shows extreme distortion of particles with little if any signs of recrystallization.

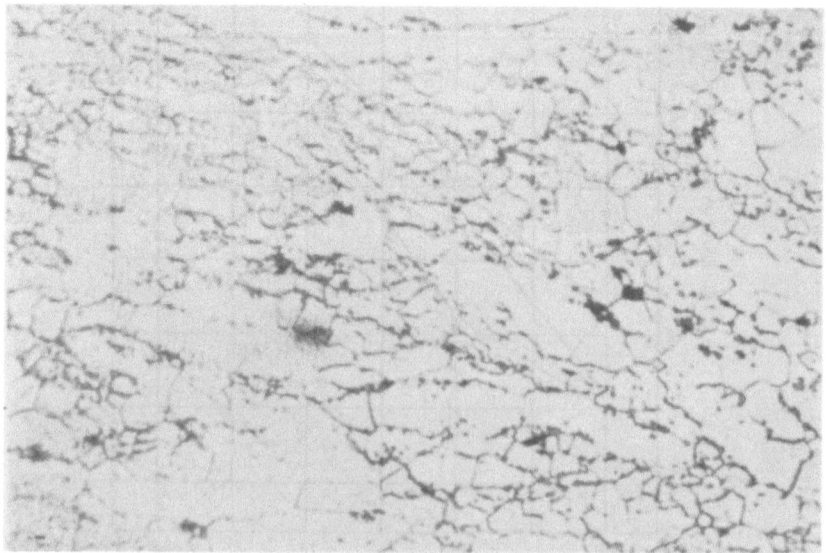

Fig. 13. Compact hot-pressed at 780°C for 450 sec at 10 tons/sq in. Original magnification 1000 x. Structure shows absence of particle boundaries and a completely recrystallized structure.

TABLE 3. Results

Temper-ature, Deg. C.	Time, Sec.	Den-sity, Grams per c.c.	Tensile Strength, Lb. per Sq. In.	Elon-gation, Per Cent in 1 In.	Brinell Hard-ness, 500-kg. Load, 10-mm. Ball
500	50	6.31	26,200	0	50
500	150	6.38	25,500	0	51
500	450	6.71	39,800	1.0	63
600	50	6.70	36,900	0.5	62
600	150	6.89	40,800	1.0	77
600	450	7.05	48,800	2.0	80
700	50	7.32	47,800	1.0	90
700	150	7.52	57,300	12.0	95
700	450	7.58	57,500	27.0	100
780	50	7.59	54,100	22.0	101
780	150	7.71	52,400	32.0	93
780	450	7.76	52,900	37.0	96

Discussion of Results

The results of these tests (Table 3) indicate the potent influence of time and temperature on the physical and mechanical properties of hot-pressed iron powders. These properties are dependent upon the amount and rate of diffusion, which in turn depend on the number and area of particle contacts, as well as on the amount of plastic deformation to which the compact is subjected. Increasing temperature and time at temperature under sustained pressure tends to increase each of these effects individually, with cumulative results.

Density

The curves in Fig. 8 show a decided increase in density with increasing temperature and to a lesser extent with increasing time at temperature. If intermediate points were obtained, the temperature curves would probably take the form of S-curves with the maximum slope between 600° and 700°C. Within this range the effects of increased plastic deformation of the density are marked. Above 700°C, the density increases at a slower rate and for a 450-sec sustained pressure period at 780°C is 99.1% of the theoretical density of solid iron.

Hardness

The Brinell hardness curves (Fig. 11) are influence by several factors. With increasing density, higher readings are obtained, even though there may be no actual increase in the hardness of the metal because there is more metal available to resist the indentation of the ball. Since the specimens are maintained at temperature for only a short time, however, the cold-working that results from plastic deformation must account for some of the increase in hardness readings. Support for this contention is indicated in the results for specimens hot-pressed above 700°C. The metal is hot-worked and the annealing effects are indicated on the curves by the decrease in hardness.

Tensile Strength and Elongation

The curves for tensile strength (Fig. 9) and elongation (Fig. 10) clearly show the advantages that may be gained from hot-pressing. The increase in tensile strength up to 700°C is caused by increasing intraparticle bonding. The increase in contracting particle surfaces and diffusion rates with increasing temperatures under sustained pressure are mainly responsible for this bonding.

The elongation values, though nominal at the lower temperatures, increase rapidly at the higher temperatures because of the strong bonding resulting from hot-working and the complete recrystallization of the grain structure.

Conclusions

The properties that can be obtained on hot-pressed iron powders are far superior to the properties of cold-pressed and sintered iron powders. However, the time, temperature, and pressure requirements are too severe for present-day commercial equipment. With advances in processing and materials, the hot-pressing of metal powders should take its place as an important method of fabricating metal.

Some Factors Influencing the Properties of Sintered Iron

Lennart Forss

A. Johnson and Co., Inc.
Newark, New Jersey

Introduction

During experiments with development of iron powder compositions with controlled dimensional changes it was found that our knowledge of the variables which influence the properties of sintered structural materials is still quite insufficient. Some such variables are, for example, the causes of differences in sintering activity and dimensional behavior between the commercial types of iron powder and the role of phase transformations and grain growth during the sintering cycle.

The purpose of our investigation was to gain a better understanding of the sintering process in order to obtain a maximum of mechanical properties with the least costly sintering procedure.

We feel that it is necessary to understand the mechanics of sintering pure iron thoroughly before attempting to analyze multicomponent or liquid phase systems. Consequently, most of our work has been concentrated to pure iron with only minor additions of other elements.

The sintering conditions have been held to conform with American practice which usually is 10–40 min at maximum 2050°F. The linear dimensional changes have been calculated from die size, and volume changes from dimensions at 0 min sintering time which constitutes heating time only.

Comparison of Iron Powders

Figure 1 shows the relation between sintered density and tensile strength for three types of iron powder as sintered under average commercial conditions, about 30 min at 2020–2050°F in endothermic or exothermic atmosphere.

The specifications given for the three types of iron powder are as follows:

	Hydrogen Reduced	Sponge Iron	Electro-lytic
Apparent density, g./cc.	2.4	2.4	2.7
Flow rate — sec. 50 g.	39	30	29
+ 100 mesh, %	3	1	4
+ 150 mesh	13	24	11
+ 200 mesh	21	24	18
+ 250 mesh	10	12	16
+ 325 mesh	25	16	18
− 325 mesh	28	23	33
H_2 loss, %	0.4	0.70	0.3
SiO_2	0.29	0.25	0.04

These data give no indication to a reason why the slopes in Fig. 1 should show such a marked difference.

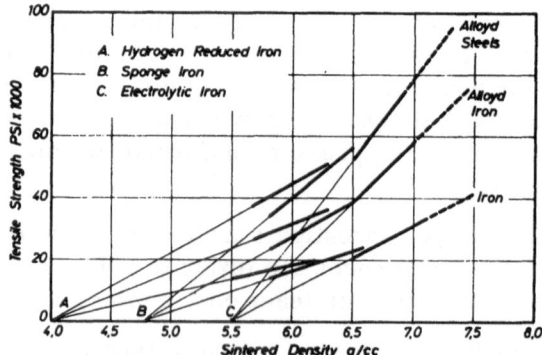

Fig. 1. Tensile strength versus sintered density for three types of compositions based on three types of iron powder.

For a closer study of the sintering behavior, a number of cylinders were pressed at 40 tsi in the direction of the axis. The die was lubricated with a zinc stearate slurry. Four types of powder were used: (1) Carbonyl iron, Inco type MCP (7.0 μ average particle size), (2) hydrogen-reduced mill scale, (3) sponge iron powder, and (4) electrolytic iron powder.

The following green densities and green expansions were found:

Powder Number	1	2	3	4
Green density, g /cc.	6.60	6.38	6.63	7.06
Green expansion, %	0.21	0.32	0.27	0.20

Microscopic examination of the compacts showed the carbonyl powder to consist of small equiaxed particles; the hydrogen-reduced Fe_2O_3 has a large number of quite small, presumably partially isolated pores; the sponge iron showed the larger pores typical of high reduction temperature, and the electrolytic iron, finally, appears as agglomerates of nearly pure iron.

The cylinders were heated for 30 min to temperatures from 1000° to 2200°F in dissociated ammonia, whereafter the radial dimensional changes from tool size and the hardness of the compacts were measured (Fig. 2).

The powder compacts show increasing densification up to the transformation temperature; above this temperature density is lower, going to a minimum at 1800–1900°F, and again increasing at higher sintering temperatures.

The shape of the curves indicates that the variation of diffusion rates with temperature (Fig. 3) [1] does not provide a full explanation of the dimensional differences in the sintering behavior between the four powders.

A third factor could be the rate of grain growth, and an approximate estimate of the average grain size of the sintered compacts is shown in Fig. 4.

Evidently the grain size of alpha iron after 30 min sintering does not change with temperature, but increases considerably at the transformation temperature, particularly in the case of electrolytic iron powder (Fig. 5a, b, c), and again becomes constant or even slightly decreased at higher temperatures.

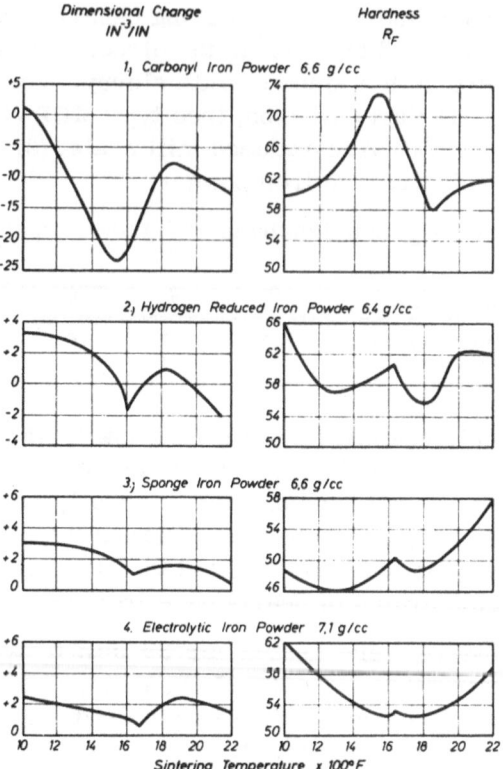

Fig. 2. Variations of dimensions and hardness with sintering temperature, 30 min in dissociated NH_3.

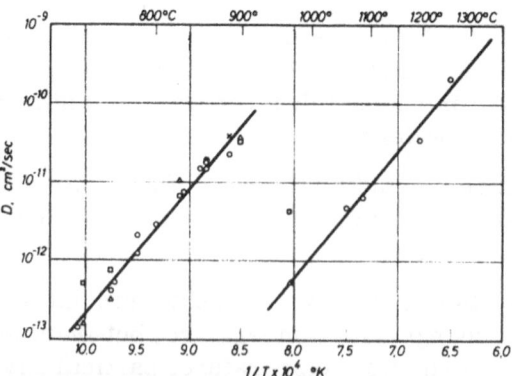

Fig. 3. Self-diffusion in alpha iron; ○ puron; △ carbonyl iron; □ commercial sheet X welded puron couple.

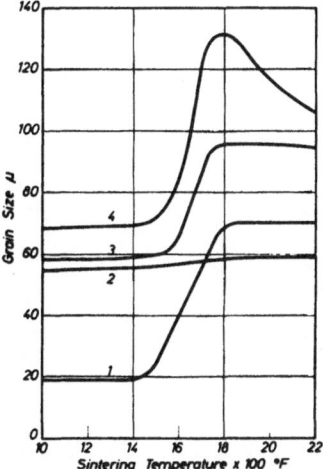

Fig. 4. Approximate average
grain size after 30 min at
indicated temperature.

Powder No. 2 is an exception and the grains do not appear
to grow larger than the original particle size, probably because of
nonmetallic inclusions.

From Fig. 5 it is evident that the number of grain boundaries
in this case is considerably decreased when the transformation
temperature is passed with the result that a substantial fraction of
the porosity previously connected with grain boundaries becomes
isolated inside the grains.

As available data on the diffusion of iron furthermore shows
that the activation energy for grain boundary diffusion is much
lower than that of lattice diffusion [2], it seems that the two
mechanisms may be regarded separately and that the shape of the
dimensional change curves in Fig. 2 should be interpreted as
effects of not only lattice and grain boundary diffusion rate but
also rate of grain growth.

In experiments with self-diffusion of nickel, W. R. Upthegrove
and M. J. Sinnott [3] found that the ratio of the grain boundary diffusion coefficient to the lattice diffusion coefficient varies from
10^3 to 10^{17} over the temperature range 1290–2010°F.

M. Tikkanen [4] studied the role of grain boundary diffusion in the
sintering of nickel but no correlation to the grain growth was made.

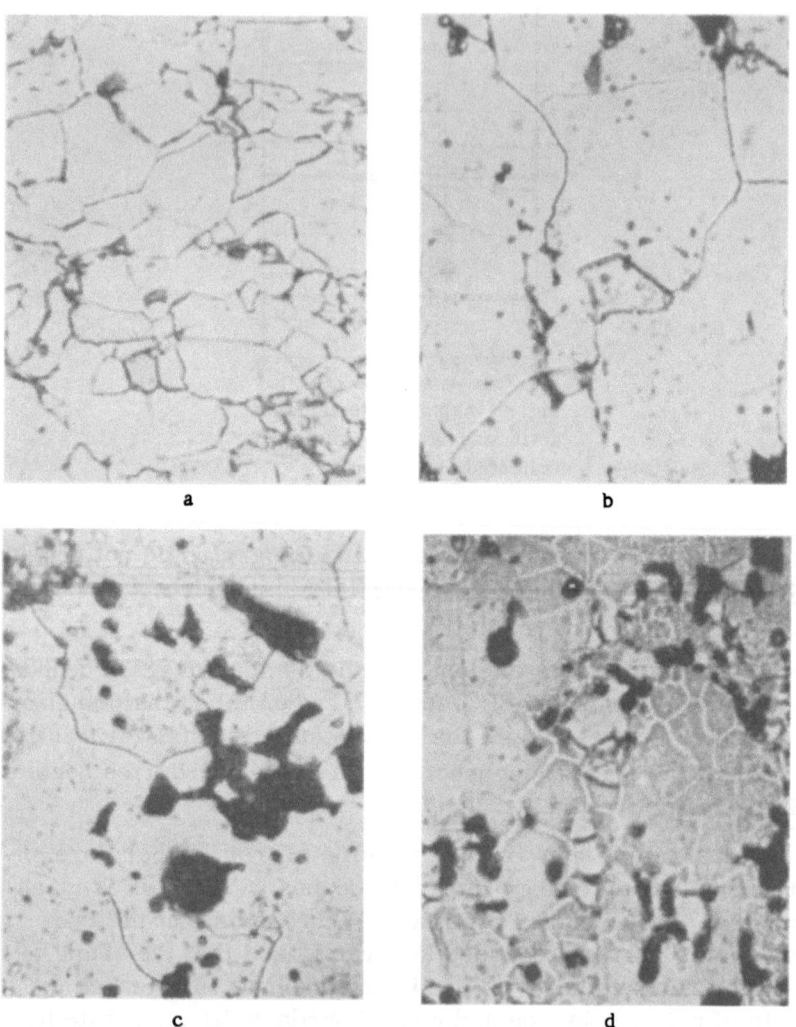

Fig. 5. Effect of sintering temperature and nickel addition on grain size of electrolytic iron powder compacts. Density: 7.0 g/cc. ×530; a) 30 min 1650°F, b) 30 min 1740°F, c) 30 min 2050°F, d) 30 min 2050°F 4% nickel added.

G. Bockstiegel [5] concludes that volume diffusion is the only mechanism of transport involved when iron powder compacts are sintered at temperatures from 1150 to 1650°F, whereas A. L. Pranatis and L. Seigle [6] assume that surface diffusion takes place when iron wire compacts are sintered at 1640°F.

As furthermore some investigators are presently finding that plastic flow plays an important role, it is evident that more basic experiments are needed before the theories of sintering are applied to practice.

If we plot the tensile strength of electrolytic iron powder pressed to 7.1 g/cc and sintered for 30 min in dissociated ammonia against sintering temperature (Fig. 6) and compare this trend with the dimensional changes in Fig. 2, we see that the strength is approximately equal at the maximum point about 1650° and the commercially customary temperature of 2050° but that the shrinkage is greater at 1650°F.

If the assumption that two different mechanisms are dominant at the two temperatures is correct, it would be of advantage to use a "duplex" cycle, i.e., sinter at both the lower and higher temperature.

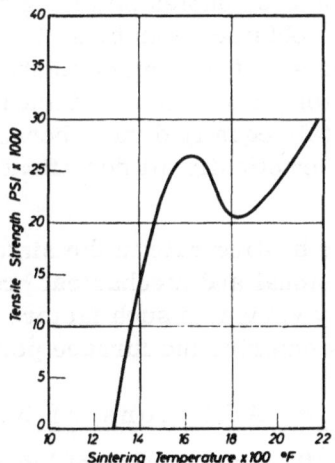

Fig. 6. Tensile strength of electrolytic iron powder, single pressed to 7.1 g/cc, sintered 30 min in dissociated NH_3 at temperatures from 1000 to 2200°F.

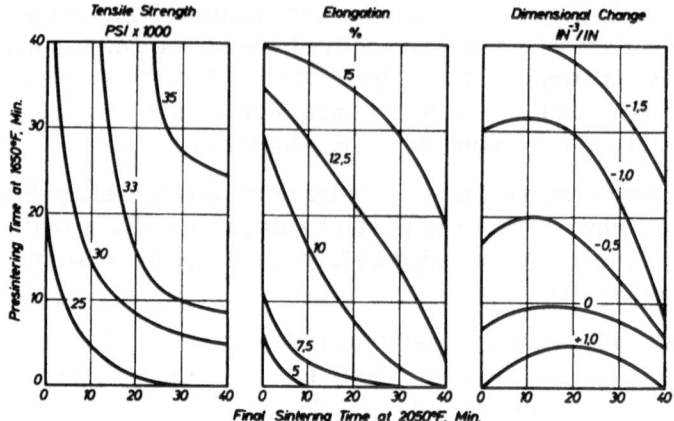

Fig. 7. Tensile strength, elongation, and dimensional changes
versus sintering time—temperature for electrolytic powder.
Green density: 7.1 g/cc; presinter: 0–40 min at 1650°F; final
sinter: 0–40 min at 2050°F dissociated NH_3.

In order to study the effect of such cycles on properties,
similar bars were presintered 0–40 min at 1650°F and given
a final sinter also 0–40 min at 2050°F. The resulting properties in Fig. 7 indicate that by duplex sintering a considerably
greater strength can be obtained than by sintering at any one temperature alone. It seems that maximum mechanical properties
such as strength and elongation can be obtained with available sintering time approximately equally divided between 1650° and 2050°F
whereas, as expected, practically all densification takes place at
the lower temperature.

It follows that the heating rate in the sintering cycle has an
influence on the dimensional and mechanical properties of a part
which may consequently vary with such factors as the energy input
and the weight of parts entering the furnace per time unit.

Effect of Additions of Nickel

In experiments with various types of lubricants added to iron
powder, it was found that 1% nickel stearate increases both tensile
strength and elongation of sintered iron powder compacts.

Nickel-containing powders were, therefore, prepared by mixing electrolytic iron powder with amounts of nickel stearate cor-

responding to 0.1 and 0.2% and subsequently burning off the organic matter at 1350°F in dissociated ammonia, simultaneously reducing the nickel oxide to metallic nickel.

The resulting powders were lubricated with 0.75% zinc stearate, pressed to 7.0 g/cc, and sintered 30 min at 2050°F in dissociated ammonia. The following properties were obtained:

	Tensile Strength, psi	Elongation %	Dimensional Change %
0 % Ni:	20,400	6.9	+ 0.0017
0.1% Ni:	27,200	11.2	+ 0.0003
0.2% Ni:	32,100	14.7	− 0.0023

The volume shrinkages of compacts of electrolytic iron with varying amounts of carbonyl nickel of average particle size of 3μ added were also studied. The percentages of nickel were 0, 0.25, 1.0, and 4.0; only die lubrication with zinc stearate slurry was used and the cylinders were sintered 15−480 min at 1650° and 2050°F in dissociated ammonia. The resulting volume shrinkages are listed in Table 1.

Addition of 4% nickel causes approximately equal densification rates at 1650°F and 2050°F.

The photomicrographs in Fig. 5(c and d) reveal that the nickel-rich phase tends to inhibit the grain growth whereby a greater fraction of the voids are connected with grain boundaries, and thus the rate of sintering is increased also in the gamma range.

TABLE 1. Volume Shrinkages in Percent Versus Percentage Nickel Addition and Sintering Conditions

Sintering		Nickel Addition			
Temp.	Time	0%	0.25%	1.0%	4.0%
1650	15	0.72	0.72	0.78	0.72
1650	30	1.07	1.07	1.01	0.91
1650	60	1.07	1.23	1.32	1.26
1650	120	1.04	1.44	1.68	1.73
1650	480	1.05	1.82	2.28	2.44
2050	15	0.07	0.17	0.44	0.68
2050	30	0.11	0.25	0.59	0.93
2050	60	0.12	0.40	0.80	1.23
2050	120	0.00	0.55	0.99	1.48
2050	480	−0.23	0.83	1.26	2.40

Conclusions

There is strong reason to believe that the sintering of iron powder compacts at moderate temperatures is a result of at least two mechanisms which should be regarded separately: grain boundary diffusion which is rapid and takes place under shrinkage as long as the grain boundaries are connected with pores, and lattice diffusion which is much slower and apparently has a strengthening effect, independent of presence of grain boundaries.

Compacts of pure iron powder show a considerable grain growth when the transformation temperature is passed, and the rate of sintering can be accelerated either by sintering in both the alpha and gamma range or by small additions of, e.g., fine nickel.

It seems that future work on the theory of sintering by the metal physicists would benefit from more data on grain growth and grain boundary diffusion rates in connection with dimensional behavior.

References

1. Birchenall, L. E., and A. F. Mehl, Trans. Am. Inst. Met. Eng. 188:144-49 (1950).
2. Smithells, C. J., Metal Reference Book, 3rd Ed. Vol. 2, p. 593.
3. Upthegrove, W. R., and M. J. Sinnott, Trans. Am. Soc. Met. 50:1030-46 (1958).
4. Tikkanen, M., Contribution to the Theory of Sintering, Jernkontorets Powder Metallurgy Research Panel, Stockholm, 1963.
5. Bockstiegel, G., J. Metals 8:580-85 (1956).
6. Pranatis, A. L. and L. Seigle, in: Powder Metallurgy, Interscience, New York, 1961, pp. 53-73.
7. Ljungberg, I., Unpublished report from Husqvarna Vapenfabriks AB.
8. Bockstiegel, G., Powder Met. No. 10, London, 1962, pp. 114-18.

Study of the Mechanical Properties
of Porous Iron in Tension and Torsion

G.S. Pisarenko, V.T. Troshchenko,

and A. Ya. Krasovskii

Institute of Materials Problems
Academy of Sciences of the Ukrainian SSR

Constructional elements of modern machines generally operate under conditions of a complex stressed state, and for engineering calculations of the strength of such elements, it is necessary to have information on the ability of materials being employed to resist fracture in the presence of various combinations of the principal stresses. Porous materials prepared by powder metallurgy methods are no exception in this respect. The range of parts made from such materials is growing rapidly, covering new fields of application, e.g., and now in contact with various corrosive environments and at high or low temperatures, also includes load-carrying constructional elements operating under different types of stress. So far, however, the ability of porous materials (particularly those based on ductile metals) to resist a complex stressed state has not been studied at all, and no experimental investigations of the properties of these materials under conditions of a complex stressed state have been carried out. (An exception is to be found in tests involving shear and nonuniform compression from all directions;

in view of the shortcomings of tests of this type, their results can hardly provide the information required for this purpose.)

In this respect, the most promising and simple are torsion tests, the results of which permit a relatively uncomplicated analysis of the stressed state. Although in torsion the material is subjected to only one combination of the principal normal stresses, in which σ_1 and σ_3 are equal in magnitude and oppositely directed, while σ_2 is missing, nevertheless, an analysis of the resistance of a material to torsion may provide some information on its operation under conditions of a complex stressed state. Many authors compared tensile and torsional test results in studies of the operation of ductile materials under conditions of a complex stressed state [1-3], and it has been established that the ratio of the yield stress in torsion τ_S to the yield stress in tension σ_S for such materials as iron and low-carbon steels lies in the range 0.55-0.75 [4]. In the present investigation, a study was made of the following main problems:

1. What is the relationship between the mechanical properties of iron specimens of different porosity, as determined in tensile and torsion tests on specimens from the same material?

2. Which of the existing theories of strength is the most appropriate for describing the operation of porous iron under the conditions of a complex stressed state?

3. What is the relationship between the elastic properties of porous iron, as determined in tension and torsion?

4. How does the type of stressed state influence the ductility of porous iron?

All batches of specimens of different porosity were made from reduced iron powder of the APZhM grade (fine of the following composition (in %):

Fe	C	S	Mn	P	Insoluble residue	Balance (oxides)
98.2	0.09	0.01	0.3	0.02	0.3	1.08

The particle-size distribution of the iron powder is shown in Table 1. Using a specially designed and constructed die assembly, compacts $150 \times 30 \times 30$ mm in size were obtained by double-

TABLE 1

Fraction, mm	Above 0.14	From 0.14 to 0.09	From 0.09 to 0.071	From 0.071 to 0.053	Below 0.053
Amount of fraction, %	0.5	16.85	39.10	23.15	20.40

Density of freely poured powder 1.7 g/cc.

action pressing "to the support." As a result of using large amounts of powder and this method of compaction, a high uniformity of porosity was secured within each batch of specimens (porosity scatter did not exceed a fraction of one percent). After pressing, the compacts were sintered for 2 h in a hydrogen atmosphere at 1473°K. From the sintered compacts, cylindrical specimens were machined according to the drawing in Fig. 1. The gauge section of the specimens came from the core portion of the compacts, thereby eliminating the nonuniformity of density over the compact cross section, which is present to some extent in all sintered parts.

In view of the danger of the mechanical properties being disturbed by the work hardening and breaking up of the surface layer during machining, the turning of all specimens was followed by polishing with abrasive cloths of four grades in the order of decreasing abrasive particle size, during which the specimen diameter was reduced by 0.4-0.8 mm depending on porosity. It should be noted that the undesirable effect of the surface layer manifests itself in torsion tests even more strongly than in tensile tests, because the resisting moment of the section is proportional to the third power of the radius, and consequently, special attention was

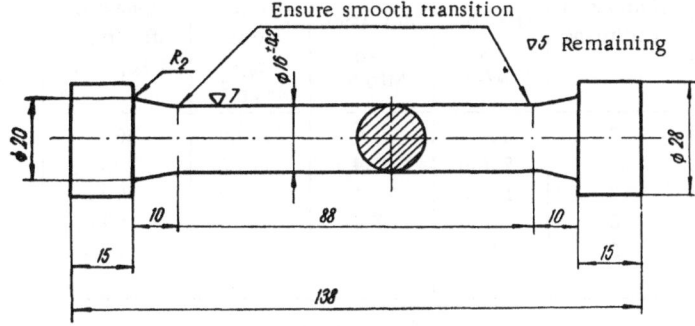

Fig. 1. Drawing of test specimen (all dimensions in mm).

TABLE 2

Specimen number in batch	From 1M-10 to 20M-10	From 1M-20 to 5M-20	From 1M-30 to 20M-30	From 1M-40 to 5M-40	From 1M-50 to 20M-50
Porosity of gauge part, %	12	21	31	37	43
Number of specimens	20	5	20	5	20

given to this circumstance in the present investigation. In order to eliminate the work hardening caused by polishing and reduce the oxides formed in the material between the end of sintering and the end of polishing, the specimens were annealed for 1 h in hydrogen at 1173°K.

The porosity of sintered compacts was checked by calculating density by weight and by volume, as a result of which the total porosity could be determined. In addition, depending on the batch size, 1-3 specimens were selected from each batch, and their gauge portion (cylinder) was cut out. The total porosity of these cylinders was determined, and it was these values that were used in all subsequent graphs and tables. This procedure was necessary because of the above-mentioned nonuniformity of porosity over the compact cross section. The characteristics of the specimen batches prepared are shown in Table 2.

TABLE 3

Porosity, %	Number of specimens	Mean tensile strength σ_b, MN/m^2	Mean square deviation Δ_σ, MN/m^2	Mean arbitrary yield stress $\sigma_{0.2}$, MN/m^2	Mean elongation after fracture, δ, %	Mean residual reduction of area ψ, %
0*	2	330.0	—	200.0	30	—
12	15	208.0	5.4	116.5	9.2	8.75
21	3	141.8	—	94.9	5.41	5.56
31	15	82.0	2.3	58.5	2.60	2.53
37	3	49.1	—	42.4	1.50	1.48
43	11	33.5	1.6	31.9	0.80	0.79

*Results obtained in tensile tests on specimens of a low-carbon steel similar in composition to the iron powder.

Tensile and Torsion Test Procedures

Tensile tests were performed in a hydraulic press with dynamometer scale divisions of about 25 N each. Values of tensile
strength, proof stress, residual elongation after fracture, and residual reduction of area were obtained. In order to decide on the details of the test procedure, preliminary tests were made in the
same press on several specimens fitted with resistance wire strain
gauges, as a result of which it became possible to estimate and
decrease the extent to which bending was superimposed in tensile
testing. Stress-strain diagrams of these specimens were plotted.

Torsion tests were made in an MK-50 machine with grips
which were specially designed for this purpose and secured an excellent coaxiality of the specimen heads in the loading system of
the machine. The indications of the dynamometer and the dial of
the angle of twist were used for plotting general torsion diagrams
in T versus φ coordinates, where T is torque and φ is the angle of
twist.

For recording the initial portions of the strain diagram,
wire resistance strain gauges were attached to the test specimens.
The gauges were mounted in such a manner that their axes lay
along the trajectories of the principal normal stresses, on the com-
pressed and extended "fibers." The indications of the strain gauges
were recorded with an EMG 2353 "Orion" instrument. Since the
sensitivity coefficients of wire resistance strain gauges depend on
the type of strain being measured (tensile or compressive) [5], it
becomes necessary to calibrate separately gauges measuring these
strains. The calibration showed that the sensitivity coefficient of a
given strain gauge in tension was 1.2% higher than in compression.
This difference was taken into account in the processing of test
results.

Tests on all specimens made it possible to plot the initial
portion of the stress-strain diagrams (up to a strain of $\varepsilon = 2\text{-}3\%$)
of the test materials in the directions of compressive and tensile
stresses. The portion of a stress-strain curve plotted from dial
indications, after the strain gauges have ceased to be operative, is
less accurate, because it is affected by the deformation of materials
at points of contact between the specimen heads and the grips.
This deformation was largely allowed for by introducing into the
value of the angle of twist suitable corrections obtained by extra-

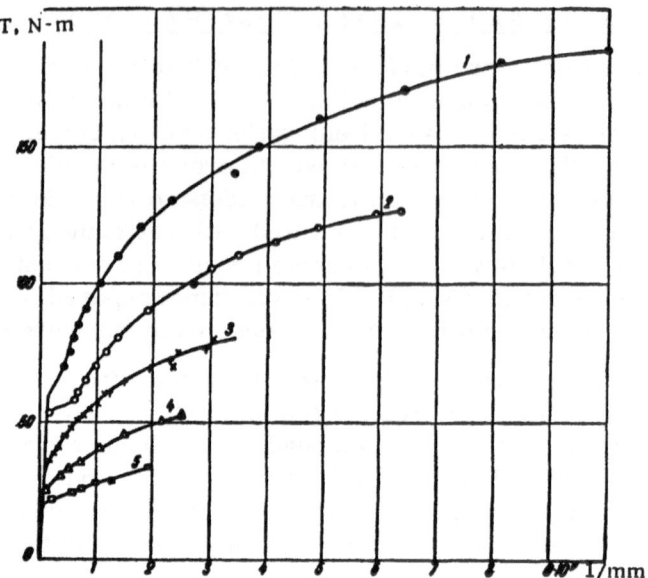

Fig. 2. Torsion diagrams of porous iron. Porosity (%): 1) 12;
2) 21; 3) 31; 4) 37; 5) 43.

polating the straight portions of T versus φ curves to the axis of abscissas. For the conclusions arrived at in this investigation, such an accuracy is entirely adequate.

Discussion of Results

The results of the tensile tests on porous iron specimens are shown in Table 3 and the torsion diagrams of similar specimens, in T versus θ coordinates, in Fig. 2. Here, $\theta = \varphi/l$ is the specific angle of twist, and l is the specimen gauge length. As in the case

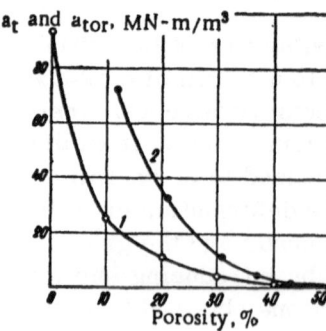

Fig. 3. Effect of porosity on specific work of rupture of porous iron in tension and tor-sion: 1) tension; 2) torsion.

of dense metals, these curves have straight line portions. Raising
the porosity of iron leads to a decrease of the breaking torque and
the maximum angle of twist. No significant changes in the specimen diameter and length were observed during the tests. It appears that, as in the case of dense metals (cf. for instance, [6–8]),
these changes are of a secondary order of magnitude compared
with the changes in tension.

With the torsion diagrams illustrated, it is possible to find
the relationship between the specific work of rupture in torsion for
porous iron and porosity. The value of this work can be obtained
with sufficient accuracy from the formula

$$a_{ton} = \frac{\int_0^{\theta_{max}} M_T \, d\theta}{\frac{\pi d^2}{4}}, \tag{1}$$

where d is the specimen diameter, and θ_{max} is the specific angle
of twist corresponding to the rupturing torque of the specimen. The
numerator in this formula is given by the area under the torsion
curve of a material of the corresponding porosity. A comparison
of this work with the specific work of rupture in tension, as determined in [9], shows that the energy capacity of a material in torsion
is much greater (Fig. 3). It will be shown later that the ductility
of porous iron in torsion is substantially higher than in tension,
and this largely explains the marked difference between the values
of the work of rupture.

For processing the diagram obtained, use was made of
Ludwick–Nadai's formula [1, 6, 10]:

$$\tau_{max} = \frac{4}{\pi d^3} \left(3M_T + \frac{dM_T}{d\theta} \theta \right), \tag{2}$$

where τ_{max} is tangential stress on the periphery of the transverse
cross section of the specimen. Using this formula τ_{max} versus
θ curves were plotted.

The results obtained by processing the torsion diagrams
are shown in Fig. 4. All the curves exhibit yield arrests and even
small stress drops at the point of transition between the elastic
and plastic zones. Similar yield arrests were observed in the tensile tests.

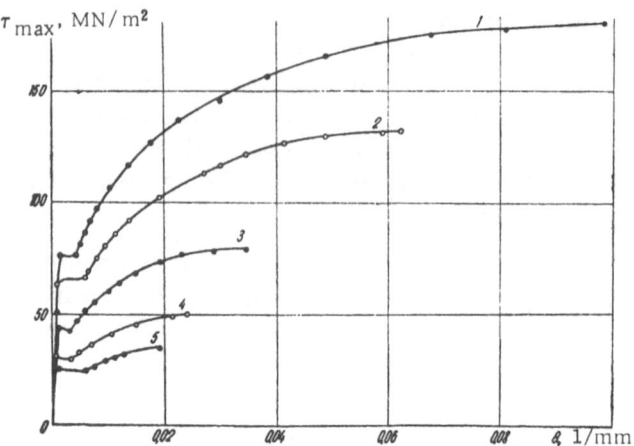

Fig. 4. Strain curves of porous iron in torsion. Porosity (%):
1) 12; 2) 21; 3) 31; 4) 37; 5) 43.

The validity of the straight radii hypothesis in the torsion of porous iron was also verified. For this purpose in specimens of 10, 30, and 50% porosity (specimens of 8-mm diameter, employed in other tests), holes of 0.5-mm diameter were drilled at three points on the length of the gauge portion of the specimens, in a diametrical direction. Copper wires were forced into these holes, and the specimens were then subjected to torsion tests to rupture. Next, the specimen cross sections holding the wires were used for preparing microsections, which were examined under the microscope. No deviation from straight lines was found with any of the porosities investigated (Fig. 5). On the basis of this experiment, it may be concluded that the straight radii hypothesis remains valid also for the case of torsion of porous materials (for specimens of cylindrical shape). The principal results of the torsion tests on specimens are shown in Table 4.

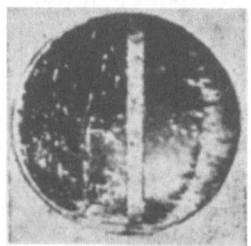

Fig. 5. Transverse cross section of specimen with inserted copper wire, subjected to torsion to rupture. No bending of wire has taken place.

TABLE 4

Porosity, %	Number of specimens	Mean ultimate strength τ_B, MN/m^2	Max. angle of twist φ, deg	Mean yield limit $\tau_{0.3}$, MN/m^2	Shear modulus $G \cdot 10^{-5}$ MN/m^2
0*	—		—	180-220	0.8015
12	2	180.0	572	76.2	0.619
21	1	132.0	360	65.1	0.434
31	2	78.3	198	43.5	0.335
37	1	49.5	138	30.6	0.171
43	2	32.0	102	24.3	0.138

*Data for steel 10 (0.1% C) obtained from [11].

References

1. P. Ludwick and R. Scheu, Stahl und Eisen, 45:373 (1925).
2. G. Sachs, VDI-Zeitschrift, 72(22):734 (1928).
3. H. L. Cox and D. G. Sopwith, Proc. Phys. Soc., London, 49:134 (1937).
4. I. A. Oding, Permissible Stresses and Cyclic Strength in Engineering, Moscow, Mashgiz (1962).
5. Z. Ruzga, Electrical Resistance Strain Gauges, Moscow, MIR (1964).
6. A. Nadai, Plasticity and Fracture of Solid Bodies, Moscow, IL (1954).
7. Ya. B. Fridman, Mechanical Properties of Metals, Oborongiz (1952).
8. Ya. B. Fridman, T. K. Zilova, and N. I. Demina, Study of Plastic Strain and Fracture by the Method of Engraved Grids, Moscow, Oborongiz (1962).
9. A. Ya. Krasovskii, Poroshkovaya Met., 4:3 (1964).
10. Yu. N. Rabotnov, Strength of Materials, Moscow (1962).
11. NAMI-VNITOMASH Handbook, Automobile Constructional Steels, Moscow, Mashgiz (1951).

The Effect of Minor Additions of Sulfur on Pore Closure and Case-Hardenability of Sintered Iron

Gerhard Bockstiegel

Höganäs AB
Höganäs, Sweden

I. Introduction

It is well known that sintered iron structural parts, especially in the low- and medium-density range (i.e., approx. 6.4-6.8 g/cc), often behave trickily during gas- or pack-carburizing treatment. This is mainly due to the fact that the carburizing gases, via interconnected pores, can easily penetrate the sintered parts to a greater depth. If the density of sintered iron parts is less than ~ 7.0 g/cc, the carburizing action is usually so fast that it becomes difficult to control case depth and to prevent carburization of the core. This can be especially inconvenient if larger charges are involved, or if a charge contains parts of varying size or density. Then, because time is too short to achieve an even temperature distribution throughout the whole charge, some parts may become through-carburized before others have developed a thin carburized case.

It is obvious that the lowest possible interconnected porosity is required, if satisfactory case-hardening results are to be obtained. To achieve this it would seem necessary, at a first glance, to produce the sintered iron parts to a very high density, i.e., preferably > 7.0 g/cc. In practice, many sintered structural parts must in any event for strength reasons be made with high densities.

On the other hand, there are many applications where high strength is not required but wear-resistance is of great importance. To give these parts a higher density only to make them more amenable to case-hardening treatment may, however, not always be economical.

Therefore, it is natural to seek simpler and less expensive means of reducing interconnected porosity. There is, for instance, the possiblity of filling the pores by infiltrating the sintered iron parts with copper or copper alloy during or after sintering. This method, which is practiced to a certain extent, though more for strength reasons than to improve case-hardenability, can hardly be regarded as less expensive or less intricate than making high-density parts by applying high pressures and using double-pressing and double-sintering techniques.

It might be anticipated that the addition of a few percent of copper powder to the iron powder would reduce interconnected porosity, the reason in this case being that a liquid copper phase appears in sintering which might be expected to be sucked into and to clog the capillary channels between the pores. In actual fact, however, the pore-closing effect of this process is rather ineffective, since the liquid copper appears only transiently and at a rather late phase of sintering, where it quickly diffuses into the solid iron, leaving new cavities behind. As a consequence, the sintered parts swell and the total porosity is increased rather than decreased. The number of channels that become closed or clogged when the liquid copper phase appears is partly or completely outbalanced by the number of cavities newly created when the liquid phase disappears again.

II. Possible Advantages of Sulfur Additions

The unsatisfactory pore-closing effect obtained with copper additions makes it desirable to seek more effective and less expensive additives. In this respect, sulfur appears to be a most promising substance, for the following reasons:

Sulfur is inexpensive; it can easily be blended with iron powder and has no detrimental effect on the green properties of the iron powder.

Sulfur melts at 119°C, and, because of its low viscosity and excellent wettability for iron in the temperature range between ~120 and 200°C, it can be expected to become distributed quickly over the whole internal surface of the iron powder compact, concentrating especially in capillaries and at the many contact faces between adjacent powder particles.

In the temperature range between ~200 and 400°C, the viscosity of sulfur suddenly increases to substantially higher values. This circumstance practically eliminates the risk that with rising temperatures the sulfur will drain out of the compacts before having had a chance to react with the iron. The formation of iron sulfide, FeS, for instance, begins to take place at ~200–250°C, i.e., a temperature well below 400°C, where the sulfur just before reaching its boiling point (444°C) becomes very liquid again.

The iron sulfide formed, i.e., FeS, has a melting point of 1190°C, and a eutectic exists between iron and FeS at 988°C. The wettability between liquid FeS and solid iron is extremely good, while the solid solubility of sulfur in iron, even at higher temperatures, is very limited (max. 0.005 wt%).

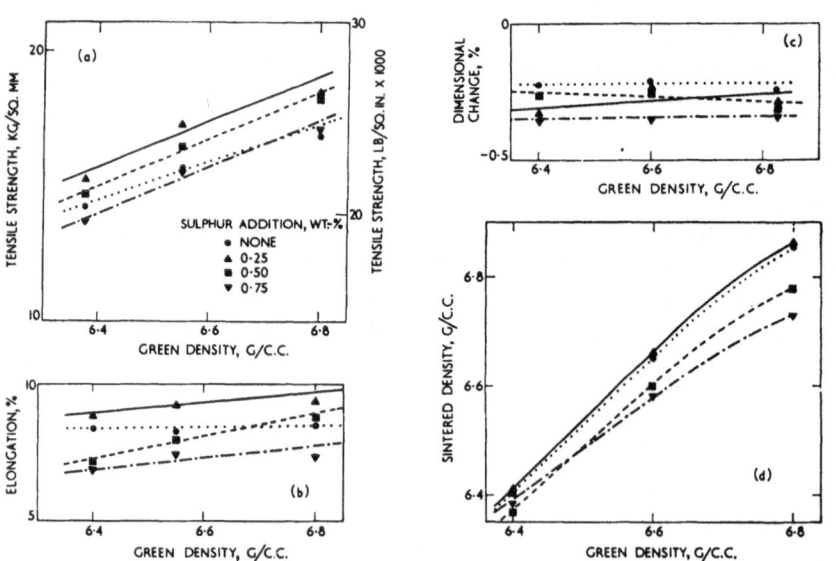

Fig. 1. The influence of sulfur additions on the mechanical properties of sintered iron, plotted as a function of green density: (a) tensile strength, (b) elongation, (c) dimensional change, (d) sintered density. Specimens sintered for 1 h at 1150°C in hydrogen.

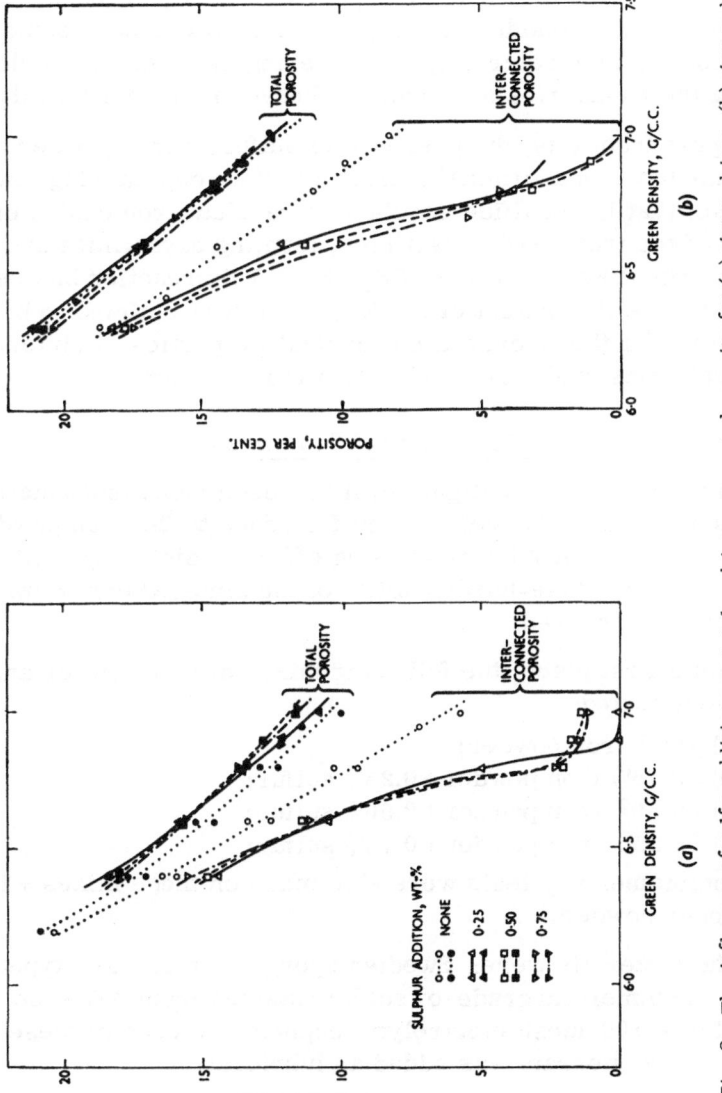

Fig. 2. The influence of sulfur additions on total and interconnected porosity for (a) sintered iron, (b) sintered iron+copper. Specimens sintered for 1 h at 1150°C in hydrogen.

In these circumstances it can be expected that in the sintering of compacts made from iron powder with sulfur additions a liquid phase will appear above 988°C and will persist, even at higher temperatures sustained over long periods. This will occur just at the very locations inside the compact where it can be assumed to be most effective in accelerating the sintering process and in closing or clogging the capillaries and channels between interconnected pores.

Apart from this, the presence of sulfur in iron powder compacts may have an additional activating effect on sintering, owing to the fact that iron sulfide reacts with residual iron oxides on the surface of the iron powder particles, forming oxysulfides and metallic iron ($2 FeS + FeO \rightleftharpoons Fe_2OS_2 + Fe$). This reaction has been mentioned in a U.S. patent of 1957,* where it is claimed to be responsible for the improved mechanical properties of sintered structural parts made from sulfatized iron powders.

III. Experiments

In the present investigation, it has been found that small amounts, i.e., 0.25-0.75 wt%,† of sulfur added to the iron powder, have a very pronounced pore-closing effect on sintering, with the consequence that case-hardenability of the sintered compacts is substantially improved.

In the first place, the following mixes of iron powder and sulfur were tested:

    (a) 100% iron powder;
    (b) 99.75% iron powder + 0.25% sulfur;
    (c) 99.50% iron powder + 0.50% sulfur;
    (d) 99.25% iron powder + 0.75% sulfur;

Some complementary tests were also made on theee mixes + 2.5 or 7.5% copper powder.

The materials were: Swedish sponge iron powder, type MH 100.24; a commercial grade of sulfur obtained from AB Kebo, Stockholm; < 100-mesh electrolytic copper. To each of these mixes 1% zinc stearate was added as lubricant.

Mixing was carried out in a 5000-cc laboratory shovel-mixer of type Lödige- M4E, in such a way that first the iron powder and

---

*R. E. Blue to Chrysler Corp., U.S.A., U.S. Patent No. 2,942,344, Jan. 18, 1957, June 28, 1960.
†All compositions are given in weight percent, unless otherwise stated.

the sulfur were intimately mixed for 5 min, and then the zinc stearate was added and mixing continued for a further 5 min. Standard test bars (MPI-nr 10-50) and cylindrical compacts of 1 in. diameter and $\sim ^3/_4$ in. high were then made from the mixes, with green densities varying from ~ 6.2 to 7.0 g/cc.

All compacts were sintered for 1 h at 1150°C in dry hydrogen. Before sintering, the lubricant was distilled off under hydrogen for 15 min at ~ 500°C in the preheating zone of the laboratory sintering furnace. During sintering the compacts were placed on a thin layer of -60 mesh alumina in covered sintering boats approximately $120 \times 75 \times 25$ mm, made from 1-mm iron plate. Only compacts made from the same powder mix were sintered in the same boat.

After sintering the change in length was measured, and the sintered density of the test bars was determined by the hydrostatic method. The specimens were then pulled to determine tensile strength and elongation. The results are presented in Fig. 1.

The sintered cylinders were used for the determination of total and interconnected porosity. The procedure was as follows. First, the interconnected porosity of the sintered cylinders was

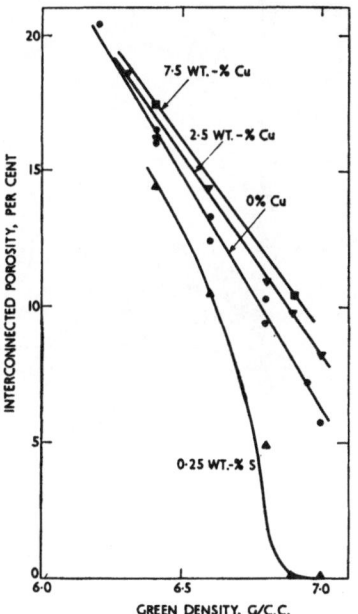

Fig. 3. The influence of copper and sulfur additions on interconnected porosity in sintered iron. Specimens sintered for 1 h at 1150°C in hydrogen.

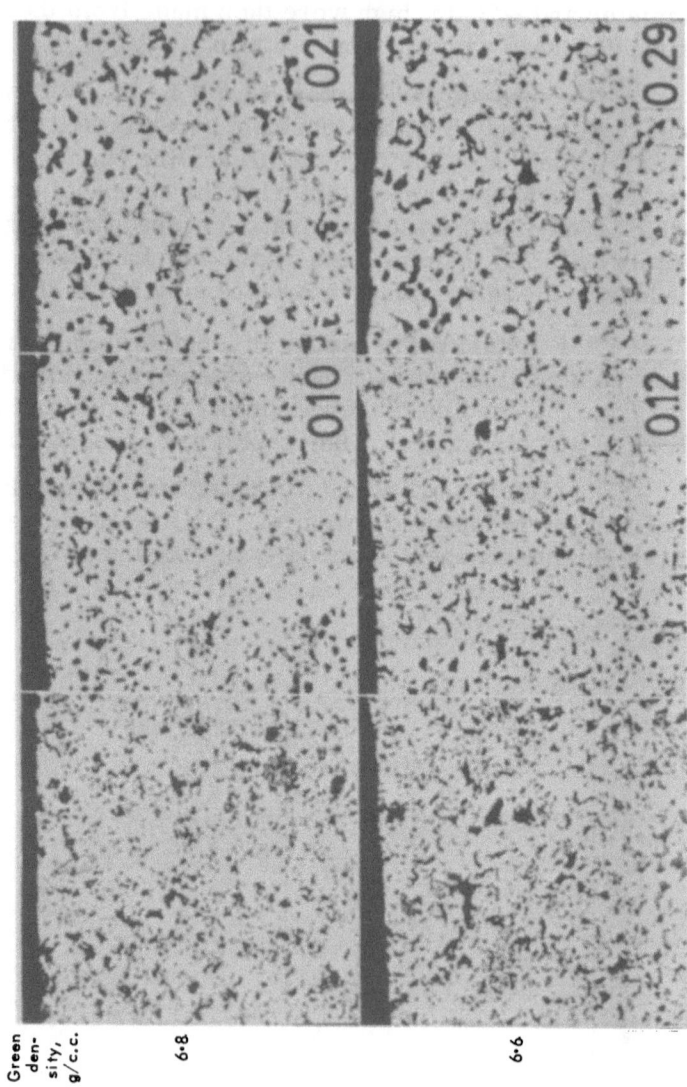

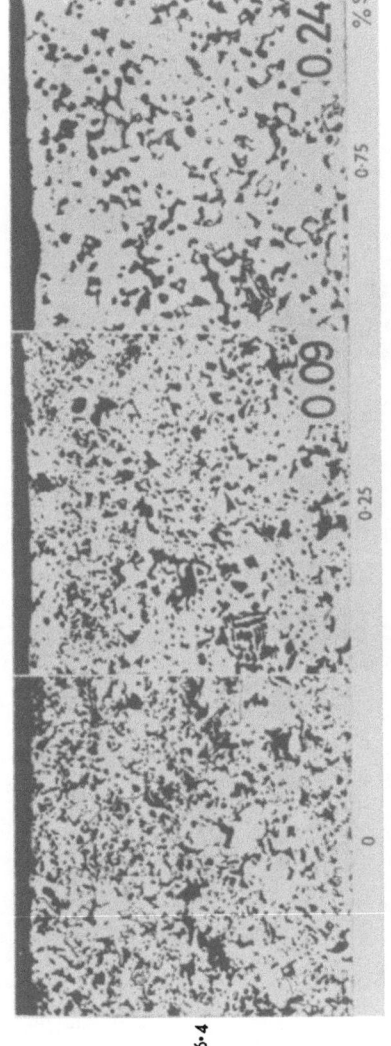

Fig. 4. The influence of sulfur additions and green density on the pore structure of sintered iron. The figures on the photomicrographs indicate the wt% of residual sulfur after sintering. Specimens sintered for 1 h at 1150°C in hydrogen. Etched in Nital. x75.

Fig. 5. The influence of sulfur additions and green density on case depth after pack-carburizing for 1 h at 900°C and quenching in water. Etched in Nital. × ~4.

measured by the method developed by Notari. This method works on the principle of comparing the increase in gas pressure when feeding a precisely defined amount of air under strictly controlled temperature conditions into the measuring chamber of the apparatus: (1) when a solid standard specimen, and (2) when the porous test specimen is placed in the chamber. From the relative difference in gas-pressure increase between (1) and (2), the interconnected porosity can be calculated.

The apparent density of the sintered cylinders was then determined by the hydrostatic method, and the total porosity was calculated from the apparent density figures, on the assumption that the total porosity decreases to zero as the apparent density approaches the specific gravity of pure solid iron, i.e., 7.86 g/cc. The results obtained are plotted in Fig. 2a. For comparison, the results obtained with cylinders made from iron powder containing 2.5% admixed copper powder and from 0 to 0.75% sulfur are plotted in Fig. 2b. The relative effect of copper and sulfur additions is illustrated in Fig. 3. The influence of the additions of sulfur on the pore structure of compacts of different densities has been studied on a longitudinal section through the cylinders, and the respective photomicrographs are presented in Fig. 4.

Pack-carburizing experiments were carried out in the following way: The halves of the pulled test bars were placed six at a time in small platen boxes of the type used in sintering and embedded in a carburizing mix consisting of 40% $BaCO_3$ and 60% charcoal. The lids of the boxes were sealed with a heat-resisting mortar. The boxes were then placed one at a time in a laboratory chamber furnace, heated to 900°C, and held for 1 h. After this treatment, the lid was removed while the box was still inside the furnace; the box was then quickly taken out of the furnace and its contents discharged into a bucket of cold water. This quenching technique proved sufficient to protect the specimens from surface decarburization.

The shoulders of the quenched specimens, which were not distorted during tensile testing, were then sectioned and prepared for metallographic inspection and microhardness testing. The results of this carburizing test are illustrated in the macro- and micrographs of Figs. 5, 6, and 7.

Finally, the loss of sulfur when distilling off the lubricant, and during sintering, was investigated on cylindrical compacts of

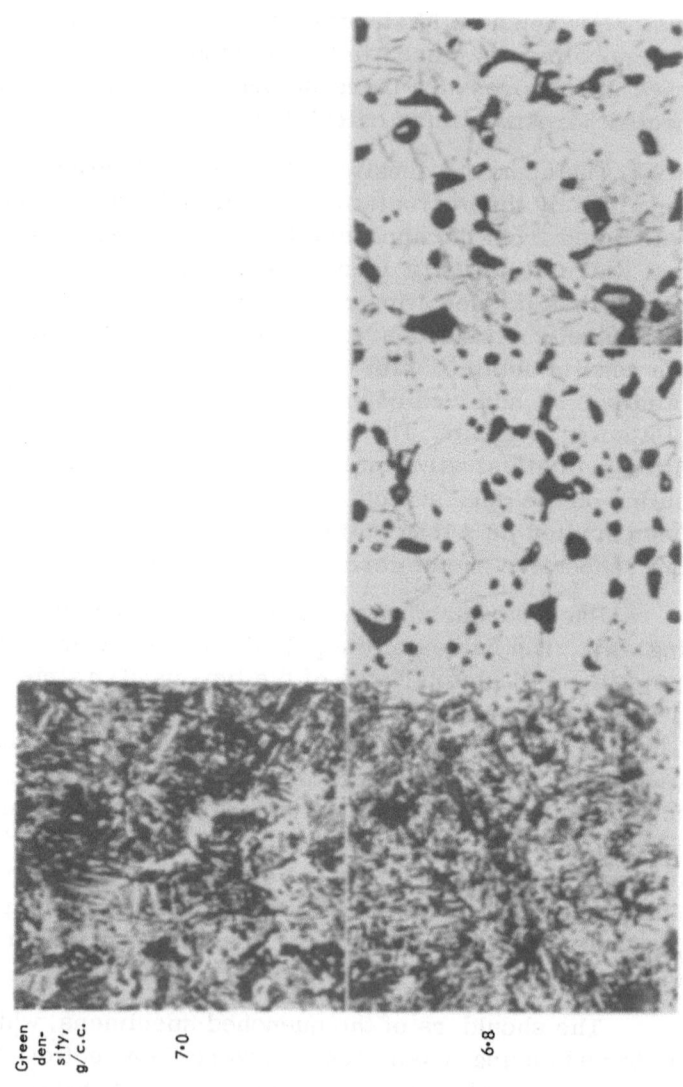

Green-
den-
sity,
g/c.c.

7·0

6·8

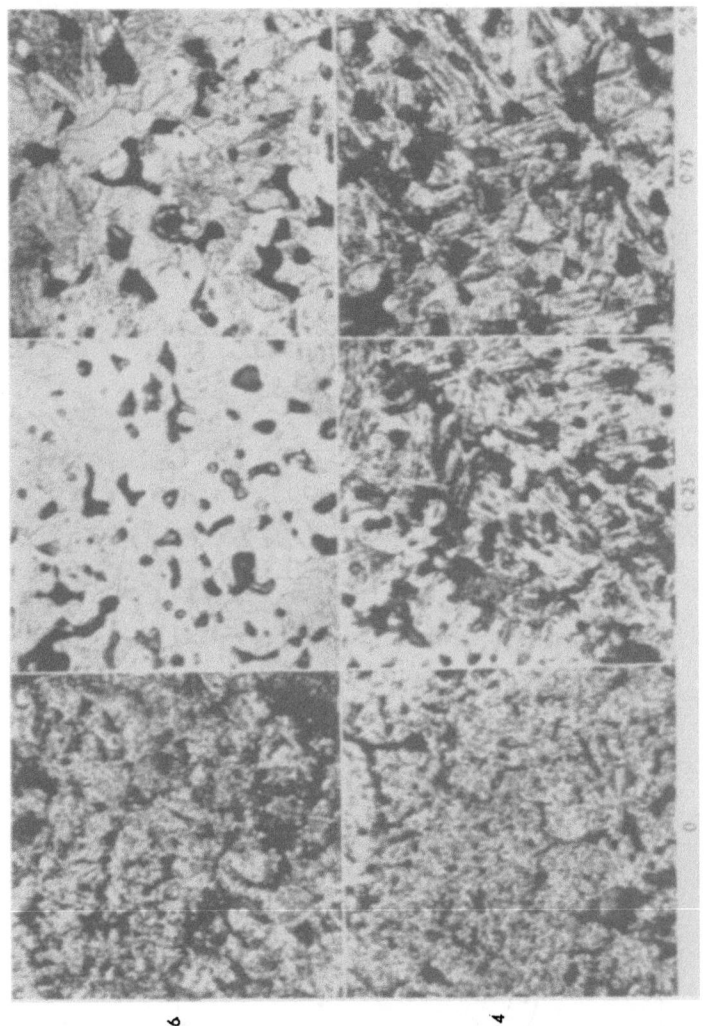

Fig. 6. Microstructures at the core of the pack-carburized sintered specimens shown in Fig. 5. Etched in Nital. x 240.

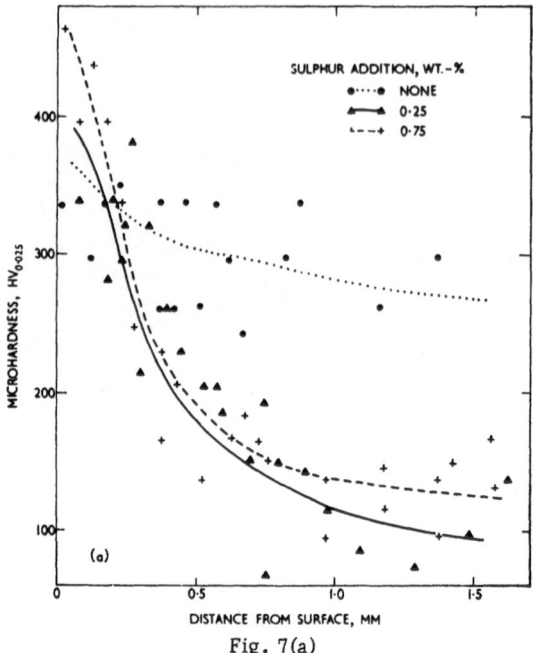

Fig. 7(a)

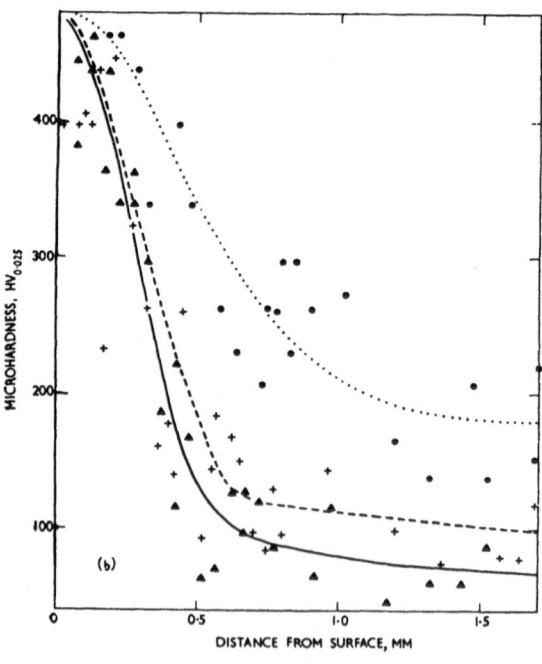

Fig. 7(b).

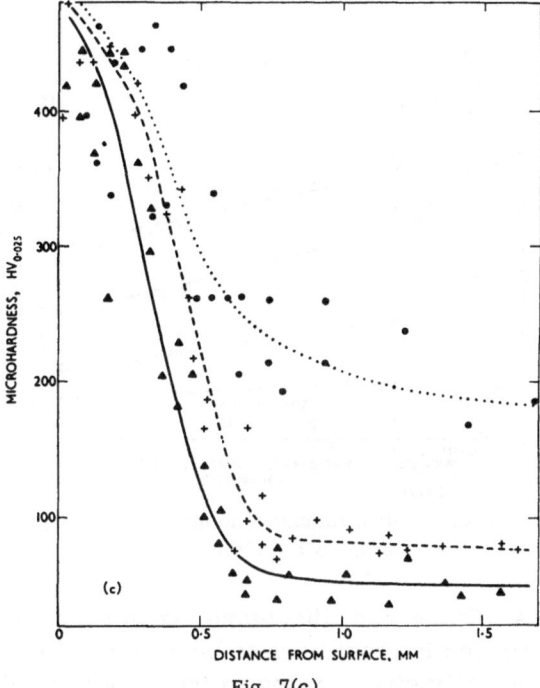

Fig. 7(c).

Fig. 7. Microhardness across the section of the pack-
carburized specimens shown in Fig. 5, for green densities of: (a) 6.4; (b) 6.6; (c) 6.8 g/cc.

various densities by chemical analysis of the specimens. The results are presented in Fig. 8.

IV. Discussion of Results

As can be seen from Fig. 1, sulfur additions produced
only a slight improvement in tensile strength and a slight increase
in shrinkage. An addition of 0.25% sulfur improves elongation,
while 0.75% decreases it somewhat. An addition of 0.5% sulfur
leads to a decrease in elongation with compacts of lower density
and a slight increase in compacts of higher density. To summarize,
under the sintering conditions involved here, the effect of these
small sulfur additions on the mechanical properties of sintered
iron powder compacts is not very significant.

A substantially better result, however, could hardly be expected since, as emerges from the curves of Figs. 2 and 3, the
total porosity is not decreased but rather slightly increased by

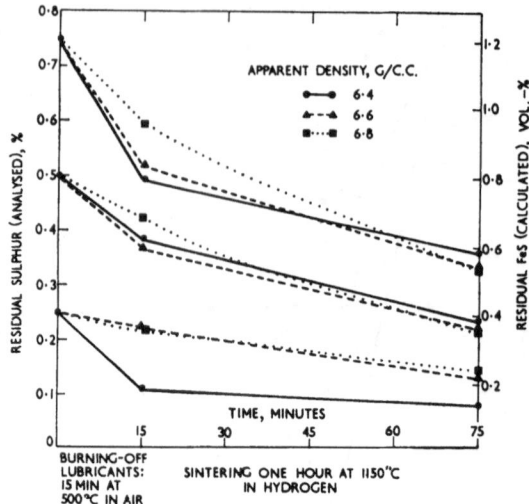

Fig. 8. Residual sulphur contents after burning off
lubricants and after sintering.

sulfur additions. One would, therefore, expect the bearing cross
section of the respective test bars to be weakened rather than
strengthed; especially since, owing to the very low solubility of
sulfur in iron, a strengthening effect through alloying cannot be expected. On the other hand, the same figures and in particular the
photomicrographs of Fig. 4, show that sulfur additions have a
pronounced pore-closing and pore-rounding effect, which may
balance or outweigh the weakening effect of the slightly increased
total porosity. A closer study of the curves in Figs. 2 and 3 reveals the following facts:

(1) If sulfur in amounts from 0.25 to 0.75% is added to the
iron powder or iron–copper powder mix, the interconnected porosity after sintering decreases much faster with increasing green
density than is the case without these additions.

(2) Even with sulfur additions as small as 0.25%, the interconnected porosity can be practically eliminated from compacts
with a green density exceeding 6.8 or 6.9 g/cc; in the absence of
sulfur additions this can be achieved only if the green density exceeds 7.5 g/cc.

(3) Additions of 2.5 or 7.5% copper powder to the iron powder retard the decrease of interconnected porosity with increasing
green density.

The results appear to be in good agreement with previous speculations regarding the role of sulfur additions in sintering. It seems, therefore, not unlikely that the mechanism of pore closing effected by small sulfur additions actually operates according to the principle outlined earlier. However, the experimental data available up to the present date are still too incomplete to allow of more serious theorizing.

The remarkable consequences of the pore-closing effect in pack-carburizing are evident from Fig. 5. It can be seen that, under the carburizing conditions involved, the case developed on a specimen with a green density of 7.0 g/cc made from iron powder without sulfur addition is hardly better than that on a specimen with a green density of 6.4 g/cc made from iron powder with 0.25 or 0.75% added sulfur. The core of both specimens is much over-carburized.

The specimens with green densities of 6.6 and 6.8 g/cc, made from iron powder with a 0.25 or 0.75% sulfur addition, have developed excellent cases and show much less carburization of the core. The best case has developed on the specimen with 6.8 g/cc green density and 0.25% sulfur added to the iron powder, and the core of this specimen is entirely free from carbon. Compare the photomicrographs in Fig. 6, which show the microstructure at the center of the core at higher magnification.

The drop in microhardness ($HV_{0.025}$) with increasing distance from the surface of the specimen can be studied in Fig. 7. These hardness figures were obtained along a transverse center-line across the respective sections shown in Fig. 5. A remarkable feature of these diagrams is not only that the drop in hardness from the surface toward the center of the specimen is steeper, but also that the hardness figures show less scatter, when the specimens are made from iron powder with sulfur additions. The decrease in scattering, as well as the steeper drop in microhardness, affords additional proof that not only a very effective pore closure and reduction of interconnected porosity, but also a substantial improvement in case-hardenability, have been achieved with the small sulfur additions in question.

An addition of only 0.25% sulfur has yielded the best case-hardenability results, and also seems preferable to larger additions for the following reasons. It emerges from Fig. 5 that increasing amounts of added sulfur seem to give rise to an increasing tendency for the formation of large single pores. This is probably due in part to inefficient mixing and in part to the development of increasing amounts of hydrogen sulfide gas which might have the effect of inflating smaller pores. Furthermore, the smaller the sulfur addition, the smaller also is the risk that the sintering gas becomes contaminated with sulfur which might be harmful to certain parts of the sintering furnace.

This brings us back to the question of what happens to the sulfur additions during sintering. It was found by chemical analysis that if the lubricants were burned off in the preheating zone of the sintering furnace the sulfur content after sintering was reduced to between one-half and one-third. If the lubricants were burned-off during a period of 15 min at 500°C in air before sintering, the residual amounts of sulfur after burning and after sintering respectively, were as indicated in Fig. 8. From this it can be seen that about one-third of the sulfur is lost during the burning-off of the lubricants, and that the residual amount of sulfur after sintering is again approximately one-half to one-third of the amount originally added to the iron powder.

It may seem surprising at first glance that substantially more sulfur is not lost during the burning-off of the lubricants. But it must be remembered that, preferably under oxidizing conditions, stable compounds such as FeS and Fe_2OS_2 are already forming at an early stage ($\sim 200\text{-}250°C$), i.e., well below the boiling point of sulfur. This effectively reduces the risk that the sulfur is distilled off together with the lubricants.

The residual iron sulfides are found mainly in the grain boundaries of the sintered structure, while both sulfides and oxysulfides, though to a smaller extent, are found clogging smaller pores. Examples are shown in the photomicrograph of Fig. 9 which was taken on a specimen with a density of ~ 6.9 g/cc made from iron powder with 0.75% added sulfur.

Finally, the question remains whether the method described here is applicable on a commercial scale, and whether and to what extent the sulfur contaminates the sintering gas and affects the sintering furnace.

At the present state of investigation, it is difficult to give a definite and general answer to this question, since results from tests on a larger scale are not yet available. It can be estimated,

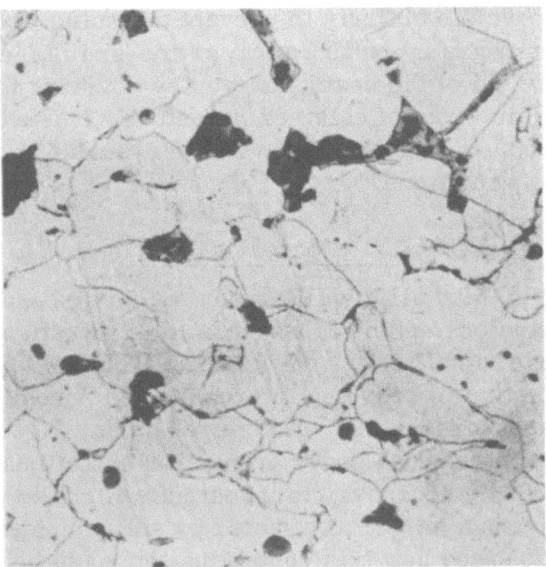

Fig. 9. Residual sulphides and oxysulphides in the microstructure of a sintered iron specimen containing 0.75% added sulphur, pressed to 6.9 g/cc. Sulphides mainly in the grain boundaries; sulphides and oxysulphides clogging the pores. Etched in Nital. ×400.

however, to what extent the sintering gas might become contaminated with H_2S. As a somewhat simplified example, let us assume an industrial sintering furnace with a capacity of 60 kg of sintered parts per hour and a gas consumption (hydrogen or cracked ammonia) of 6 m^3/h. Let us further assume that 0.25% of sulfur has been added to the iron powder from which the parts in question are made, and that the sulfur loss on sintering is 0.1%. It then follows that the sintering gas could, under adverse conditions, pick up ~10 g/m^3 of sulfur, which would correspond to approx. 0.7 vol% H_2S. This figure is, of course, much too high to be regarded as harmless in furnaces containing heating elements of the nickel–chromium type or other elements constructed from chromium or chromium–nickel steels. Thus, mesh-belt furnaces are most probably ruled out for this purpose.

On the other hand, pusher-type furnaces with molybdenum heating elements and ordinary refractory muffles will probably be much less affected by sulfur contamination in the sintering gas. Furthermore, since in this case the parts are charged in boats, it

seems not impossible entirely to prevent the sulfur from contamina-
ting the sintering gas by putting a tray of fine-grained calcined limestone under the lid of the boats, or even by covering the parts themselves with a thin layer of calcined limestone. It is well known
that this material is a most effective sulfur-binder, especially
under reducing conditions.

V.   Conclusions

Bearing in mind all the different aspects discussed above, the
following conclusions can be drawn from this investigation. On a
laboratory scale, small sulfur additions to iron powder have proved
to be most effective in reducing interconnected porosity and improving the case-hardenability of sintered parts. However, before
this method can be applied under production conditions, tests on a
larger scale will be necessary to investigate if and to what extent
the sintering furnace might be affected by sulfur contamination in
the sintering gas, and how this contamination could be prevented.
The contents of this paper may, therefore, be regarded as an
attempt to stimulate further investigation, rather than as a recipe
for immediate practical application.

Acknowledgments

The author wishes to thank his co-workers Allan Carlsson,
who carried out the sintering experiments, and Otto Struglics who
did the carburizing tests and the metallographic work. He is indebted to his colleague, Mr. Sven I. Hulthén, for stimulating this
investigation, and to Höganäs-Billesholms AB of Sweden for permission to publish the paper.

The Effect of HCl—H$_2$ Sintering on the Properties of Compacted Iron Powder

R.D. McIntyre
Richmond Heights, Ohio

The objective of this investigation was to systematically study the effects of H$_2$ and HCl – H$_2$ atmospheres on the properties of compacted iron powder sintered in the gamma-iron range (910 to 1400°C). Prior work demonstrated alpha-iron sintered readily to high densities and that strength seemed related to density [1]. It was noted that gamma-iron showed marked resistance to densification [2]. In this investigation density changes, pore characteristics, tensile strength, and percent elongation were evaluated in an attempt to determine gamma-iron sintering behavior.

TABLE 1. Characteristics of Briquettes Prior to Sintering

Pressing pressure, psi	Sample condition	Density, g/cc	Theoretical density, %	Calculated porosity, cc/g*	Measured porosity, cc/g	Closed porosity, cc/g	Largest pore diam, microns	Smallest pore diam, microns	Avg pore diam, microns
20,000	Pressed and annealed ½ hr at 200 C in H₂	4.90	62.0	0.080	0.070	0.010	5.80	0.10	3.30
50,000	Pressed and annealed ½ hr at 200 C in H₂	6.10	77.2	0.040	0.030	0.010	2.90	0.10	1.30
100,000	Pressed and annealed ½ hr at 200 C in H₂	7.10	89.9	0.010	0.010	...	0.50	0.10	0.30

*From density.

TABLE 2. Density and Pore Data for Sintered Briquettes Pressed at 20,000 psi

Temp. °C	Time, hr	Atm	Density, g/cc	Theoretical density, %	Calculated porosity, cc/g	Measured porosity, cc/g	Closed porosity, cc/g	Largest, microns	Smallest, microns	Avg, microns	Tensile strength × 10³ psi	Elong, %
950	½	H₂	5.10	64.5	0.070	0.060	0.010	5.80	0.60	3.10	6	2.0
		HCl-H₂	5.10	64.5	0.070	0.070	...	8.80	0.60	5.10	9	4.0
	1	H₂	5.20	65.8	0.060	0.060	...	8.80	0.60	2.90	7	3.0
		HCl-H₂	5.10	64.5	0.070	0.060	0.010	8.80	0.60	6.00	10	4.0
	2	H₂	5.10	64.5	0.070	0.070	...	5.80	0.60	3.80	6	2.0
		HCl-H₂	5.20	65.8	0.060	0.060	0.010	5.80	0.90	5.50	9	2.0
1100	½	H₂	5.20	65.8	0.060	0.060	...	5.80	0.90	4.60	7	2.0
		HCl-H₂	5.10	64.5	0.070	0.070	...	5.80	0.90	5.90	10	4.0
	1	H₂	5.10	64.5	0.070	0.060	...	5.80	0.70	4.40	6	2.0
		HCl-H₂	5.30	67.1	0.060	0.070	...	5.80	0.90	8.00	9	4.0
	2	H₂	5.30	67.1	0.060	0.060	...	5.80	0.70	4.20	8	3.0
		HCl-H₂	6.70	84.9	0.030	0.030	0.010	8.80	0.90	6.90	12	4.0
1375	½	H₂	6.70	84.9	0.030	0.030	...	8.80	0.70	7.10	20	6.5
		HCl-H₂	6.90	87.3	0.020	0.020	...	8.80	1.20	7.20	28	13.0
	1	H₂	6.90	87.3	0.020	0.020	...	8.80	1.20	6.50	28	10.5
		HCl-H₂	7.30	92.5	0.010	0.005	0.005	8.80	1.20	6.40	32	16.0
	2	H₂	7.50	95.0	0.010	0.000	0.010	5.90	1.20	4.20	37	24.0

TABLE 3. Density and Pore Data for Sintered Briquettes Pressed at 50,000 psi

Temp. °C	Time, hr	Atm	Density, g/cc	Density, theoretical %	Calculated porosity, cc/g	Measured porosity, cc/g	Closed porosity, cc/g	Largest, microns	Smallest, microns	Avg, microns	Tensile strength × 10³ psi	Elong, %
950	½	H₂	6.20	78.5	0.030	0.030	...	2.50	0.50	1.40	19	6
		HCl-H₂	6.30	79.8	0.030	0.030	...	2.50	0.50	1.40	23	10
	1	H₂	6.30	79.8	0.030	0.030	...	2.50	0.50	1.40	20	6
		HCl-H₂	6.40	81.0	0.030	0.030	...	2.90	0.50	1.80	24	10
	2	H₂	6.30	79.8	0.030	0.030	...	2.90	0.50	1.80	20	6
		HCl-H₂	6.30	79.8	0.030	0.030	...	2.90	0.70	2.10	23	10
1100	½	H₂	6.30	79.8	0.030	0.030	...	2.90	0.90	2.00	20	6
		HCl-H₂	6.40	81.0	0.030	0.030	...	4.40	0.60	2.80	23	10
	1	H₂	6.40	81.0	0.030	0.030	...	4.40	0.90	2.90	20	6
		HCl-H₂	6.30	79.8	0.030	0.030	...	4.40	0.90	2.80	24	10
	2	H₂	6.30	79.8	0.030	0.030	...	4.40	0.90	2.10	23	7
		HCl-H₂	7.00	88.5	0.020	0.030	0.020	4.40	0.90	3.20	28	10
1375	½	H₂	7.20	91.1	0.010	...	0.010	...	...	...	28	11
		HCl-H₂	7.30	92.5	0.010	...	0.010	...	...	...	34	20
	1	H₂	7.40	93.6	0.010	...	0.010	...	...	...	32	14
		HCl-H₂	7.50	95.0	0.010	...	0.010	...	...	...	37	23
	2	H₂	7.80	98.6	0.010	...	0.010	...	...	...	41	25

TABLE 4. Density and Pore Data for Sintered Briquettes Pressed at 100,000 psi

Temp, °C	Time, hr	Atm	Density, g/cc	Density, theoretical %	Calculated porosity, cc/g	Measured porosity, cc/g	Closed porosity, cc/g	Largest, microns	Smallest, microns	Avg, microns	Tensile strength X10³ psi	Elong, %
950	½	H₂	7.10	89.9	0.010	0.010	....	0.60	0.20	0.50	30	12
	1	HCl-H₂	7.20	91.1	0.010	0.010	....	0.70	0.20	0.40	34	20
	1	H₂	7.20	91.1	0.010	0.010	....	0.70	0.30	0.50	31	13
	2	HCl-H₂	7.20	91.1	0.010	0.010	....	0.70	0.30	0.60	34	20
	2	H₂	7.20	91.1	0.010	0.010	....	0.70	0.40	0.50	31	13
1100	½	HCl-H₂	7.20	91.1	0.010	....	0.010	0.90	0.40	0.80	34	20
	1	H₂	7.20	91.1	0.010	....	....	0.40	0.40	0.40	31	13
	1	HCl-H₂	7.20	91.1	0.010	....	....	0.40	0.30	0.40	34	20
	2	H₂	7.20	91.1	0.010	0.010	0.010	....	....	....	31	13
	2	HCl-H₂	7.20	91.1	0.010	....	0.010	....	....	....	34	20
1375	½	H₂	7.50	95.0	0.010	....	0.010	....	....	....	39	17
	½	HCl-H₂	7.60	96.2	0.003	....	0.003	....	....	....	37	24
	1	H₂	7.70	97.5	0.002	....	0.002	....	....	....	40	20
	1	HCl-H₂	7.70	97.5	0.004	....	0.004	....	....	....	38	26
	2	H₂	7.80	98.7	0.001	....	0.001	....	....	....	41	21
	2	HCl-H₂	7.90	100.0	....	....	....	....	....	....	42	28

Materials and Procedure

Electrolytic iron powder sifted -325 mesh was compacted in a single-action, stainless steel die at 20,000, 50,000, and 100,000 psi and annealed for $1/2$ h at 200°C in H_2. A typical analysis of the iron compacts gave the following result in wt%: 0.005C; 0.004P; 0.005S; 0.05O; and 0.07N; iron bal. Specimens were sintered for $1/2$, 1, and 2 h at three temperatures, 950, 1100, and 1375°C, in a quartz tube furnace. An induction coil around the quartz tube heated a tungsten suscepter block on which rested the compacted iron briquettes. Optical temperature measurements were made through an aperture at the exit end of the horizontally mounted tube. The pyrometer used was calibrated against the melting points of pure copper, nickel, and iron. A chromel–alumel thermocouple protected by a quartz tube and imbedded in the tungsten suscepter provided an additional temperature check. Temperature measurement was controlled to within ± 10°C. Pure, dry (– 67°C dew point) hydrogen and pure, anhydrous hydrogen chloride gases were passed through flowraters into the furnace via a glass T-joint. The flow rate was held constant at 3 l/min. Only a 1% HCl in hydrogen addition was used for the $HCl-H_2$ sintering. Additional HCl caused too high an iron loss.

Density and pore measurements were made on a mercury penetration porosimeter [3]. Pore sizes down to $0.050~\mu$ diameter were measurable by this technique. The limit of experimental accuracy for density determinations was ± 0.10 g/cc; for pore diameters, $± 0.10~\mu$; for pure volume, ± 0.10 cc/g. Sintered briquettes were examined microscopically to evaluate relative pore size and irregularity of pore shape.

The tensile bars had a 2-in long gauge length tapered to the center. All breaks occurred within the gauge length. Measurements were made using a 0.05 in./min strain rate. The limit of experimental accuracy for tensile strength values was ± 500 psi; for elongation, ± 2%.

Results and Discussion

Density and pore data for briquettes before and after sintering appear in Tables 1 through 4. The similarity in density for 950 and 1100°C sintering agreed with data from Hausner and King [4]. All the briquettes sintered in $HCl-H_2$ were characterized by a bright surface finish. The pore distribution was generally coarser for briquettes compacted at low and intermediate pressing pressure and sintered in HCl − H$_2$ at 950 and 1100°C. A minimum pore diameter appeared to be associated with each sintering temperature. Actual measurement of this pore diameter for 1375°C sintering was possible only for those briquettes compacted at 20,000 psi. Although pore size seemed related to densification, open porosity was not necessary for densification to occur. Whether or not densification occurred after pores closed-off appeared to be related to pore size prior to closing.

Tables 2, 3, and 4 show the tensile strength and percent elongation values for the range of sintering conditions investigated. These results agree well with those of Eudier [5]. HCl−H$_2$ sintering atmospheres clearly improved both tensile strength and elongation for all combinations of sintering variables. The largest differences in elongation appeared at the higher densities, i.e., moderate to high compaction pressures combined with moderate to high sintering temperatures. The specimens showing the highest elongation had closed pores. The larger pores for HCl−H$_2$ sintering appeared to be associated with a significant decrease in pore surface irregularity at the same density as confirmed by microscopic examination. Such a decrease was shown to increase tensile strength [6]. It is believed that changes in pore shape rather than size may be responsible for differences in the properties of HCl−H$_2$ vis-a-vis H$_2$ sintered samples at equal density as indicated in analogous studies on tungsten sintering [7].

The improvement in sintering through the addition of HCl to the dry hydrogen furnace atmosphere was probably due to the more active surfaces which result. More active surfaces may just be cleaner surfaces where there is less interference with such sintering mechanisms as gaseous transport and surface diffusion. Gaseous transport and surface diffusion are believed to be the principal mechanisms producing initial sintering.

Conclusions

1. Improvement in tensile strength and percent elongation for specimens sintered in HCl−H$_2$ relative to hydrogen sintering was observed for all combination of sintering variables.

2. The largest differences in elongation appeared at the higher densities which were produced by moderate to high compaction pressures combined with moderate to high sintering temperatures.

References

1.   A. Squire, Density Relationships of Iron Powder Compacts,
     Trans. AIME, 171:472 (1947).
2.   J. Libsch, R. Volterra, and J. Wulff, The Sintering of Iron
     Powder, Chap. 35 in Powder Metallurgy, ed. by J. Wulff,
     ASM, Metals Park, Ohio (1942).
3.   N. M. Winslow, and J. J. Shapiro, ASTM Bull. 1959, p. 39.
4.   H. H. Hausner and R. King, Effect of Powder Particle Size
     of the Sintered Material, Planseebe Pulvermet. 8(1):28
     (1960).
5.   M. Eudier, Symposium on Powder Metallurgy Institute of
     Metals, London, England 1954 p. 59.
6.   P. R. Basford and S. B. Twiss, Trans AIME, 212:124
     (1958).
7.   R. D. McIntyre, The Effect of $HCl-H_2$ Sintering Atmospheres
     on the Properties of Compacted Tungsten Powder, ASM
     Trans Quart. 56(3):468 (1963).

Controlled Oxidation Prior to Sintering of Iron Compacts

Harold T. Harrison and Carl G. Johnson

The Presmet Corporation
Worcester, Massachusetts

When we agreed to present this paper to this group, we thought we had a considerable store of knowledge on the subject. In truth, we did have considerable knowledge as to how we could influence our product by using variations of this technique; however, the whys of what happens to produce these results were more or less subject to conjecture and educated guess.

Accordingly, the last six months have been spent on additional experimentation with the idea of eliminating these educated guesses. These many experiments have led us to two very definite conclusions. First, the title of this paper should be changed to read: The Uncontrollable Oxidation Prior to Sintering of Iron Compacts." Secondly, we are still unable to make any more than educated guesses as to why we are able to improve the properties of our product with this technique. Nevertheless, we do feel that we can contribute something to the art if you will accept the subject matter as a progress report, and if this progress spurs some young, ingenious, and ardent investigator to approach the problem from a fresh point of view, and come up with some conclusions.

Several investigators, including Mr. Eudier at this meeting a few years ago, have reported on preoxidation as a means of activating the sintering mechanism. Likewise, most of you are aware that variations of this technique have been in common use in Europe for several years.

Our interest in preoxidation was the result of two factors. First, we were producing larger parts on larger presses, and we noted that these parts required excessively long sintering times to develop expected properties. Secondly, we were having difficulties in sintering iron powder to the ductility and strength levels specified in Military Specification 11073A. We did manage to meet the specification, but only with barest margin of safety. Likewise, in spite of all our precautions, we found it impossible to sinter this material for any extended period. The slightest buildup of carbon potential in the furnace resulted in parts that would not meet the ductility requirements. These slow sintering rates coupled with the excessive number of furnace setups required to sinter a given volume of parts resulted in excessive costs of sintering to the aforementioned specification.

Accordingly, we ran some quick investigations in our laboratory to determine if we could increase the tensile strength and elongation properties of straight iron by the use of some oxidation technique prior to sintering. Needless to say, these tests did confirm the sketchy but relatively accurate prior information on the subject. As a result, one of our belt conveyor furnaces was equipped with an additional zone of air heating at the charge end. This zone was completely integrated into the furnace setup, and was very carefully planned with respect to the problem at hand. However, at this later date, we would like to caution that it is not adequate in all situations. This zone is eight feet long with respect to effective heating, and is equipped with takeoff hoods at either

TABLE 1. Effect of Prior Oxidation on Sintered Properties of Iron Containing 1% Zinc Stearate

	With prior oxidation	Without prior oxidation
Density gm per cc	5.65	5.65
Ultimate tensile strength psi	13,400	8,960
Elongation, % in one inch	5	2.5
Hardness, Rockwell H	48	38

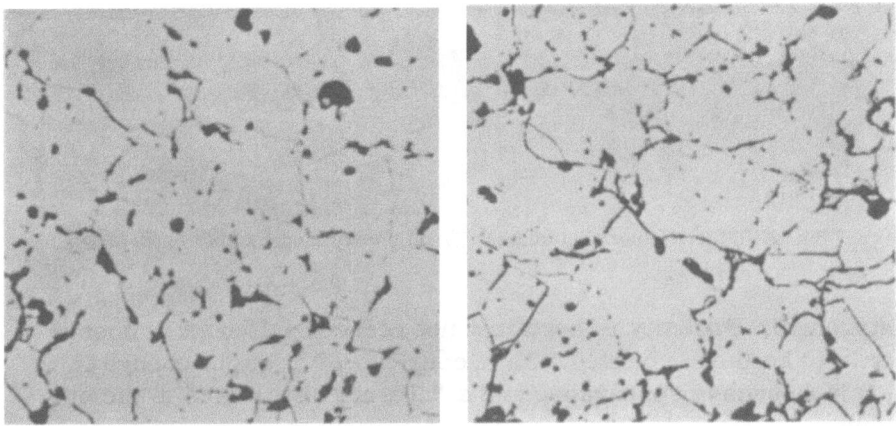

With preoxidation   *Without preoxidation*

Fig. 1. Effect of preoxidation on microstructure of sintered iron.

end to remove the gases of volatilizing stearates. It is capable of operating to 1200°F and is heated by ribbon-type nickel – chrome elements. It is equipped with a stainless steel muffle. Another very important feature is the helper belt designed to minimize belt stresses in the high temperature section of the furnace.

Table 1 illustrates the immediately successful results obtained by utilizing the added zone of air heating. The chart lists properties achieved on tensile bars sintered simultaneously with production sintering of iron rotating bands. The material is a commercially available reduced iron powder containing 1% zinc stearate as a pressing lubricant. Compact density was 5.7 g/cc. One group of bars passed through the air heating zone; this zone was by-passed in a second group by placing on the conveyor belt at a point just following this zone. You will note that bars sintered without prior oxidation fail to meet the specified property levels for Military Specification 11073A in this density range. If we were to increase the properties to acceptable levels by conventional sintering techniques, both an increase in sintering temperature and soaking time would be necessary; however, the air heating zone obviates the need for increasing the time and temperature, and the factor of safety on properties achieved is increased.

Figure 1 illustrates the microstructure of bars just discussed. The substantial increase in properties resulting from preoxidation would be quite difficult to predict from the subtle changes in micro-

Fig. 2. Effect of lubricant content (%) on preoxidized fracture appearance.

structure. Porosity is certainly not reduced, although it does appear to be more rounded. This, coupled with a slight change in grain-boundary appearance and a "cleaner" structure is the only clue.

One other fact observed at that time concerned raw materials used to produce parts to the aforementioned specification. The preoxidation technique reduced raw materials selection problems to a bare minimum. Almost any commercially available iron powder of the proper purity level could be used in the manufacture of iron rotating bands. This was in contrast to our prior experience that certain powders could not be used, and even those that could be used were selected by lot after careful testing.

Getting away from that old conversation piece and back to the subject of this paper, you will recall that I suggested that the word

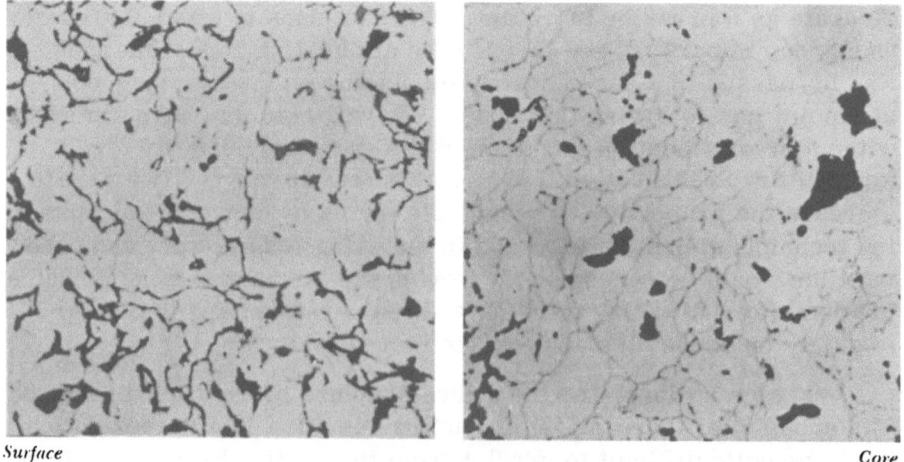

Surface   *Core*

Fig. 3. Preoxidized structure iron plus 1% zinc stearate.

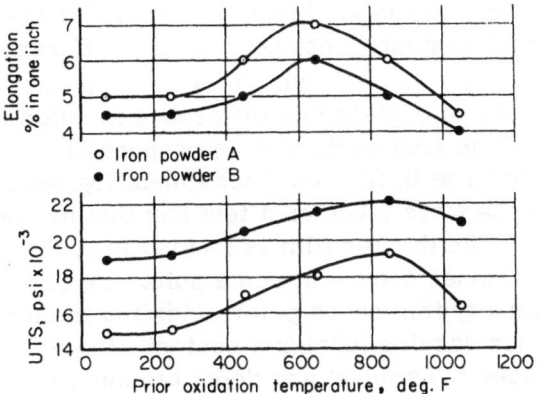

Fig. 4. Effect of varying prior oxidation temperature
on sintered properties of two iron powders containing
1% zinc stearate. Compact density 6.1 g/cc; sintered
at 2050°F for 20 min in dissociated ammonia.

uncontrollable, rather than controlled, would be more apropo. This
is actually belittling our abilities as laboratory investigators, and
is based only on our observations of applying the preoxidation
technique to production methods. We have found that the degrees
of preoxidation and the practical depth to which it penetrates our
green compacts to be very much dependent on density and the

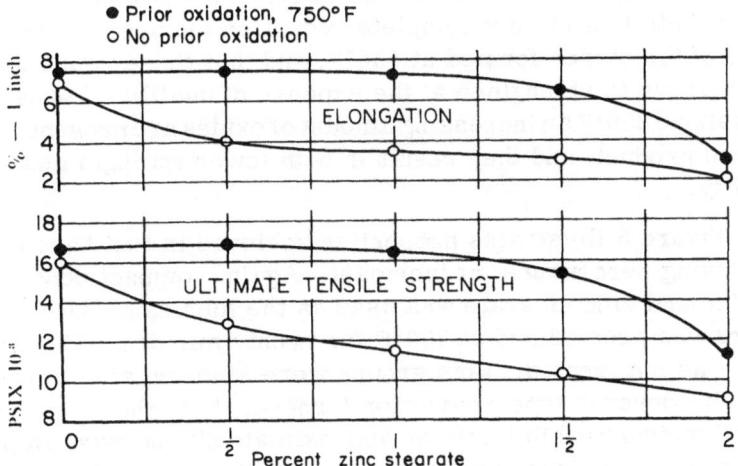

Fig. 5. Effect of varying lubricant contents on properties of sintered iron
with and without prior oxidation. Compact density 6.1 g/cc; sintered at
2020°F for 20 min in dissociated ammonia.

amount of lubricant present. Figure 2 illustrates the effect of varying lubricant content. Note not only the difference in depth of penetration, but also the different color shadings of the oxidized zone. Our efforts to explain this difference in color indicate that the thickness of the iron oxide film is responsible. Figure 3 illustrates this difference in film thickness metallographically. These photomicrographs were taken on a test bar that had been preoxidized at 750°F. The thicker film is located near the surface of the bar, while the thinner occurs at some point nearer the core. Varying compact density influences prior oxidation in much the same manner noted for varying lubricant content. Increasing density gradually reduces the amount and depth of oxidation, and the strengthening effect to be realized gradually diminishes.

These variables of density, lubricant content, and compact mass, pointed out to us that the word "controlled" was a misnomer when applied to our attempts at incorporating preoxidation with present production techniques.

Figure 4 illustrates the effect of varying preoxidation temperature on two iron powders containing 1% zinc stearate as a pressing lubricant. Standard ASTM tensile bars pressed to 6.1 g/cc were used in this investigation. Of interest here is the fact that prior oxidation affects both powders in similar manners. Elongation reaches a maximum at a preoxidation temperature of 650°F, whereas ultimate tensile strength peaks at about 850°F. We attribute this to an incomplete reduction (during the sintering cycle) of the oxides formed at 850°F, and that those remaining oxides serve to strengthen at the expense of ductility. At temperatures above 850°F an increasing amount of oxides are present in the finished product, and they result in both lower strength and lower ductility.

Figure 5 illustrates properties achieved in test bars containing varying percentages of lubricant. Again, compact density was 6.1 g/cc and zinc stearate was used as the lubricant. One group of bars was preoxidized at 750°F for twenty minutes. The other group was not treated. Both groups were sintered simultaneously in a belt conveyor type production furnace. Note that bars containing no lubricant exhibit only an approximate 5% increase in properties when preoxidized, whereas bars with the normal $1/2$% to 1%

zinc stearate are improved about 90% in elongation and 35% in ultimate tensile strength.

These facts are interesting in that they point out graphically the great handicaps that we impose on ourselves by our present techniques for incorporating pressing lubricants. They also indicate that only a small percentage of the increase in properties achieved by preoxidation can be attributed to activated sintering – if activated sintering is considered in the light of surface energies and the other theoretical mechanisms involved. Conversely, some very great advantage that is related strictly to the mechanics of lubricant removed can be attributed to the preoxidation technique.

During the course of these investigations, we did arrive at a theory to explain some of the results. One of the graduate students working with us made the statement, "We wouldn't be working on this project if we sintered from the inside out instead of the outside in." This is quite a profound statement when analyzed, and as a result we came up with a rather simple explanation that we call the "Inside Out Theory." Preoxidation of lubricated compacts results in a constantly diminishing oxide concentration as you approach the inside – or core. Therefore, in the all important initial stages of heating, the high concentration of near surface oxides retards sintering; as a result, porous oxides and partially reduced oxides are still present at that time the core is passing through the critical initial stage of sintering. The resulting permeability of the structure nearer the surface allows expanding lubricant volatiles from the core to escape the confines of the compact. To put it more simply, preoxidation allows us to approach that condition of sintering from "the inside out."

Densification and Grain Growth in the Later Stages of Sintering of Alpha-Iron

H.F. Fischmeister
Chalmers Tekniska Högskola
Goeteborg, Sweden

Introduction

Only few studies have been made of the mechanism of sintering of body-centered cubic metals. On the author's part, interest in the sintering of α-iron is enhanced by the observation [1, 2] of a dominant contribution of surface diffusion during the initial stage of sintering of wire spool models in the α range. Together with the empirical fact that fine iron powders do exhibit considerable shrinkage on sintering in the alpha range (which is incompatible with a surface diffusion mechanism) these wire spool experiments indicated that the mechanism of material transport might depend on the particle size of the sintering system. This stimulated a systematic study of particle size dependence which is still in progress. As a first step it was decided to study the sintering of carbonyl iron powder. This powder can be sintered to nearly full theoretical density below the transformation temperature, which makes it very suitable for a study of shrinkage mechanism.

A theory for the shrinkage of powder compacts during the intermediate and final stages of sintering has been developed by

Coble [3]. To date, it has been verified only for a ceramic powder (α-Al$_2$O$_3$) at quite high temperatures. The theory is based on the concept of the dissolution of the pores in the metal lattice, in the form of individual vacancies, and of the diffusion of the vacancies thus formed to the surrounding grain boundaries where they are annihilated. It predicts a time law for the shrinkage of a compact, the form of which depends on the kinetics of the grain growth that accompanies sintering. To provide data for a comparison with this theory, grain growth was studied concurrently with shrinkage, using a semiautomatic linear analyzer which allows a very detailed characterization of both grain and pore structure of partly sintered compacts.

Experimental

Carbonyl iron powder from Vakuumschmelze Hanau AG (Hanau, Germany) was used. The producer's batch analysis states the following main impurities: Ni 0.03%, Mn 0%, Si 0.003%, Cu 0.002%, C 0.01%, P trace, S 0%. The weight loss on hydrogen reduction [4] was 0.23%.

The particle size of the powder is given in Fig. 1. The band in the figure contains the results of double runs with two sedimentation techniques, a sedimentation balance (Sartorius AG) and a turbidimetric apparatus (Evans Electroselenium Ltd), carried out in a 1:1 mixture of ethylene glycol and water. Turbidimetric results obtained in cyclohexane also fall within the band. The average particle size by air permeability, as determined with a Fisher Subsieve Sizer, was 4.065 μ. The microstructure of the powder is reproduced in Fig. 2. Only few particles exhibited the typical "onion structure," the greater part having recrystallized.

The powder was compacted into standard ("MPI") tensile test bars [5], of density (as pressed) 5.75 ± 0.05 g/cc. A complementary series was pressed to a density of 6.50 ± 0.05 g/cc, to provide data for a study of the influence of green density on the shrinkage process. Camphor was added as lubricant in an amount of 0.5%. It was preferred to other lubricants because of its volatility which made it possible to remove it completely by preheating to a temperature safely below the onset of shrinkage.

Sintering was carried out at 823.5°C in hydrogen dried over activated alumina and P$_2$O$_5$ to a dew point of -20°C in a furnace

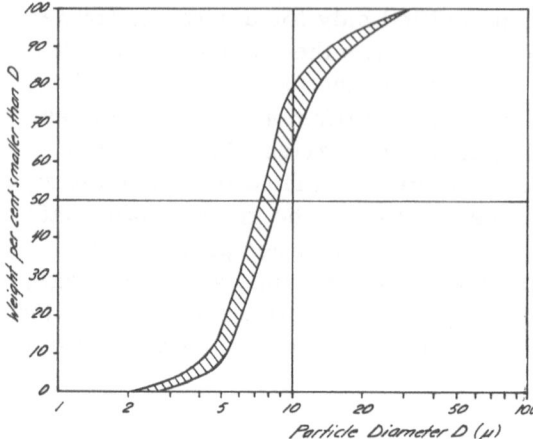

Fig. 1. Particle size distribution of carbonyl iron powder as
determined by sedimentation
balanced and turbidimetric
analysis.

equipped with a single stainless steel tube for the hot zone, the preheating zone, and the cooler. Loading and unloading was carried out
with the furnace and preheating zone at temperature and with nitrogen as a protective atmosphere, to avoid burning off hydrogen at
the loading port, which would have resulted in condensation of
moisture in the cooler. After loading, the compacts were flushed
with hydrogen for 30 min while in the cooler, then pushed into the
preheating zone which was kept at 380°C, and left there for 1 h to
remove the camphor and surface oxide. It had been ascertained
by thermogravimetric measurements that complete removal of
both impurities occurred under these preheating conditions. There-

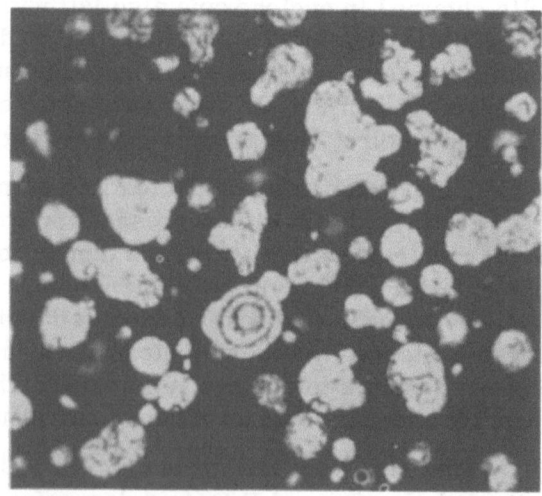

Fig. 2. Microstructure of carbonyl iron powder. Etched in
1% nital, 2000 ×.

after, the boat was pushed on into the hot zone thus avoiding the risk of reoxidation before sintering and finally withdrawn into the water mantled cooler.

The temperature of the charge was monitored by a thermocouple inside the hollow push rod, the joint of which was situated in the center of the boat. Experiments with a thermocouple drilled into one of the bars had shown close correspondence between the heating and cooling curves of the bars themselves and of the monitoring couple. The contribution of the heating and cooling periods to sintering could be neglected thanks to their shortness except for the shortest sintering times of 12 and 48 min. For these runs, corrections were avoided by sintering only two bars at a time in a boat of very small heat capacity, which decreased the heating and cooling times still further, and by raising the furnace temperature 5° above its nominal value, which made the average holding temperature equivalent to the nominal.

The loading port and the back end of the furnace were rubber stoppered, and gas-tight stop boxes were used to feed through the push rod, and at the back end of the furnace, a fixed thermocouple was used to control the temperature of the hot zone. The preheating temperature was controlled by a simmerstat relay, less accuracy being required there. The precautions against contamination of the sintering atmosphere with air or moisture inherent in the operating procedure and furnace construction are believed to have effectively guaranteed an atmosphere of pure hydrogen.

Measurements

The density of the sintered tensile bars was determined by weighing in air and water, open pores having been closed by dipping the bars in a solution of 1% paraffin in benzene containing 10% of silicone oil. Each sintering batch consisted of ten bars. The standard deviation of the individual measurements indicates a limit of error (at 95% confidence level) of $\pm$ 0.022 g/cc in the density, corresponding to an error of $\pm$ 0.3% units in porosity. To facilitate comparison with Coble's theory, the data were plotted as porosity rather than density values. In calculating the porosity from the expression

$$P = \left(1 - \frac{\rho_s}{\rho_m}\right) \cdot 100\% \tag{1}$$

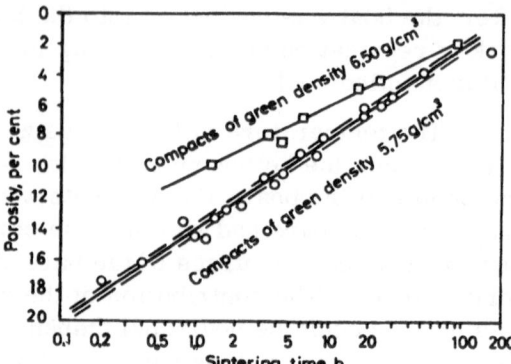

Fig. 3. Porosity of carbonyl iron powder compacts sintered isothermally at 823.5 °C in hydrogen. Values based on archimedic density determinations.

the theoretical density ρ_m of the fully compacted metal was assumed to be 7.780 g/cc. Figure 3 shows the mean values of porosity thus obtained for all sintering runs.

The data can be fitted by the expression

$$P = 0.1366 \ \ln \frac{t}{t_f} \qquad (t_f = 235 \ h)$$

(2)

The numerical factor in the above equation is adapted to the definition of P as a volume fraction, not as percentage. To comply with equation (1), it should be multiplied with 100. The band around the line in Fig. 3 delineates the range of uncertainty of the data points at 95% confidence level.

An alternative to gravimetric density measurement is to calculate the sintered density ρ_s from the green density ρ_p and the

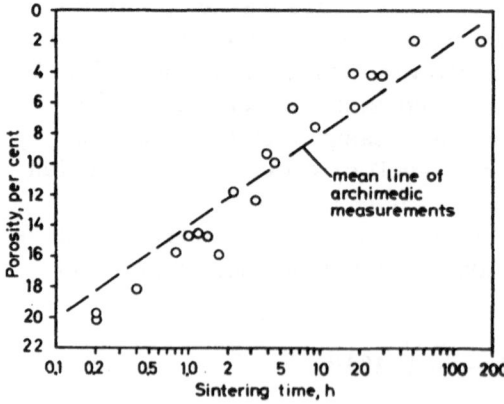

Fig. 4. Porosity values calculated from green density and linear shrinkage [equation (3)].

shrinkage, which was measured with a screw micrometer:

$$\rho_s = \frac{L_p W_p T_p}{L_s W_s T_s} \cdot \rho_b \tag{3}$$

where L = length, W = width, T = thickness of bar; index p = pressed, and index s = sintered state.

The results obtained in this way are plotted in Fig. 4. Beside greater scatter, the points indicate a slope somewhat steeper than that of the line through the Archimedic data points. This might be caused by a porosity-dependent weighing error, such as leakage of water into the pores of highly porous specimens, or by deviations from the overall average of the dimensional change at the points where the bars were measured. The latter explanation is preferred since checks with mercury instead of water ruled out the penetration of liquid into the pores as a source of error. It should be noted that small errors in green density – which are difficult to avoid – affect calculated porosity most at low values, where the deviations from Archimedic values are indeed largest.

The degree of interconnected porosity was assessed qualitatively by measuring the air permeability of cylindrical specimens 1 in. in diameter by $^1/_2$ in. high in a constant pressure apparatus

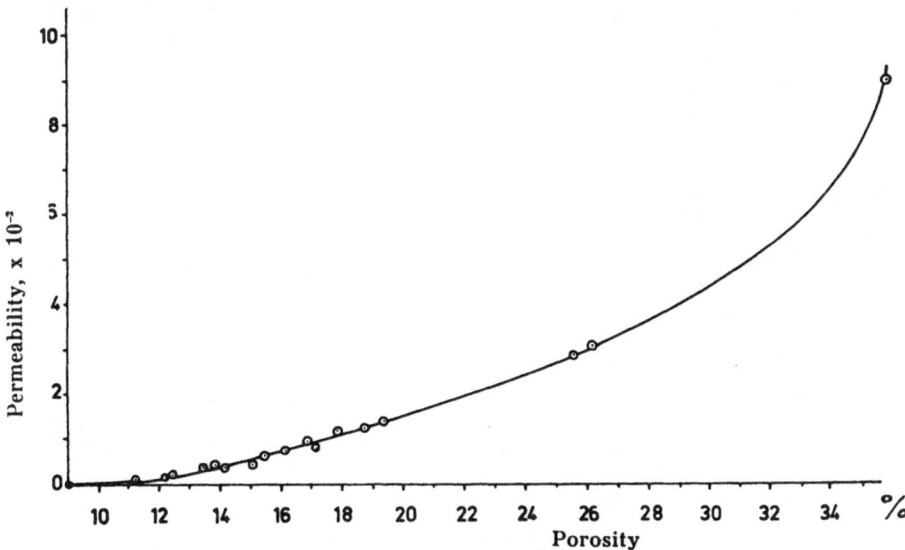

Fig. 5. Air permeability of sintered specimens.

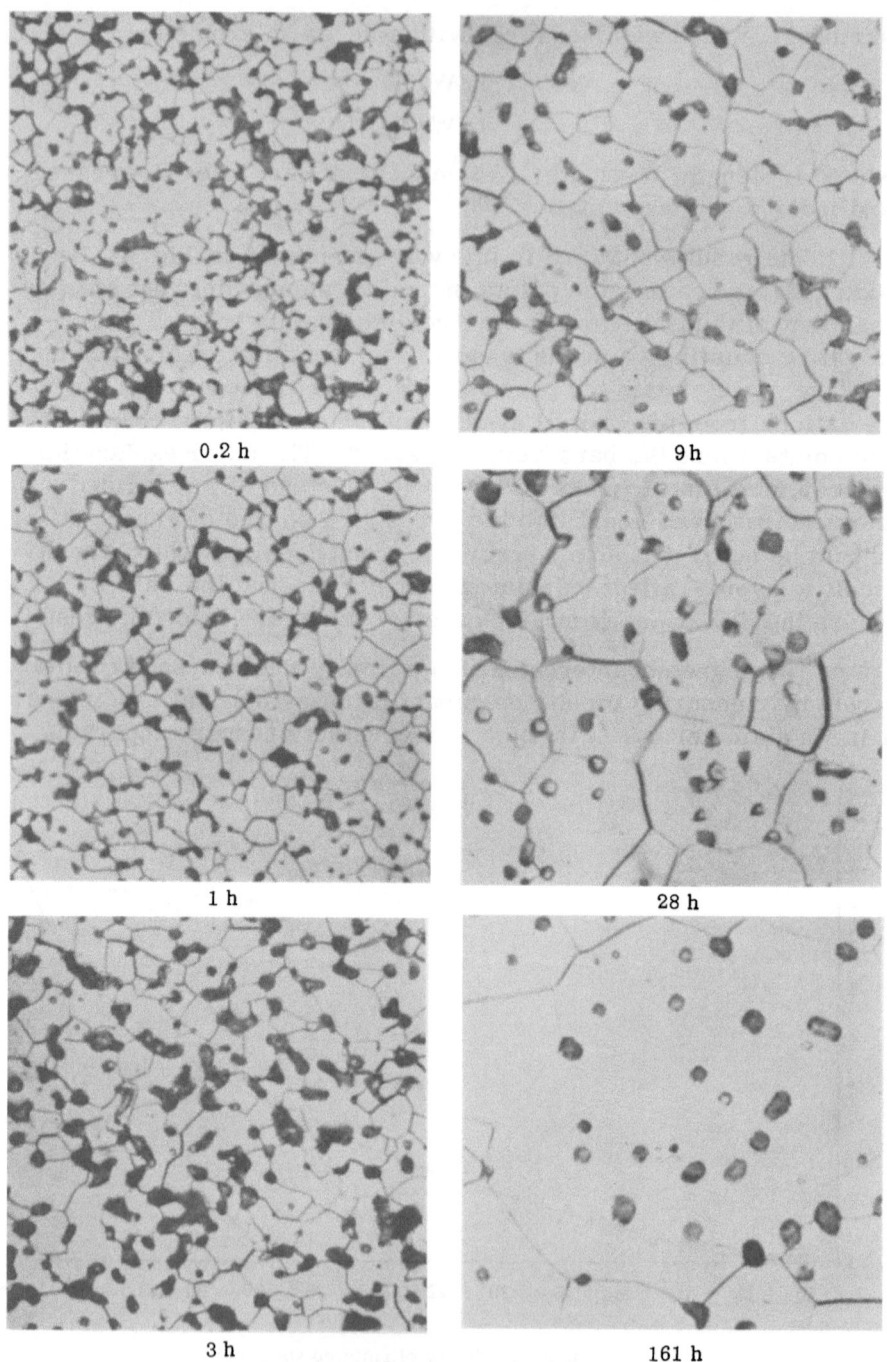

0.2 h   9 h

1 h   28 h

3 h   161 h

Fig. 6. Microstructure of compacts sintered for stated times at 823.5°C in H_2. 600 ×.
Etched in 1% nital.

similar to the one described by Astbury et al. [6]. Permeability is
defined as

$$\pi = \frac{V\,h}{t\,A\,p}\quad [cm^4 \cdot g^{-1} \cdot sec^{-1}] \tag{4}$$

where V = volume of air passed through the specimen, t = time of
passage, h = specimen height (apparent length of passage), A = specimen cross section, and p = pressure difference between top and
bottom surface of specimen.

Permeability becomes zero at a porosity of about 10% (cf.
Fig. 5) which corresponds to sintering times around 5 h at
823.5°C. Below this level of porosity, no throughgoing pore channels exist. The transition from channel- to bubble-shape of the
pores must begin at porosities somewhere still further below.

The micrographs in Fig. 6 indicate a gradual transition of
pore structure from irregular, oblong shapes to regular, round
ones at sintering times in the range 8-16 h.

Metallographic measurements of the following quantities
were made by linear analysis:

mean grain size of metal phase

$$\bar{L}_m = \frac{l_m}{n_{mm}} \tag{5}$$

grain-size distribution of metal phase

$$N\ (L_m)$$

mean size of pores

$$\bar{L}_p = \frac{l_p}{n_p} \tag{6}$$

size distribution of pore intercept

$$N\ (L_p)$$

specific inner surface (i.e., amount of metal-pore interface
in cm^2 per unit volume of compact)

$$S_m = \frac{2\,n_{mp}}{l_m + l_p} \tag{7}$$

The symbols in the above definitions have the following significance (cf. Fig. 7): l_m is the total length of the scanning line traversed
in the metal phase $= \Sigma L_m$; L_m is the intercept of the scanning line
with the individual metal grain; l_p is the total length of the scanning line
traversed within the pore phase $= \Sigma L_p$; L_p is the intercept of the scanning line with the individual pore or portion of pore; n_{mm} is the
number of grain boundaries in the metal phase intersected by the
scanning line of length $l = l_m + l_p$; n_p is the number of pores intersected by the scanning line of length $l = l_p + l_m$; and n_{mp} is the number of metal-pore phase boundaries intersected by the scanning
line of length $l = l_m + l_p$. Note: For each pore intercept, only the
first boundary (i.e., the one encountered on leaving the metal and
entering the pore phase) is counted. The opposite pore-metal
boundary belonging to the same pore is not counted.

It will be recognized from Fig. 7 that n_p is different from
n_{mp} when pores of irregular shape give more than one intercept
with the scanning line. The number of such "superfluous intersections," $n_{mp} - n_p$, reflects the amount of reentrant contours in the
pore surface and thus the tortuosity of the pore surface. It decreases with sintering time as the pores become more and more
spherical.

A short description of the linear analyzer [7] may be appropriate at this point (cf. Fig. 8). It consists of a motor-driven
traveling stage attached to an ordinary metallographic microscope.
A slotted disc attached to the drive interrupts a light beam falling
on a photo transistor, producing exactly 10 pulses for each micron of

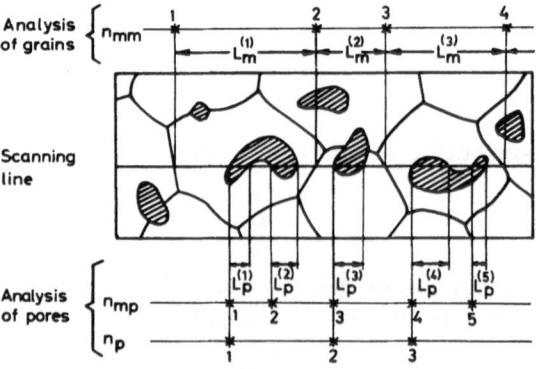

Fig. 7. Principles of linear analysis of porous structures.

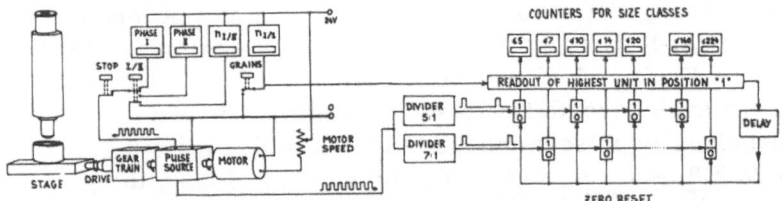

Fig. 8. Linear analyzer (schematic).

stage travel. These pulses are fed to electronic counting and classifying devices.

As the stage advances, the center of a cross hair eyepiece scans a straight line in the specimen plane. The operator marks intersections with grain boundaries by pushing a "grain boundary key" which actuates the grain size analyzer. Another key is pressed during the periods of travel through the pore phase. This deflects the pulses from counter I (Fig. 8) – which records the length of scan over the metal phase, l_m – to counter II, recording the length of the pore phase intercept, l_p. The ratio $l_p/(l_m + l_p)$ is equivalent to the volume fraction of the pore phase. Each depression of the pore key is recorded in counter n_m giving n_{mp}. A further key (not shown) is used to record $n_{mp} - n_p$ when reentrant pore contours occur.

The grain size analyzer consists of two chains of binary counting elements (flip-flop circuits), which record the number of pulses received between two operations of the grain boundary key.

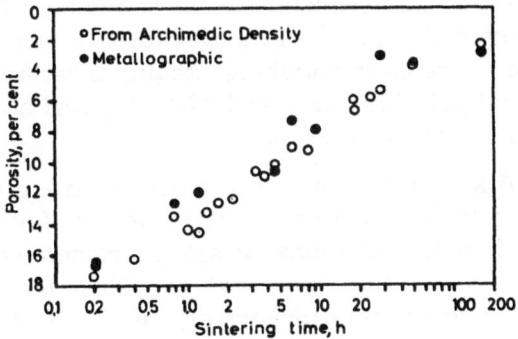

Fig. 9. Metallographic and Archimedic porosity values.

One such chain will record numbers (i.e., intercept lengths) on a binary scale, 1, 2, 4, 8,.... The two parallel chains are fed by a frequency divider which transmits each fifth pulse to the first chain and each seventh to the second chain. The two chains together then classify pulse numbers in terms of the scale 0.5, 0.7, 1.0, 1.4, 2.0, 2.8, 4.0,...μ, which resembles very closely the geometric scale $\frac{1}{2}\sqrt{2^0}$, $\frac{1}{2}\sqrt{2^1}$, $\frac{1}{2}\sqrt{2^2}$,...,$\frac{1}{2}\sqrt{2^i}$ commonly used for the classification of grain size [7, 8]. When the grain boundary key is pressed, the highest digit in the binary chain causes a recording in the appropriate counter. Thus for a grain 17 pulses (= 1.7 μ) long, counter "< 20" is actuated, the lower counters being blocked by an interlocking network. The special suitability of the $\sqrt{2^i}$ scale is due to the fact that size distributions of metallographic features (grains, particles, inclusions, and pores) as a rule follow the logarithmic normal distribution [9].

Metallographic specimen preparation had to be carefully standardized to yield true and reproducible values of pore sizes. Exploratory tests showed that, while too short and too long etching times as well as too short polishing times falsified the porosity values determined by linear analysis, there was a safe range within which apparent metallographic porosity was independent of polishing and etching treatment and corresponded fairly closely to Archimedic porosity values (Fig. 9). The technique finally adopted consisted in vacuum impregnation of the sectioned, etched, and dried bodies with methylmetacrylate, hardening at 50°C for 24 h, and grinding and diamond polishing, as preparatory treatment. The etching after sectioning was to open the pores of the sectioned surface for impregnation; it was performed on the sawed and emery-ground sectioning surface without polishing. The final treatment involved diamond polishing with 1 μ diamond paste, followed by two successive polishing-etching treatments consisting in 2 minutes' polishing on a soft Al_2O_3 lap and etching for 40 sec at room temperature in 1% nital.

The results of the grain size analysis are plotted in Fig. 10, giving the complete intercept distributions, and Fig. 11, which shows the median and mean grain intercepts as a function of sintering time at constant temperature. Since the intercept distributions are approximately lognormal, i.e., skewed on a linear scale of intercept length, there is a consistent difference between the median and the mean, the average ratio being $\overline{L}_m / L_m^{50} = 1.25$.

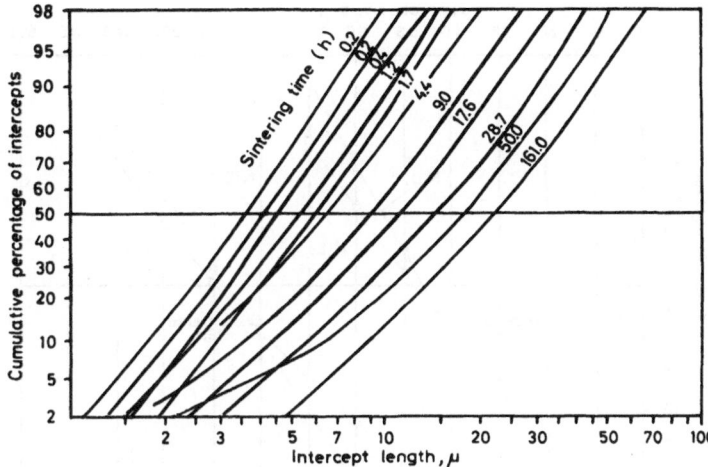

Fig. 10. Intercept distribution of grain sizes of compacts sintered for sintering time.

In metals containing particulate inclusions, the rate of grain growth is often controlled by the coalescence of the particles through lattice diffusion of the atoms from small inclusions to larger ones. This gives rise to a theoretical time law of grain growth [10]

$$X^3 - X_o{}^3 = k \cdot t \qquad (8)$$

where X is some linear measure of grain size. For later stages of growth, where the initial grain size X_0 becomes negligible, this reduces to

$$X^3 = k \cdot t \qquad (9)$$

In Fig. 11, lines of slope 1/3 give a fair link between the data points, indicating that both $L_m{}^{50}$ and $\overline{L}_m$ follow the cubic time

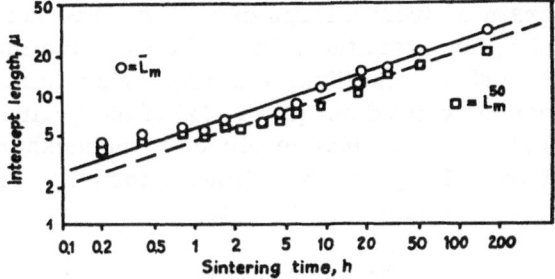

Fig. 11. Median and mean grain intercepts as a function of sintering time.

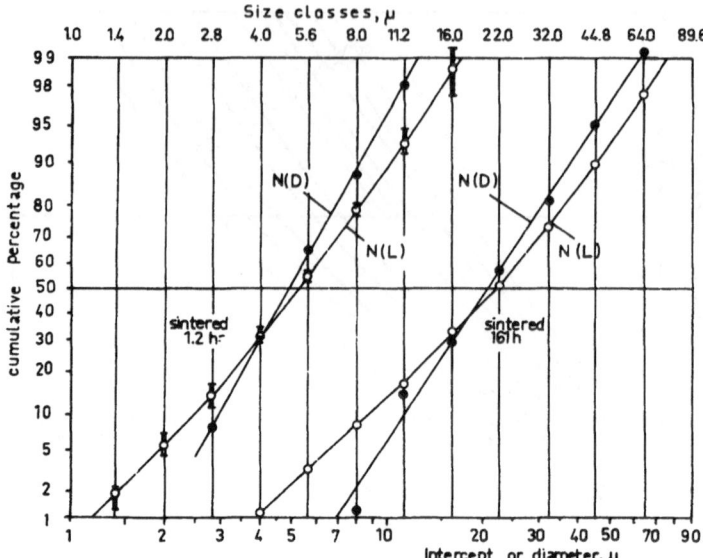

Fig. 12. Intercept and diameter distributions of grains in compacts
sintered for short and long times.

law to satisfactory approximation. There is a deviation at short
sintering times which is believed to be due to the loss of the finest
grains resulting from limited optical resolution and etching difficulties.

From Fig. 12, the time laws of grain growth can be written

$$\overline{L}_m = 0.352 \cdot 10^{-4} \cdot t^{1/3} \ (t \text{ in sec}, \overline{L}_m \text{ in cm}) \tag{10}$$

$$L_m^{50} = 0.281 \cdot 10^{-4} \cdot t^{1/3} \ (t \text{ in sec}, L_m^{50} \text{ in cm}) \tag{11}$$

The linear intercept of an individual grain is not uniquely related to its true size in space. A small intercept may result from
a truly small grain or from a large one intersected near its periphery. Statistically, however, the relation between the intercept distribution, N(D), and the distribution of true grain diameters in
space, N(D), can be worked out [9, 11, 12] if the grains may be
treated as spherical. A method developed by Schückher [9] and
simplified by Exner [13] gives the following formula for the conversion from intercept to diameter distribution:

$$P_i = 2p_{i+1} - p_{i+2} \tag{12}$$

where P_i stands for the percentage of diameters in size class i (defined as the interval $\sqrt{2^{i-1}}\ldots\sqrt{2^i}$) and p_i for the percentage of intercepts in the same interval.

Calculation has to start with the highest size class, proceeding towards the fine grain end of the distribution.

The relation between intercept and diameter distribution is illustrated in Fig. 12 for a fine grained specimen and for the coarsest one produced in this study. It is interesting to note that, while the intercept distributions are only approximately lognormal, the diameter distributions are truly so.

Figure 12 also illustrates the reproducibility of intercept distributions obtained with the linear analyzer. Five runs, comprising together 2318 grains, were made on a specimen sintered for 1.2 h. Their maximum deviation from the common mean is indicated in the figure. The comparatively large deviations at extreme values are due partly to the relative scarcity of the largest and smallest intercepts, and partly to the extension of the probability–percentage scale. The usual run comprised about 100 grains.

The diameter distributions of all specimens as calculated from formula (12) are given in Fig. 13.

The change in pore structure during sintering can also be described by parameters obtained from linear analysis. In fact,

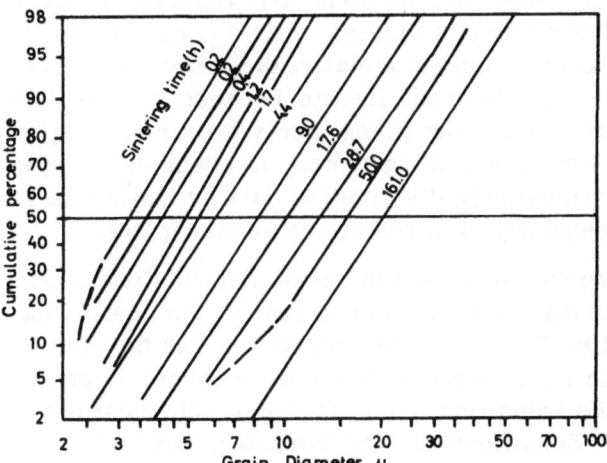

Fig. 13. Grain diameter distributions calculated from the intercept distributions of Fig. 10 (smoothed).

TABLE 1. Pore Size Measurements

Sintering time t, h	Mean pore intercept L_p, μ	Intercept distribution		
		Lower decile L_p^{10}, μ	Median L_p^{50}, μ	Upper decile L_p^{90}, μ
0.2	1.93	0.66	1.41	3.89
0.2	1.91	0.58	1.38	3.98
0.8	1.89	0.75	1.41	3.55
1.2	2.10	0.83	1.72	3.80
4.4	2.45	1.02	2.19	4.17
6	2.35	1.02	2.04	4.37
9	2.72	1.29	2.37	4.73
28.7	2.44	1.03	2.19	3.98
50	2.46	0.93	2.29	4.37
161	3.74	1.32	3.39	5.25

the analyzer described above yields these data at the same time as those for grain size characterization.

Mean pore intercepts together with the median and the top and bottom decile of the intercept distributions are collected in Table 1.

The variation of pore parameters with sintering time is surprisingly small in comparison to the range of grain size. Some complete pore size distributions are reported in Fig. 14 in terms of intercept frequency, a method which brings out the small differences more clearly than cumulative diagrams. Conversion to diameter distributions is inadmissible because the pore shapes, at least at shorter sintering times, are very far from spherical (cf. Fig. 6). At longer sintering times, it becomes increasingly difficult to obtain reliable distribution data by lineal analysis, owing to the low frequency of pores on the scanning line.

It is worth noting that the pore distributions become distinctly narrower during sintering, in contrast to the broadening of the grain size distributions. This narrowing and the rather small increase of pore size throughout the wide range of densities studied, are due to the transition from tubular to spherical pore shape that accompanies densification. Cylindrical pores will give a broader distribution and a larger mean intercept than spherical ones of the same radius, the size of the maximum intercept being limited only by the length of straight segments of pore channel.

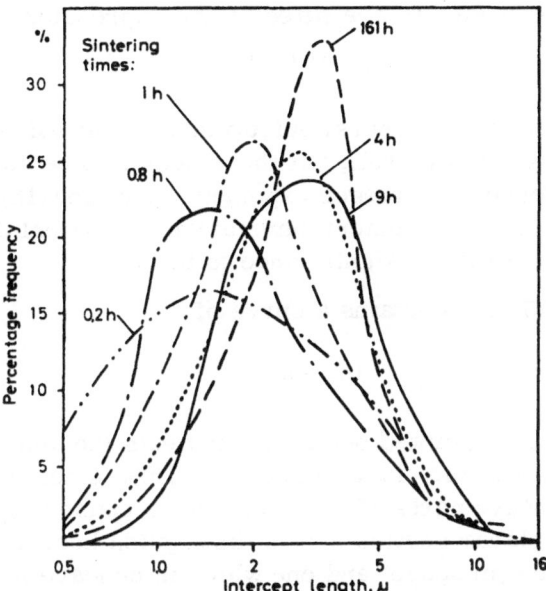

Fig. 14. Intercept distribution of pores (smoothed).

Finally, Fig. 15 shows the decrease of total pore surface
per unit volume of compact with sintering time. The measurement
of a surface in space by the number of its intercepts with a randomly oriented line is based on equation (7) which has the character of
a general geometrical theorem and is independent of shape and
other special circumstances [14]. Therefore, this measurement
of inner surface of a compact reflects directly and without approxi-
mations the excess energy associated with inner surfaces, which
is the thermodynamical driving force of densification.

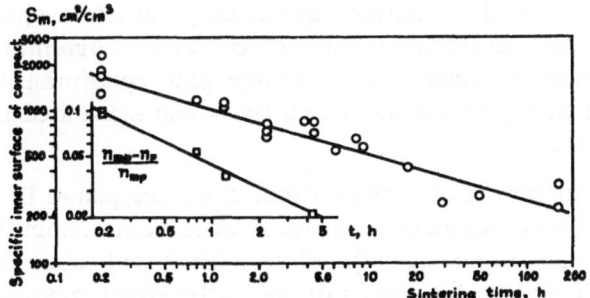

Fig. 15. Inner surface of compacts (area of the pore-metal
interface) as a function of sintering time.

The data in Fig. 15 are fitted by the expression

$$S_m = 1020 \cdot t^{-0.303} \tag{13}$$

with S_m in cm^2/cc and t in hours. A truly cubic expression, $S_m = k \cdot t^{-1/3}$, would fit almost equally well, but there seems to be no theoretical basis for preferring an integer exponent. Equation (13) should be viewed merely as a convenient summary of empirical data, without attaching theoretical significance to its form.

Figure 15 also contains a curve of

$$(n_{mp} - n_p)/n_{mp}$$

which, as has been pointed out above, reflects the amount of reentrant contours in the plane of section. This is not an exact measure of the concave parts of pore surface, because data obtained from a single planar section cannot distinguish between a truly concave surface (in space) and one which is concave in one direction and convex in the other. However, this quantity will indicate in a qualitative fashion the degree of complexity of the pore surface, being greatest for a tortuous surface and becoming zero for a purely convex one. By this criterion, the absence of an appreciable amount of concave surface parts is indicated after 8-h sintering time.

Discussion

A theoretical model of the later stages of densification has been developed by Coble [3]. It visualizes a compact as consisting of polyhedral grains of equal size. The shape of the grains is represented by a regular tetrakaidecahedron, which satisfies the requirements that the idealized grain shape must be space-filling and that its intersections with planes of random orientation should resemble grain sections in a metallographic specimen, i.e., a collection of polygons among which five- and six-sided ones are most frequent.

In the intermediate stage where the pore phase is continuous through the whole compact or at least over some short distance within it, the pores are visualized as cylindrical tubes of equal diameter situated on the edges of the polyhedral grains. In the final stage, where the pore phase has become completely discontinuous, the pores are described as isolated, spherical voids of equal size, placed at the corners of the polyhedra. In each case, porosity (as volume fraction of pores) is given by the radius of the tubular or spherical pores together with the edge length of the polyhedra which determines the length of the pore channels per unit volume, or the number of spherical pores per unit volume. Thus, for cylindrical pores,

$$P = \frac{3\pi}{4\sqrt{2}} \cdot \frac{r_c^2}{\lambda^2}$$

and for spherical ones   (14)

$$P = \frac{\pi}{\sqrt{2}} \cdot \frac{r_s^3}{\lambda^3}$$

where $r_{c.s}$ = radius of spherical or cylindrical pore and λ = edge length of regular polyhedral grains.

Each pore emits a flux of vacancies, by virtue of the increase in vacancy concentration under a curved surface,

$$\frac{\Delta c}{C_0} = \frac{\gamma \Omega}{kT} \cdot \frac{1}{r_c} \text{ (for cylindrical pores)} \tag{15}$$

where c_0 is the equilibrium concentration of vacancies under a plane surface (or at a grain boundary), γ the surface energy of the metal-pore interface, Ω the atomic volume of the metal, k the Boltzmann constant, and T temperature.

The vacancy flux emerging from unit length of a cylindrical pore is

$$j = 4\pi D_0 \Omega \Delta c = \frac{4\pi D \gamma \Omega}{k T} \cdot \frac{1}{r_c}$$

$$(D = D_0 C_0 \Omega) \tag{16}$$

On the other hand, the vacancy flux is directly proportional to the volume change of the pore from which it originates. Knowing Δc and the geometrical expression for the pore volume, one can now calculate the rate of change of the pore volume, which is obtained as

$$\frac{dP}{dt} = -N\frac{D\gamma\Omega}{kT\lambda^3} \qquad (17)$$

The numerical factor N has the value 10 for the intermediate stage (cylindrical pores) and the value $6\pi/\sqrt{2}$ for the final stage (spherical pores).

Equation (17) still contains λ, the edge length of the grains. This, however, can be expressed as a pure function of time if the rate law of grain growth is known. If, as in the present case [equation (10)], grain growth follows a cubic time law, the rate law of porosity becomes especially simple:

$$\frac{dP}{dt} = -\frac{N}{A} \cdot \frac{D\gamma\Omega}{kT} \cdot \frac{1}{t} \qquad (18)$$

where A stands for the third power of the rate constant in the time law of grain growth, equation (10). The rate law of porosity can now be integrated between the moment of observation, t, and the time of final densification, t_f, for which $P = 0$:

$$P = \frac{N}{A} \cdot \frac{D\gamma\Omega}{kT} \cdot \ln\frac{t}{t_f} \qquad (19)$$

The form of this rate law corresponds with the time–dependence of porosity observed in the present study [equation (2)]. The change in N, from 10 to $6\pi/\sqrt{2} = 13.3$, which accompanies the breakup of the pore phase into isolated units, can hardly be expected to make itself felt, being small and spread out over a long period of time.

The assumptions of the model concerning pore structure can be put to a test by means of the data obtained on the inner surface of the compacts. The specific surface (i.e., surface-to-volume ratio) of an array of geometrical bodies is known to be very sensitive to changes in the size and shape of the bodies. The specific surface of the pore phase can be obtained from that of the compacts by division with the volume fraction of pores:

$$S_p = \frac{S_m}{P} \qquad \left(P = 1 - \frac{\rho_s}{\rho_m}\right) \qquad (20)$$

On the other hand, the model of cylindrical pores in the intermediate stage and of spherical pores in the final stage of sintering gives

$$S_p = \frac{2r_c\pi\lambda}{r_c^2\pi\lambda} = \frac{2}{r_c} \text{ for cylindrical pores} \tag{21}$$

and

$$S = \frac{4r_s^2\pi}{4r_s^3\pi/3} = \frac{3}{r_s} \text{ for spherical pores} \tag{22}$$

Thus, the model predicts that the change of pore configuration will be performed in such a way that the specific surface of the whole pore system remains almost unaffected, unless the pore radius changes at the same time. As has been pointed out above, pore size parameters change only insignificantly during densification (cf. Table 1). Calculation of S_p from the smoothed experimental values of P and S_m (Figs. 3 and 15) shows indeed that only very small changes occur in S_p, except at the very beginning of sintering and at its very end (Table 2).

A very shallow minimum occurs in S_p in the range of sintering times where microscopic observation and permeability measurements indicate the beginning of the breakdown of the pore phase. The "equivalent pore radii" listed in the table are those required to give the observed specific surface in a model made up of pores of equal size throughout. They will, therefore, be larger than the real average pore radii, since the real system contains pores of widely varying size, the smaller ones of which contribute greatly to specific surface.

TABLE 2

Sintering time t, h	Specific pore surface S_p, cm²/cc	Equivalent pore radii	
		cylindrical $r_c = 2/S_p$	spherical $r_s = 3/S_p$
0	54,700*	—	—
0.1	10,250	$2.16 \cdot 10^{-4}$	—
1	7,280	$2.74 \cdot 10^{-4}$	—
3	6,670	$3.00 \cdot 10^{-4}$	—
10	6,430	$3.11 \cdot 10^{-4}$	$4.68 \cdot 10^{-4}$
30	7,020	$2.85 \cdot 10^{-4}$	$4.28 \cdot 10^{-4}$
100	11,830	—	$2.54 \cdot 10^{-4}$

*Calculated from specific surface of powder as given by its Fisher Subsieve Sizer value and from the porosity of the unsintered compacts (density 5.75 g/cc).

The measured pore parameters (Table 1) agree fairly well with these equivalent radii. Qualitatively, the description of the behavior of the pore phase during densification resulting from this study is similar to that given by Rhines, Birchenall, and Hughes [15] for copper specimens. A more detailed analysis could be carried out if it were possible to convert the intercept data to length and diameter values. This presupposes knowledge of the statistical pore shape in space. At present the only way to obtain such information is the study of a sufficient number of pores through a series of consecutive sections, which would have exceeded the scope of this investigation by far.

Acknowledgments

It is a pleasure to acknowledge the skillful and accurate work of Miss C. Müller-Snyders who performed all of the metallographic work and most of the other measurements reported here. The air permeability measurements and the construction of the apparatus for them, were carried out by civil engineer G. Lindelöf. The author is indebted to the sixteen companies which sponsored the Jernkontoret Laboratory for Powder Metallurgy during the period of this research in particular to Höganäs – Billesholms AB and to Dr. G. Bockstiegel of that company, who provided the pressed tensile bars. Financial support was also given to the project by the Swedish Technical Research Council (Statens Tekniska Forskningsråd).

References

1.   A. L. Pranatis, L. S. Castleman, and L. Seigel, Rep. SEP-250, Sylvania Research Lab. Bayside, L. I. (1958).
2.   H. Fischmeister and R. Zahn, Abh. Deut. Akad. Wiss., Kl. Math., Physik Tech., (1962), p. 93.
3.   R. L. Coble, J. Appl. Phys., 32:787, 793 (1961).
4.   MPA Standard 2-48 (Metal Powder Industries Fed., New York 17).
5.   ASTM Standard E 8 - 61 T, Fig. 19a.
6.   N. F. Astbury et al., Trans. Brit. Ceram. Soc. 60:658 (1961).
7.   H. Fischmeister, 1st Intern. Congress for Stereology, Vienna 1963. Proceedings, paper 27 (Congressprint, Vienna VI).
8,   H. Fischmeister, Chapt. XIII, "Quantitative Metallography," Ed. F. N. Rhines and R. De Hoff, McGraw-Hill (in press).

9.   F. H. Schückher, Chapt. VIII, "Quantitative Metallography,"
      Ed. F. N. Rhines and R. De Hoff, McGraw-Hill (in press).
10.   M. Hullert, Acta Met. 13 1965 p. 227.
11.   G. W. Lord and T. F. Willis, ASTM Bull. 177:56–61 (1951).
12.   J. W. Cahn and R. L. Fullman, Trans. AIME 206:610 (1956).
13.   E. Exner, Dissertation at Mont. Hochschule Leoben and
      Jernkontorets Lab. f. Pulvermetallurgi, Stockholm, Stock-
      holm 1964.
14.   C. S. Smith and L. Guttman, Trans. AIME 197:81 (1953).
15.   F. N. Rhines, C. E. Birchenall, and L. A. Hughes, Trans.
      AIME 188:378 (1950).

Effect of Powder Particle Size
on the Grain Size
of the Sintered Material

H.H. Hausner

Consulting Engineer
New York, New York

and

R. King
Sylvania Corning Nuclear Corp.
Bayside, N. Y.

Introduction

For many years powder metallurgy has been known as a
method for fabrication of sintered materials with special physical
properties not obtainable by any other more conventional metallurgical method. Reference is made, for example, to materials with
controlled porosity, combinations of metals which do not form
alloys, and sintered products with special electrical properties.
Little attention, however, has been given heretofore to the grain
structure of sintered materials, and hardly anything more is known
with respect to the structure than the general fact that the grain
size in sintered metals is usually smaller than that of the corresponding cast material heat-treated in a similar way.

There are many factors which determine the grain size in a sintered metal, such as sintering temperature and time, atmosphere, compacting pressure, shape, size and surface condition of powder particles, and others. This study is concerned exclusively with the effect of particle size of electrolytic iron powder on the grain size of sintered iron made therefrom.

It has been shown [1] that the well-known equations for grain growth in solid metals are not correct for the grain growth in metal powder compacts during sintering. In solid metals isothermal grain growth occurs, and the average grain diameter D is given by

$$D = Kt^n \tag{1}$$

where K and n = constants at constant temperature and t = time of heating. K can be defined by the usual Arrhenius relation

$$K = Ae^{-H/RT} \tag{2}$$

where A = a constant, H = activation energy, R = gas constant, and T = absolute temperature.

For solid metals the continuous grain growth from a grain size D_1 to D_2 occurring by heating from temperature T_1 to T_2 during the time t_1 to t_2 can be expressed by

$$\ln \frac{t_1}{t_2} = \frac{H}{R} \left(\frac{1}{T_1} - \frac{1}{T_2} \right) \tag{3}$$

The average rate of linear grain growth dD/dt is a function of the average grain diameter D and can be expressed according to P. A. Beck [2]:

$$\frac{dD}{dt} = \frac{K}{D^m} \tag{4}$$

According to D. Harker and E. R. Parker [3], grain growth should continue as long as the grain face is curved. Zener [4], however, has stated that inclusions will inhibit the grain-boundary migration and growth will cease when the inhibiting force is equal to the growth force due to the boundary curvature. Assuming that the growth force is inversely proportional to the radius of curvature in the grain boundary, Zener developed the following equation:

$$\frac{R}{r} = \frac{1}{f} \tag{5}$$

where R = average radius of curvature of the grain boundary, r =

average radius of spherical inclusions, and f = volume fraction of inclusions in specimen.

Equations (1) to (4) cannot be applied without modifications of grain growth during sintering. The reason for this lies in the porosity of the green metal powder compacts and in the fact that the pores or voids hinder the grain-boundary movement in a manner similar to that of impurities in metals made by more conventional methods.

It has been suggested [1] that for grain growth during sintering, equation (1) should take care of the pores in the specimen, and should be modified to the form

$$D = Kt^n - f(P\ r) \qquad\qquad (6)$$

where P = total pore volume and r = pore size.

Equation (6) permits the prediction that compacts with lower green density, and therefore larger total pore volume, will result in a sintered material of a smaller grain size than compacts with greater green density and smaller pore volume, and that the pore size also affects the grain size in a rather nebulous way. Recent observations, however, have shown that equation (6), even in its very indefinite form, is not generally applicable, and that conditions may exist where compacts with low green and sintered densities, and therefore a larger pore volume, show considerably larger grains after sintering than compacts of higher density. This increased rate of grain growth seems to be closely connected with the particle sizes of the powder from which the compacts are prepared.

Experimental Data

In order to study the effect of powder particle size on the grain size of sintered products, tests were made with electrolytic iron powder of -40 mesh size. The powder was screened into 7 mesh size fractions, so that the particles in each fraction did not vary more in diameter than 1 to 2. Microscopic inspection indicated that the dimensions of the crystal grains within each powder particle varied strongly and that some grains were approximately 10 times as large as the smallest ones. Another observation, however, was of greater significance: the finer powder particles in

TABLE 1. Minimum and Maximum Diameter Grains within Annealed Electrolytic Iron Powder Particles

Powder particle size in U.S. sieve mesh	in microns	Grain size in microns Min.	Max.
— 40 + 60	250 — 420	7	70
— 60 + 100	149 — 250	5	60
—100 + 140	105 — 149	5	50
—140 + 200	74 — 105	5	45
—200 + 230	62 — 74	5	32
—230 + 325	44 — 62	5	28
—325	—44	2	23

general contained smaller grains than the larger ones. The respective data on minimum and maximum grain size in the powder particles of various size are shown in Table 1.

In order to treat the iron powder particles of all fractions as uniformly as possible during the compacting and sintering procedure, a compacting pressure of 40 tsi was uniformly applied to all specimens of the first test series, and the green compacts were sintered in purified hydrogen atmosphere for 4 h at 900° and 1100°C respectively. Densities were measured before and after sintering and minimum and maximum grain size determined in each sintered specimen. The results are shown in Table 2, and the data listed

TABLE 2. Effect of Sintering Temperature on the Sintered Density and Grain Size of Electrolytic Iron Powder Compacts Compacted at 40 tsi and Sintered in Purified Hydrogen for 4 h at 900° and 1100°C, Respectively

Particle sieve size	Pressed density g/cc.	Sintering temp., ° C.	Sintered density g/cc.	Grain size microns
— 60 + 100	7.297	900	7.331	10 — 125
—100 + 140	7.228		7.278	10 — 100
—140 + 200	7.176		7.252	10 — 85
—200 + 230	7.119		7.223	10 — 75
—230 + 325	7.123		7.203	7 — 60
—325	6.961		7.089	7 — 45
— 60 + 100	7.296	1100	7.299	25 — 220
—100 + 140	7.253		7.272	20 — 160
—140 + 200	7.190		7.214	15 — 140
—200 + 230	7.160		7.197	15 — 120
—230 + 325	7.109		7.159	10 — 100
—325	6.967		7.012	10 — 85

therein indicate the following:

a) The green densities decrease with decreasing powder particle size on account of the greater friction caused by the smaller particles.

b) The sintered densities also decrease with decreasing particle size.

c) The densities obtained by sintering at 900° are considerably higher than those of the 1100°C sintered specimens, due to the variation in crystal structure below and above the phase transformation temperature.

d) The grain size of sintered material is slightly greater than that of the powder particles from which the compacts were made, but follows the same trend as shown in Table 1 for the grain size within the particle: sintered compacts pressed from large particles show larger grains than those compacted from smaller particles.

e) The grain size of the 1100° sintered compacts is greater than that of specimens sintered at 900°C, although the densities of these compacts are lower.

However, the results of this test series with respect to the grain size are inconslusive on account of the fact that the compacts in the "green" as well as in the sintered state differ in density according to the pressure losses which are different for each particle size fraction. In order to compensate to a certain extent for the variation in pressure losses, it was decided to compact the fine particles at higher pressures and decrease the compacting pressure with increasing particle size.

For the specimens of the two following test series, the compacting pressure was, therefore, varied from 42.4 tsi for -60 +100 mesh size particles to approximately 53.6 tsi for the -325 mesh size fraction, so that the sintered densities of the compacts were fairly uniform.

For the first series of specimens, the sintering temperature was 900°C and the sintering times were 4, 8, 16, and 32 h. It took one hour to reach the sintering temperature. Densities were measured before and after sintering, and hardness and minimum and maximum grain size of the sintered specimens were determined. The test results are shown in Table 3, and the data indi-

TABLE 3. Effect of Sintering Time on The Density, Hardness, and Grain Size of Compacted Electrolytic Iron Powder Compacts Sintered at 900°C in Purified Hydrogen

Particle sieve size	Compacting pressure, tsi	Pressed density, g/cc.	Sintering time, hrs.	Sintered density, g/cc.	Rockwell H hardness, Top	Bottom	Grain size, micron
− 60 + 100	42.4	7.383	4	7.426	84	88	10 − 85
− 100 + 140	44.7	7.366		7.410	88	89	7 − 60
− 140 + 200	46.9	7.382		7.422	89	88	8 − 50
− 200 + 230	49.1	7.392		7.449	90	90	7 − 40
− 230 + 325	51.3	7.379		7.449	92	91	7 − 35
− 325	53.6	7.297		7.446	95	95	5 − 25
− 60 + 100	42.4	7.403	8	7.439	85	84	12 − 100
− 100 + 140	44.7	7.356		7.396	88	89	10 − 70
− 140 + 200	46.9	7.358		7.436	85	87	8 − 60
− 200 + 230	49.1	7.387		7.470	89	89	8 − 45
− 230 + 325	51.3	7.383		7.459	91	93	7 − 40
− 325	53.6	7.288		7.474	94	95	6 − 25
− 60 + 100	42.4	7.356	16	7.413	79	81	15 − 130
− 100 + 140	44.7	7.379		7.429	81	81	10 − 80
− 140 + 200	46.9	7.358		7.454	82	82	10 − 70
− 200 + 230	49.1	7.386		7.474	84	84	9 − 55
− 230 + 325	51.3	7.352		7.480	90	90	8 − 50
− 325	53.6	7.295		7.520	95	94	7 − 30
− 60 + 100	42.4	7.358	32	7.409	80	80	15 − 150
− 100 + 140	44.7	7.358		7.432	81	82	10 − 100
− 140 + 200	46.9	7.356		7.463	82	82	10 − 90
− 200 + 230	49.1	7.377		7.503	83	83	9 − 70
− 230 + 325	51.3	7.381		7.529	84	84	10 − 60
− 325	53.6	7.281		7.504	89	89	7 − 35

cate the following:

a) The green densities after compacting at 42.4 to 53.6 tsi are fairly uniform with the exception of the −325 mesh size powder compacts, which showed slightly lesser density than the compacts made from coarser powder particles.

b) The densities of the specimens sintered for 4 h were fairly uniform for all compacts.

c) After sintering for 8 h the compacts pressed from the smallest (−325) particle-size fraction showed a slightly higher density than the coarser particle-size compacts.

d) Densification of the fine particle compacts increased considerably after sintering for 16 h.

e) The grain size of all specimens increased slightly with sintering time, increasing from 4 to 32 h.

f) The grain size was smallest for compacts made from finest powders and increased with the particle size of the powders from which the specimens were compacted.

It seems that continuous grain growth takes place at a more delayed rate in the fine powder particle compacts (with a greater number of pores) than in the compacts made from coarse powders.

The whole series was repeated under similar conditions, with the exception of a change in the sintering temperature, which was increased to 1100°C ($1\frac{1}{4}$ h to reach this temperature). The test data for the second series are shown in Table 4 and indicate the following results:

a) The green densities were fairly uniform with the exception of the compacts pressed from the finest powder-size fraction, whose density was the lowest.

b) The sintered density of the finest powder size compacts in general was slightly lower than that of the compacts made from larger particles.

TABLE 4. Effect of Sintering Time on the Density, Hardness, and Grain Size of Compacted Electrolytic Iron Powder Compacts Sintered at 1100°C in Purified Hydrogen

Particle sieve size	Compacting pressure, tsi	Pressed density, g/cc.	Sintering time, hrs.	Sintered density g/cc.	Rockwell H hardness,		Grain size, micron
					Top	Bottom	
− 60 + 100	42.4	7.388	4	7.404	79	81	15 — 300
− 100 + 140	44.7	7.346		7.343	85	84	15 — 250
− 140 + 200	46.9	7.365		7.388	80	82	15 — 200
− 200 + 230	49.1	7.345		7.388	83	87	10 — 180
− 230 + 325	51.3	7.375		7.401	83	84	10 — 150
− 325	53.6	7.304		7.358	89	91	7 — 150
− 60 + 100	42.4	7.348	8	7.367	80	82	20 — 400
− 100 + 140	44.7	7.360		7.387	79	80	20 — 400
− 140 + 200	46.9	7.276		7.311	83	82	15 — 300
− 200 + 230	49.1	7.357		7.379	85	82	12 — 250
− 230 + 325	51.3	7.390		7.426	85	84	10 — 200
− 325	53.6	7.302		7.384	89	88	7 — 200
− 60 + 100	42.4	7.357	16	7.331	76	74	20 — 500
− 100 + 140	44.7	7.351		7.344	75	75	20 — 500
− 140 + 200	46.9	7.302		7.290	82	80	15 — 450
− 200 + 230	49.1	7.349		7.333	80	80	15 — 400
− 230 + 325	51.3	7.353		7.358	80	82	12 — 350
− 325	53.6	7.276		7.321	85	89	10 — 600
− 60 + 100	42.4	7.353	32	7.372	74	79	20 — 3000
− 100 + 140	44.7	7.355		7.373	79	78	20 — 2000
− 140 + 200	46.9	7.360		7.377	80	78	20 — 900
− 200 + 230	49.1	7.377		7.438	82	81	15 — 200
− 230 + 325	51.3	7.359		7.388	79	79	15 — 2700
− 325	53.6	7.278		7.345	84	83	10 — 4000

c) Increase in sintering time from 4 to 32 h hardly changed the densities.

d) The grain size of the 4- and 8-h sintered specimens followed the same pattern as in all the above-described series: fine powder compacts showed finer grains than compacts made from coarser powders.

e) Sintering for 16 h resulted in rapid growth in the compacts pressed from finest powders, so that the grain size of these specimens was greater than that of any specimens made from coarser powders, although the density of these specimens was on the low side.

f) Sintering for 32 h showed the phenomenon that very rapid grain growth took place in the compacts pressed from the -325 mesh and -230 +325 mesh size powders, and also, but to a lesser degree, in the -60 +100 mesh size powder compacts. Minimum grain size is obtained in compacts made from medium size powders.

g) The hardness follows neither density nor grain size of the sintered compact, but the original particle size, from which the specimens were compacted.

The interesting and new item in this test series is the rapid grain growth of the specimens made from fine powders at prolonged sintering time and the complete independence of grain size from the density or total porosity of the specimens.

Microscopic examination of the specimens indicated that a larger number of the voids in the 16- and 32-h sintered specimens which were compacted from the finer particles were considerably more round and spheroidized than the voids of the 4- and 8-h sintered specimens.

Discussion of Results

From the results of the test series listed in Table 4, one does not find any correlation between percent density and grain size of the sintered compact, such as had been observed previously for other metals [5]. However, one has to consider sintered density in terms of the porosity of the compact in order to understand the observed phenomena. The total porosity of the compact depends on the number and average size of the pores:

$$P = \frac{W}{d_{th} - d_s} = nv \tag{7}$$

where P = total pore volume in 1 cc, W = weight of specimen in g,

d_{th} = theoretical density of the material, g/cc, d_S = sintered density of the compact, g/cc, n = total number of pores in 1 cc, and v = average volume of one pore in 1 cc.

The test results have indicated that, up to a certain sintering temperature and time, compacts with a larger number n_1, of fine pores (v_1) result in a fine grain structure, whereas compacts with a smaller number, n_2, of coarser pores (v_2) show coarser grains, even in the case when

$$v_1 \, n_1 = v_2 \, n_2$$

and the total porosity of both specimens is equal.

Equation (6) is correct only for compacts with irregular or cusp-shaped voids, but does not seem to be applicable for spheroidized voids.

Compacts made from small powder particles contain a large number of small pores which are characterized by a considerably larger surface energy than the larger pores of the coarse particle

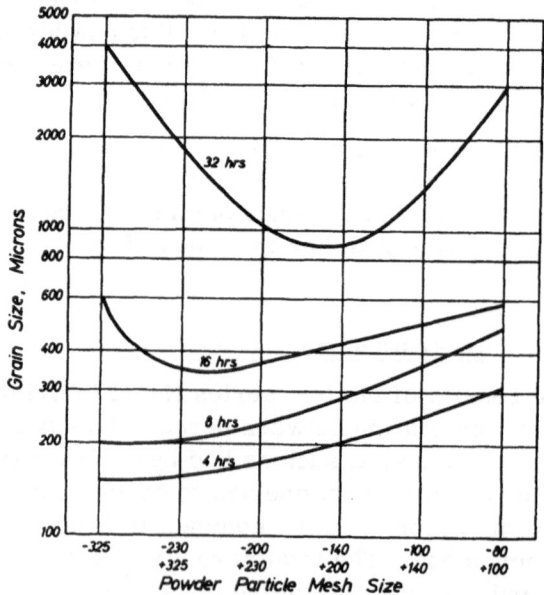

Fig. 1. Effect of iron powder particle size on the grain size of sintered iron. (Sintering temp. 1100°C; Sinterting time, 4, 8, 16, 32 h).

TABLE 5. Effect of Sintering Time on the Grain
Size of Sintered Silver (in microns)

Time of sintering at 900° C., hrs.	Powder particle size in microns				
	1.5	5	9	38	125
2	17	14	15	34	65
5	60	30	32	34	65
10	65	30	32	34	65
20	65	30	32	34	65
50	65	30	32	34	65

compacts. On account of this larger surface energy, the fine pores
reach the spherical shape earlier than the larger pores in the
coarse particle compacts. When the spherical pore shape is
reached, porosity no longer seems the grain-growth determining
factor, and a considerably more rapid grain growth takes place.
It, therefore, can be assumed that, in fine powder compacts with
faster spheroidization of the fine voids, the rapidly increasing grain
growth will start at lower sintering temperature and require less
sintering time than in coarse particle compacts with larger voids
which spheroidize only at higher temperature or prolonged sintering time.

The results of the above-described tests have shown that
this assumption is correct: rapid grain growth in sintered powder
compacts starts earlier when fine powder particles are used than
when compacting coarse particles. This is shown in a schematic
way in the diagram, Fig. 1.

The results of our experiments are in accord also with tests
described by K. Ogawa, G. Matsumura, and D. Okubo [6] who studied
the effect of silver powder particles of various sizes on the grain
size of sintered silver compacts. Their results, shown in Table 5,
indicate also the more rapid grain growth of fine silver powder
compacts than of coarser ones during sintering at 900°C for 2 to
50 h. The results, however, are inconclusive, inasmuch as the
authors used various methods for preparing their silver powder of
various sizes, and did not indicate the densities of the respective
green and sintered compacts.

Conclusions

1. The density of the sintered compact is not necessarily the
criterion for grain size of the compact.

2. During the earlier stage of sintering, the grain size in a sintered material depends on the original grains within the powder particle (continuous grain growth) and on the number and size of voids between the particles.

3. During a later stage of sintering, with progressive spheroidization of the pores, a more rapid grain growth occurs.

4. Compacts made from fine powders contain smaller pores which spheroidize faster than the coarse pores between coarse particles; the rapid grain growth in the fine powder compacts, therefore, starts earlier than in the coarse powder compacts.

References

1.   H. H. Hausner, Symposium on Powder Metallurgy, Spec. Rep. No. 58, Iron and Steel Inst., London 1954.
2.   P. A. Beck et al., Metals Technol., Vol. 14 AIMME Techn. Publ. No. 2280 (1947).
3.   D. Harker, and E. R. Parker, Trans. Am. Soc. Met., 34:156 (1945).
4.   J. E. Burke, Metals Technol., Vol. 15 AIMME Techn. Publ. No. 2472 (1948).
5.   H. H. Hausner and H. S. Kalish, J. Metals 3:625 (1951).
6.   K. Ogawa, G. Matsumura, and D. Okubo, Mh. Chem., 85:1281 (1954).

Structure Formation in
Iron—Graphite Compositions

V.A. Dymchenko and Yu. F. Morozov
NIIPTMASH Institute
Kramatorsk, USSR

Depending on the intended application of parts made from iron—graphite compositions, their structure must meet different requirements. Thus, for instance, antifriction materials must contain graphite but no structurally-free cementite; in constructional materials, the presence of graphite is highly undesirable, while cementite is permissible or, sometimes, even indispensable.

It has been established that the pearlitic-graphitic structure is the most desirable in antifriction materials, but, as shown by experience, such a structure is not very easy to secure. The difficulty is due to the fact that, at the sintering temperature of components (1050-1100°C), austenite is capable of dissolving much more carbon than is required for the formation of the pearlitic structure. At the same time, the high rates of cooling in sintering furnaces are responsible for the metastable decomposition of austenite and for the inevitable appearance of secondary cementite.

The available literature information on the control of structure-formation processes is restricted to recommendations of slow cooling [1] and holding in the eutectoid temperature zone [2]. The authors are of the opinion that, because of the presence of pores,

which eliminate the limiting influence of the self-diffusion of iron,
and of remnants of undissolved graphite, which act as ready crys-
tallization centers, the graphitization process in iron—graphite
alloys takes place at a very high rate.

These factors undoubtedly influence the kinetics of graphitization, but nevertheless, possibly because of the low silicon content, this process requires a very long time. In our experiments,
changing the rate of cooling from 15 to 1°C/min produced no
change of phase composition. Methods based on the dissolution of
the cementite which has already formed or on the decomposition of
austenite in a stable process involve complex sintering procedures
and can hardly be regarded as acceptable in powder metallurgical
practice.

According to the Fe—C phase diagram, the limiting solubility
of carbon in austenite is 2%. At a temperature of 1050°C, which is
usual for the sintering of iron—graphite compositions, the amount
of carbon which can dissolve is 1.7%. When an austenite of this
composition decomposes during cooling by a metastable process,
the resultant structure consists of pearlite and about 15% of cementite. Consequently, in order to secure a pearlitic-graphitic
structure, it is necessary, irrespective of the sintering temperature and the amount of graphite in the charge, to halt the dissolution of carbon at a certain level, in this case 0.8%.

A diametrically opposite problem arises in the production
of parts from constructional materials, in which graphite inclusions weaken the cross section and adversely affect strength. The
production technology of constructional parts must ensure the full-
est possible dissolution of the graphite introduced into the charge.

It has thus been shown that the final structure of iron—graphite alloys is predetermined by the amount of carbon dissolved in
the austenite. However, in spite of the great importance of the
kinetics of graphite dissolution, this problem has received practically no attention in the literature. In the present investigation,
an attempt was made to study and describe in a general outline
some regularities of the graphite dissolution processes in sintered
iron—graphite compositions.

The graphite dissolution mechanism in cast metal may be
schematically described in the following manner. The structural
changes taking place in the iron matrix during the heating of an

alloy to temperatures slightly exceeding the eutectoid one disturb
the equilibrium of the iron–graphite system and bring about the
dissolution of carbon. The appearance of carbon on the austenite
interface releases the diffusional transport of carbon atoms into
the iron matrix, as a result of which the austenite composition becomes equalized, and further dissolution of graphite is ensured.

Together with the transfer of carbon atoms from a graphite
inclusion toward austenite and the diffusion of carbon in austenite,
iron atoms undergo self-diffusion toward this front. The space
evacuated by the dissolution of graphite is filled by iron atoms,
and iron–graphite contact is thereby secured. The dissolution of
graphite continues until either the latter is completely exhausted or
the system reaches a state of equilibrium. This process is relatively rapid [3], and, at 1000-1050°C, saturation of austenite to the
equilibrium concentration is reached in a few minutes. The rate
of dissolution of graphite is determined by the rate of the two
slowest processes, namely, the migration of carbon atoms though
the austenite–graphite interface, which is associated with the
severance of strong covalent bonds, and the self-diffusion of iron.

In contrast to cast metals, in which graphite inclusions are
packed in a dense matrix, powder compacts contain rough, irregularly shaped particles stacked in an unordered fashion, and are,
therefore, incapable of securing full contact between graphite and
iron. Because of its irregular shape, the void containing a graphite
inclusion has a much greater volume and area than the inclusion
(Fig. 1).

The kinetics of graphite dissolution in the porous iron matrix
are decisively affected by the circumstance that the diffusion processes taking place during sintering cannot fully restore and main-

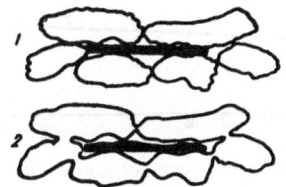

Fig. 1. Diagrammatic arrangement
of graphite inclusions among iron
powder particles: 1) before sintering;
2) during sintering.

tain the contact between iron and graphite, which is being disturbed by the decreasing volume of the graphite particles undergoing dissolution. The iron atoms supplied through self-diffusion become distributed over the whole void surface and tend to occupy positions with the least energy content, by filling micro- and macropores and cracks, rather than projecting areas which could provide contacts.

The dissolution of projections and smoothing out of relief take place also as a result of the surface diffusion of iron atoms, the propagation of which, particularly during the initial period of sintering, is activated by the reduction of the oxide film on the powder. The development of these processes leads to a reduction of the contract surface area in iron−graphite material, which in turn reduces the dissolution of graphite, often only to a very slight extent. Thus, the graphite dissolution process in iron−graphite compositions may be arbitrarily divided into two periods.

1) Initial period. Carbon atom transfer is carried out through contact zones established as a result of the iron particles and graphite having been brought closer together during compacting. The rate of dissolution is high.

2) Contact zones are established through the self-diffusion of iron. The rate of carbon dissolution is lower than in the first period. Transport of carbon through the gaseous phase is possible.

Figure 2 shows curves of the saturation of iron by carbons as a function of isothermal holding time and of the graphite content of the charge. The investigation was made on specimens of 20% porosity, $10 \times 10 \times 55$ mm in size, made from as-supplied powder and sintered in a cracked ammonia atmosphere. The graphite was

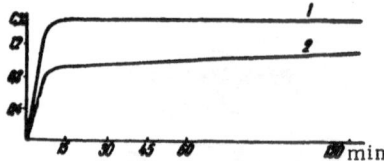

Fig. 2. Change in amount of dissolved carbon as function of duration of isothermal holding during sintering and graphite content of charge: 1) 3%; 2) 1.5%. Sintering temperature 1050°C. Graphite of 0.063-mm fraction.

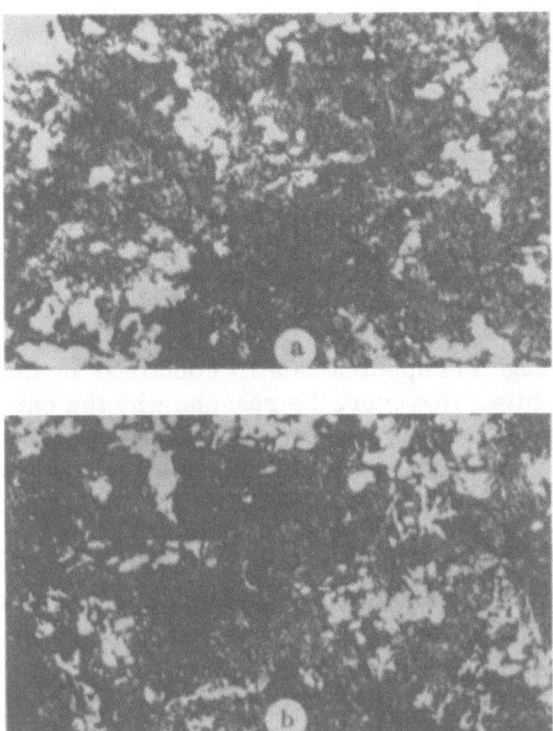

Fig. 3. Structure of specimens sintered for 15 (a) and
120 (b) min. Graphite content of charge 3%, fraction
smaller than 0.063 mm. Etched. Pearlite + cementite.
× 300.

of the KLZ grade and was passed through Nos. 04, 0315, 025, 01,
and 0063 sieves. In order to minimize the effect of heating time,
the specimens were placed in a furnace preheated to the isothermal
holding temperature (1050°C).

A study of the microstructure of specimens demonstrated
that the dissolution of carbon during the initial period of sintering
proceeds very rapidly and, in the case of finely-ground graphite
(less than 0.063 mm), is practically completed within a few minutes.
This period is also sufficient for the completion of diffusion processes, which tend to equalize the carbon content throughout the
whole specimen volume (Fig. 3).

It follows from Fig. 2 that carbon dissolution in our experiments was completed or strongly inhibited in the tenth minute of

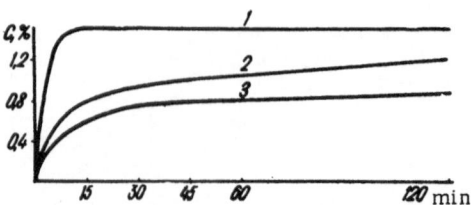

Fig. 4. Change in amount of dissolved carbon depending on sintering time and particle size of graphite: 1) less than 0.63 mm; 2) 0.1-0.25 mm; 3) 0.315-0.4 mm.

sintering, although the specimens still contained an adequate amount of graphite. However, the reasons why the carbon dissolution processes became arrested in the compositions with 3 and 1.5% graphite were different. In the former case, saturation of austenite with carbon proceeded particularly vigorously. The composition had much more graphite than it could dissolve, the existing iron–graphite contact zones were able to convey into austenite the required amount of carbon, and the system soon attained a state of equilibrium.

In the latter case (1.5% C), the rate of graphite dissolution was also very substantial during the initial period of sintering but sharply dropped when the iron–graphite contacts became disturbed. And, although carbon dissolution continued during the whole period of isothermal holding, only the first 5-10 min determined the structure. During that time 85% of carbon (out of the total amount assimilated) dissolved, and during the remaining 110 min only 15%.

If this outline of the graphite dissolution mechanism is correct, control of the structure formation process must consist in

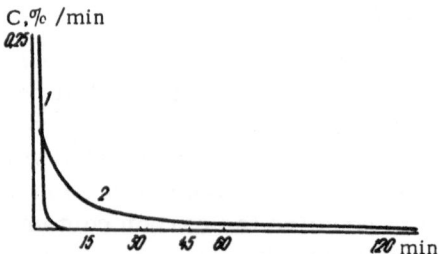

Fig. 5. Dissolution rate of graphite of different particle size: 1) less than 0.063 mm; 2) 0.1-0.25 mm.

altering, depending on the required degree of graphite dissolution, the iron−graphite surface in the compact. The easiest way of achieving this is by choosing a graphite of an appropriate particle-size distribution. The finer the graphite powder, the greater are its specific and, consequently, contact surfaces.

Our experiments confirmed this hypothesis. Figures 4 and 5 show the results of a structural analysis of specimens containing 3% graphite of different particle size, sintered for different isothermal holding periods. With decreasing graphite particle size, the rate and degree of graphite dissolution increase within very wide limits. A study of these specimens revealed that the use of coarser graphite does not slow down the equalization of austenite composition. Irrespective of the particle size of graphite, the structure of the composition remains the same (Figs. 3 and 6) throughout the whole period of isothermal holding (after the first 5-10 min).

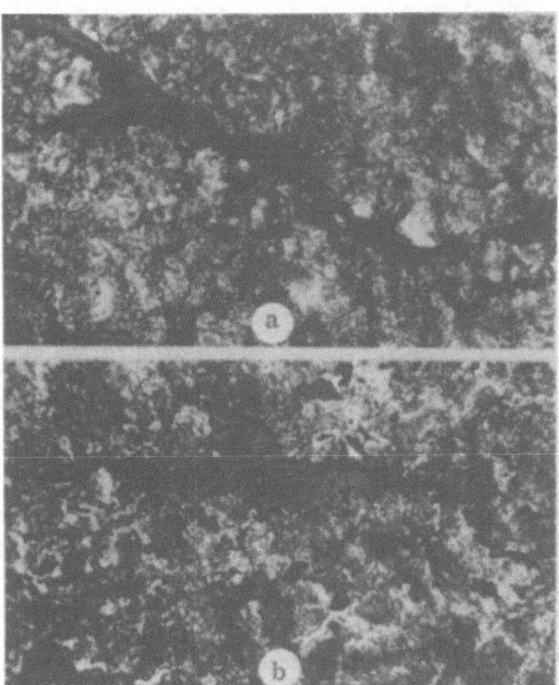

Fig. 6. Structure of specimens sintered for 15 min, pearlite + ferrite (a) and 120 min, pearlite + cemenite (b). Graphite content of charge 3%, fraction 0.25-0.315 mm. Etched. × 300.

The carbon content of austenite will also depend on the composition of the furnace atmosphere. Since the latter is only rarely strictly balanced, the composition undergoing sintering may become either carburized or decarburized as a result of reaction with the atmosphere. The final austenite composition will be determined by the summation of two component processes: graphite dissolution and reaction with the atmosphere. The kinetics of these two processes will be influenced to a greater or smaller extent by a number of factors, such as the degree of oxidation of the powder, shape and size of the iron powder particles, and porosity of the part. It is evident that quantitative evaluation of the influence exerted by graphite particle size on the extent of graphite dissolution and, especially, on the carbon content of austenite can only be carried out individually for each specific case.

The proposed scheme of the mechanism by which graphite particle size influences the extent of graphite dissolution has undoubtedly been simplified. A change of the specific surface of a graphite powder affects not only the graphite-iron contact surface, but also the graphite-atmosphere contact surface, powder activity, and the uniformity of graphite distribution in the charge. In subsequent investigations, an attempt will be made to discover all the processes responsible for the transport of carbon from graphite to iron and to establish their correlation with different technological factors.

Summary

1. The rate of graphite dissolution in iron−graphite compositions is determined by the size of the iron−graphite contact surface, which in turn depends on the particle size of the graphite.

2. The structure of iron−graphite compositions may be changed by choosing graphite powders of different particle size. The use of coarser graphite fractions enables structurally-free cementite to be completely eliminated or its amount to be substantially reduced.

3. The time of isothermal holding required for the structure formation processes (the actual sintering process is not considered here) decreases with increasing fineness of the graphite fraction employed, and may be limited to 10-15 min.

4. At a 3% graphite content of the charge, it is possible, by suitably changing the sintering time and the particle size of the graphite, to obtain materials with any required structure, ranging between pearlitic-ferritic and pearlitic-cementite ones.

5. For the preparation of constructional materials, graphite of the highest possible degree of dispersion should be used.

References

1. P. I. Bebnev, Collection: Powder Metallurgy, p. 90. Yaro-slavl', NTO Mashprom (1965).
2. V. I. Likhtman and A. N. Smirnov, Collection: Powder Met-allurgy, Moscow, Metallurgizdat (1954).
3. M. N. Kunyavskii, Collection: Metallography and Present Day Methods of Heat Treatment of Cast Iron, p. 58, Moscow, Mashgiz (1955).

The System Iron—Carbon
in Powder Metallurgy*

P. Ulf Gummeson

Höganäs Corporation

Riverton, New Jersey

The iron–carbon system is not generally utilized to the extent it ought to be in the powder metallurgy industry. It is true that there are simpler combinations of materials that will give as good or perhaps better physical properties and dimensional control as iron–carbon, but in almost every case the raw materials for other combinations are more expensive. It is not the intention of this paper to try to pinpoint the potential savings, if any, in using iron–carbon, but to point out some of the advantages from a technical standpoint and at the same time to mention some difficulties and how they can be avoided or overcome.

Figure 1 shows the tensile strength of some common powder metallurgy mixes. At the bottom of the graph is shown the composition. For each composition three densities are reported. The iron is a sponge iron, the copper is of the reduced type, and the carbon is Bavarian 20-μ graphite. Sintering has been done at 2070°F (1132°C) for 40 min in a roller hearth furnace in purified exogas with roughly 70% nitrogen, 21% hydrogen, and 9% carbon monoxide.

*Previously published by Höganäs Corporation, Riverton, N. J.

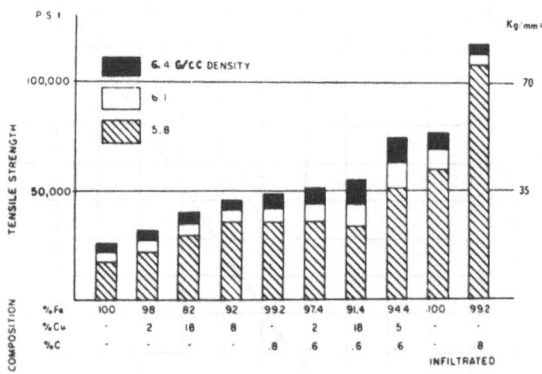

Fig. 1. Tensile strength of iron powder parts.

May we point out, as shall be proved later, that we consider this atmosphere rather on the lean side for sintering iron—carbon. Nevertheless we find that iron—carbon in this case is stronger than iron and copper at practically any copper content and one will have to consider 5% copper and carbon or infiltrated parts to find a combination that is considerably stronger than iron—carbon.

It should be mentioned also that with the aid of a very rich sintering atmosphere, the strength of iron—carbon under the time—temperature conditions pictured in the diagram can be made to surpass anything but the infiltrated parts.

The following are factors which should be kept in mind and controlled thoroughly in order to accomplish the best results when using iron—carbon.

1. Quality of the iron powder (reactivity, hydrogen loss).

2. Quality of the graphite.

3. The atmosphere composition.

Quality of the Iron Powder

The nature of the property "reactivity" is somewhat obscure and it does not seem to be possible to define reactivity in terms of chemistry or physical properties of the elements. For example, it is easily proved that the cleanliness of the iron particle surface is of importance. An iron powder can be rendered nonreactive to

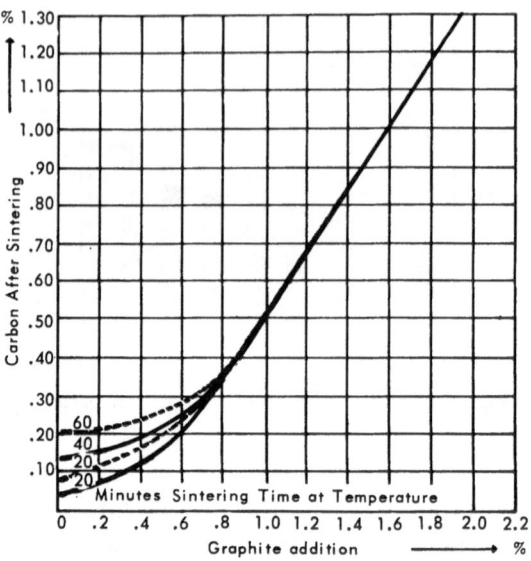

Fig. 2. Carbon content after sintering versus graphite addition. Iron: Ancor 80 (0.86% H_2 loss); graphite: Ceylon -325 mesh; sintering: 2050°F (1125°C); atmosphere: 28% CO +72% H_2; test bars: 1.250 × 0.500 × 0.250 in.; density: 5.5 g/cc (dashed curve), 6.2 g/cc (solid curve).

carbon pickup by merely mixing in very minute amounts of mineral dusts, amounts that would not be considered excessive as impurities.

It is also an established fact that small amounts of alloying elements like manganese have a beneficial effect on the reactivity with carbon. This is rather easy to accept since we know that manganese has a high affinity for carbon. However, it is by no means necessary or always desirable to have any alloying elements since these will often have detrimental influences in other respects. One property, however, is of considerable importance; i.e., oxide content. The major portion of the oxides in an iron particle will have to be reduced by the graphite and the sintering atmospheres before the combination of iron and graphite takes place at any considerable rate. It is not necessary that the powder has a low hydrogen loss value although it is helpful. It is merely important to know how much, since the proper amount of graphite has to vary accordingly. It is usually desirable that the sintered part have a combined carbon content corresponding to eutectoid composition of around 0.8-0.9% and it is, therefore, also obvious that the graphite addition will have to be higher than this figure.

Figure 2 shows in simple form percent combined carbon after sintering versus percent graphite addition for an iron powder with 0.86% hydrogen loss. In this particular case we find that to obtain 0.99% combined carbon, an addition of about 1.5% graphite should be made. Normally 1.25% is enough.

Figure 3 shows strength versus percent combined carbon. Maximum strength is obtained at eutectoid composition.

Quality of the Graphite

Althouth this is one of the more confusing aspects of the system iron—carbon, we can establish again one property, i.e., reactivity, being of importance. Of course, it can generally be said that the finer the particle size of any material, the higher is its reactivity, but not even this seems to hold true, except for one and the same graphite. Some of the types combine easily and freely under rather adverse conditions with iron powders. Others refuse to do so except to a very limited degree. Unfortunately we cannot state today exactly what properties in the graphite make it suitable and what makes it unsuitable. However, a few points have been cleared up.

In Table 1 are listed a number of different grades of graphites and one iron sintered simultaneously at two different

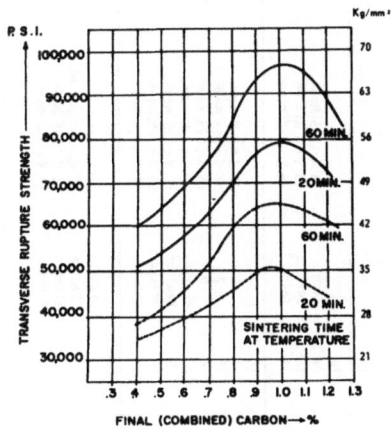

Fig. 3. Transverse rupture strength versus final carbon content (combined). Iron: Ancor 80 (0.86% H_2 loss); graphite: Ceylon -325 mesh; sintering 2050°F (1125°C); atmosphere: 28% CO +72% H_2; test bars: 1.250 x 0.500 x 0.250 in.; density: 5.5 g/cc (dashed curves), 6.2 g/cc (solid curves).

TABLE 1. Sintered Properties of Iron—Graphite Composition Influence of Graphite Quality

M. P. I. F. transverse rupture test bars 1.250"x.500" were sintered at 2070 F(1132 C)for 35 minutes in purified exogas atmosphere in a production furnace

		Graphite Addition %	Ash	Dimensional Change % 5.8 1.1	5.8 1.4	6.4 1.1	6.4 1.4	Hardness R_B 5.8 1.1	5.8 1.4	6.4 1.1	6.4 1.4	Modulus of Rupture P.S.I. 5.8 1.1	5.8 1.4	6.4 1.1	6.4 1.4
Mfg. A															
(1)	Ceylon	5 Micron Average	2.8	.06	.10	.18	.19	27	22	53	52	51,400	53,300	72,200	83,000
(2)	Ceylon	10 Micron Average	2.2	.10	.06	.18	.21	24	4	51	47	46,600	44,100	70,100	80,600
(3)	Ceylon	97 %<325 Mesh	5.2	.06	.06	.10	.15	17	16	41	49	39,300	43,900	63,700	65,300
(4)	Ceylon	97 %<325 Mesh	3.5	.03	.18	.14	.22	-5	20	34	42	33,600	34,800	49,500	55,700
Mfg. B															
(1)	Ceylon	5 Micron Average	1.7	.07	.14	.14	.23	22	28	50	55	49,600	56,700	74,100	81,900
(2)	Ceylon	20 Micron Maximum	1.6	.06	.14	.14	.26	17	24	45	55	42,100	51,800	72,500	80,800
(3)	Ceylon	20 Micron Maximum	2.9	.02	.18	.11	.30	17	36	46	62	47,100	53,800	70,400	81,200
(4)	Ceylon	98 %<325 Mesh	1.5	.06	.21	.26	.14	8	37	44	57	36,500	53,700	72,000	80,000
(5)	Ceylon	100 Mesh, 60-75%<325 Mesh	2.1	.10	.22	.24	.31	27	33	50	62	47,700	53,000	75,100	80,900
(6)	Madagascar	25 Micron Maximum	2.7	-.02	.13	.09	.28	14	37	46	61	48,100	53,100	70,000	83,300
(7)	Bavaria	20 Micron Maximum	3.8	-.02	.09	.14	.17	22	33	53	57	48,800	54,500	72,900	88,900
(8)	Bavaria	20 Micron Maximum	4.2	.06	.10	.15	.23	21	35	51	61	50,300	58,900	72,700	89,300
Mfg. C															
(1)	Texas	98 %<325 Mesh	3.6	.16	.10	.10	.24	-8	15	22	56	34,100	39,000	56,100	73,000
(2)	Texas	95 %<8 Micron	2.9	.07	.14	.09	.32	16	38	43	63	49,300	58,100	72,600	89,500
(3)	Texas	85 %<2 Micron	3.5	.06	.22	.14	.22	23	37	52	62	46,900	46,900	73,500	91,800
Mfg. D															
(1)	?	60 %<8 Micron	1.8	.06	.18	.14	.26	24	35	52	63	48,800	54,700	74,700	85,400
(2)	?	99 %<325 Mesh	4.8	.02	.03	.13	.14	20	17	51	48	47,800	51,400	72,000	79,600
(3)	?	98 %<325 Mesh	2.3	.05	.10	.10	.18	6	28	38	54	41,000	48,800	61,900	69,800
(4)	Mexican	70 %<8 Micron	17.6	-.02	-.01	.10	.17	-24	-24	14	10	30,300	26,600	39,000	32,100
(5)	Mexican	99 %<325 Mesh	18.9	-.10	-.18	.23	.39	-34	-32	6	6	21,700	19,400	32,800	25,400
(6)	Mexican	98 %<200 Mesh	18.2	-.23	-.34	.42	.58	-49	-53	-6	-15	15,800	12,800	22,800	17,300
(7)	Mexican	98 %<200 Mesh	18.9	.31	.47	.47	.58	-58	-64	-11	-19	14,000	11,100	20,000	15,600
Synthetic															
(1)	Electro	Colloidal	.9	-.02	-.02	.10	-.06	-34	-24	6	12	20,600	-	31,300	-
(2)	Electro	25 Micron Maximum	1.0	-.81	1.18	.62	.86	-22	-17	28	35	11,700	10,500	23,400	25,000
(3)	Electro	98 %<325 Mesh	1.7	-.81	1.11	.58	.82	-14	-4	35	42	13,000	12,500	27,200	27,700
(4)	Electro	98 %<200 Mesh	1.6	1.00	1.43	.78	1.07	-24	-27	24	32	9,100	9,300	22,000	22,400

densities and at two different graphite additions. Reported are, on the graphites, the origin, particle size, and ash content and on the resulting bars, dimensions, hardness, and strength. Sintering was done at 2070°F (1132°C) for 40 min in an atmosphere of purified exogas in an actual production furnace. The actual values of dimensions, hardness, and strength are immaterial in this case. Only the comparison between the various graphites is of interest at this point. We find that the synthetic graphites can be ruled out as unsuitable in spite of the fact that they are all of a particles size comparable to the natural graphites and with very low ash contents. The reason for this is not clear.

We know that natural and electrographites have the same crystal structure. One of the important differences for our purpose is, however, that the ashes in electrographite are mainly silicon carbide and in natural graphite might be mica, clay, etc. Therefore, one-half of one percent of ashes in electrographite might be worse from powder metallurgy standpoint than 5% of ashes in natural graphites.

Graphite can be rendered inactive in the presence of a molten phase. It is entirely possible that the ash components in the graphite can form this molten phase and, thereby, influence its reactivity.

It is further known that carbon black or lamp black is not very reactive in spite of a particle size ranging from $0.5-0.005\ \mu$.

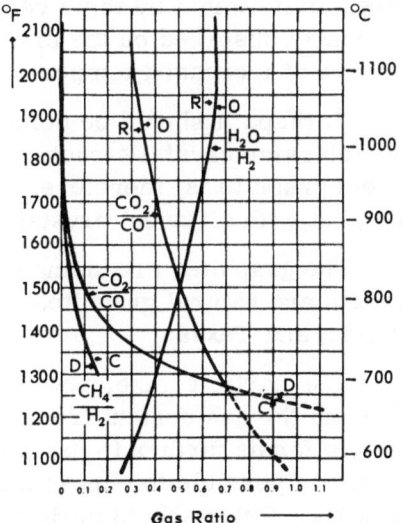

Fig. 4. Schematic illustration to show how mixtures of CO_2-CO, H_2O-H_2, and CH_4-H_2 are reducing, oxidizing, decarburizing, and carburizing at different temperatures versus gas ratio. R — reducing; O — oxidizing; C — carburizing; D — decarburizing.

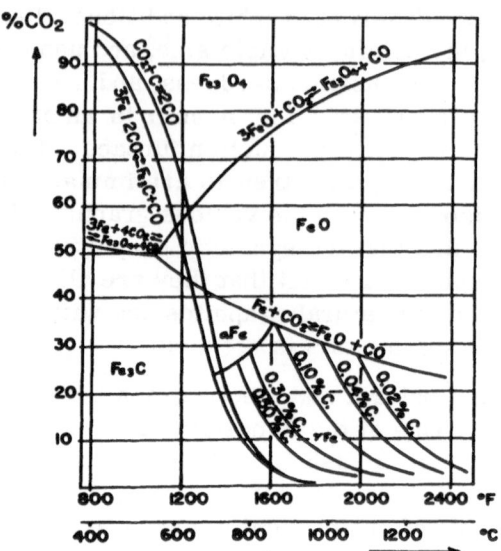

Fig. 5. Equilibrium diagram for the system Fe−
O−C. % CO_2 of $[CO + CO_2]$ versus temperature.
$CO + CO_2 = 100\%$.

They are also completely detrimental to compressibility. These
factors seem to depend on the fact that carbon black, lamp
black, and related products have a highly disordered lattice and
have an extremely fine crystal structure. One reason for their
nonreactivity might be their very high affinity for oxygen. Oxygen
is strongly absorbed on the crystal's surface with a bond that gets
stronger at increased temperatures.

Perhaps it should be pointed out that all graphites as such
have the same crystal structure. To speak about flake, amorphous,
or vein graphite is, therefore, not entirely proper. They are all
flaky, but with different crystal sizes.

The graphites producing good results for hardness and
strength are natural graphites with origins in Ceylon, Madagascar,
Bavaria, and Texas.

The particle size, however, ranges from 5 μ to -100
mesh and the ash content from 1.7% to 4.8%. It is at this stage impossible for us further to explain this phenomenon. However, it
is entirely possible to choose a suitable graphite among the many
that work. Price should perhaps also be mentioned as a factor.

Before leaving the subject of graphite quality, we would like to mention these considerations:

(a) For mixing, an easily distributed graphite that gives good flow and widest range of mixed densities is needed. For this a relatively coarse, flaky graphite with good lubricity is preferred, not forgetting that graphite aids lubrication in pressing.

(b) For pressing, good lubricity, minimum briquetting pressure and lowest green expansion are required. These considerations rule out the synthetic graphites, soot, and all the high ash graphites.

(c) For proper sintering, the previously mentioned high reactivity for fast reduction of oxides and fast carburization is needed.

Therefore the choice is narrowed to natural graphites of low ash content, less than 5%, and fine particle size and a graphite that is below 100 mesh, of natural origin, flaky structure, and with low ash content is most desirable.

The Atmosphere Composition

Two things are important:

(1) Equilibrium conditions (carbon potential).
(2) Reaction velocities (rich versus lean gas).

1. Equilibrium Conditions (carbon potential)

One reason that the atmosphere plays such an important part is, of course, the risk of decarburization.

The reactions to be in equilibrium are:

1. $CO_2 + C \rightleftharpoons 2CO$
2. $CH_4 \rightleftharpoons C + 2H_2$
3. $H_2 + CO_2 \rightleftharpoons H_2O + CO$

It is almost impossible to predict exactly if a given atmosphere is going to be carburizing, neutral, or decarburizing in the furnace for the simple reason that the composition will change with temperature, load of the furnace, and gases expelled from the parts being sintered and consequently might be entirely different in composition and in effect in various parts of the furnace. It is, however, helpful to know the characteristics of the individual equilibrium reactions. This is in simplified form pictured in Fig. 4

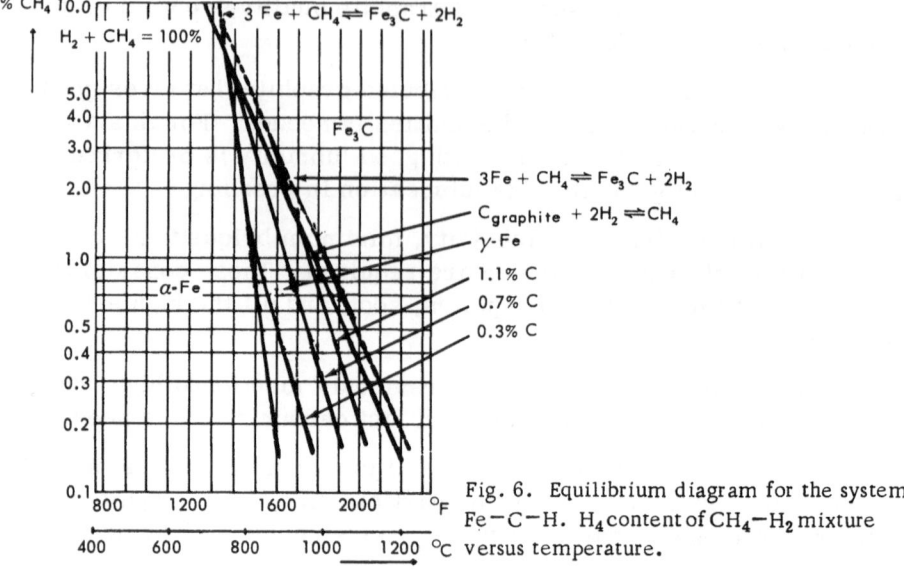

Fig. 6. Equilibrium diagram for the system Fe−C−H. H_4 content of CH_4−H_2 mixture versus temperature.

where the reducing, oxidizing, carburizing, or decarburizing effects are plotted for various temperatures and gas ratios. It is interesting to note that the ratio CO_2 to CO can be fairly high at all temperatures without causing oxidation of the iron. On the other hand the ratio must at all practical sintering temperatures be very low to prevent decarburization.

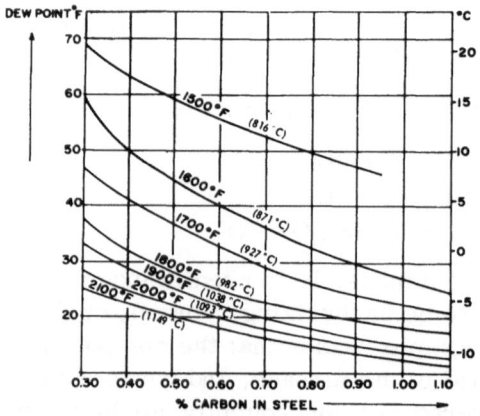

Fig. 7. Equilibrium between normal endogas and carbon in steel at different temperatures. CO, 20%; H, 40%; CH, 1%; N, balance.

At lower temperatures the CO_2:CO ratio can be permitted to increase considerably without causing either oxidation or decarburization. In other words, the CO constituent decreases its carburizing effect with increasing temperatures. This is contrary to the CH_4:H_2 ratio. As can be seen, the CH_4 increases its carburizing effect with increasing temperatures. This means in any event and at any practical temperature that it takes only small amounts of CH_4 to cause carburization.

The equilibrium diagrams for the individual systems in combination with iron oxides are shown separately.

In Fig. 5 we see that at practical sintering temperatures the CO_2 content as mentioned before will have to be kept very low to permit the formation of iron carbide.

In Fig. 6 we see the diagram for the system Fe – C – H. It shows that the equilibriums are strongly towards carburization (carbide formation) at all practical temperatures for CH_4 contents as low as 1%.

Finally regarding the influence of the dewpoint, this is shown in Fig. 7 for a normal endogas. It will be seen that the moisture content is more strongly decarburizing the higher the temperature is.

As mentioned before, when all the gases are present in the furnace, as they always are, it is impossible to put the ultimate equilibrium conditions in a diagram form. However, we have made

TABLE 2. Sintering Atmospheres (Composition of gas mixtures in equilibrium with eutectoid steel at 2050°F, in volume percent)

N_2	H_2	CO	CO_2	CH_4	H_2O
0	0	99,56	0,44	0	0
0	20	79,554	0,28	0,006	0,16
0	40	59,58	0,16	0,02	0,24
0	60	39,64	0,07	0,05	0,24
0	80	19,72	0,02	0,10	0,16
0	99,85	0	0	0,15	0
40	0	59,84	0,16	0	0
40	20	39,84	0,07	0,01	0,08
40	40	19,88	0,02	0,02	0,08
40	59,95	0	0	0,05	0
80	0	19,98	0,02	0	0
80	10	9,984	0,004	0,002	0,01
80	19,994	0	0	0,006	0

The following compositions will theoretically create the desired conditions when heated to 2050° F

N_2 = 0%	N_2 = 40%	N_2 = 80%
H_2 = 88%	H_2 = 54%	H_2 = 12%
CO = 12%	CO = 6%	CO = 2%

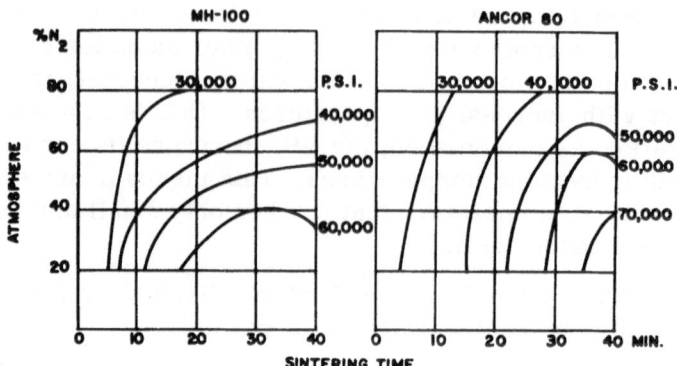

Fig. 8. Transverse rupture strength versus sintering time and atmosphere.

some theoretical calculations of various compositions or gas mixtures that would be in equilibrium with eutectoid steel at 2050°F (1121°C). The calculations have been made for 0-40-80% nitrogen and for zero and up percent of hydrogen, carbon monoxide, and carbon dioxide. This is shown in Table 2.

In all practical gas compositions we find that the methane content and the dewpoint will have to be kept low. At the bottom there are listed three additional compositions which are practical to use for laboratory work and that will create the desired conditions when heated to 2050°F (1121°C). We would like to point out one interesting feature in this calculation. Cracked ammonia, of course, consists of 75% hydrogen and 25% nitrogen. Forgetting the inert nitrogen for the moment and considering hydrogen as 100%, we

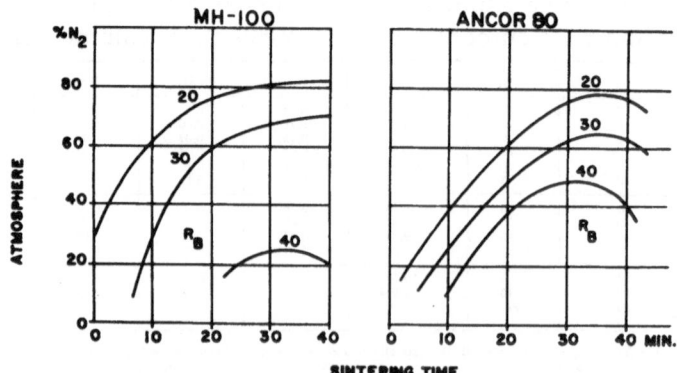

Fig. 9. Hardness versus sintering time and atmosphere.

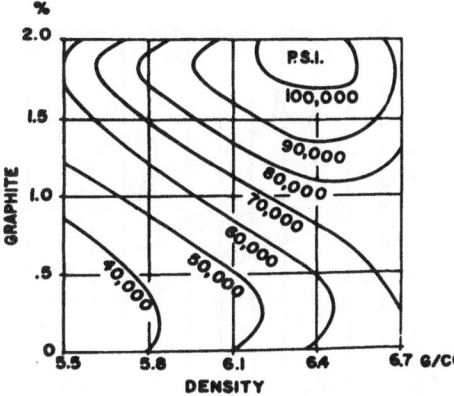

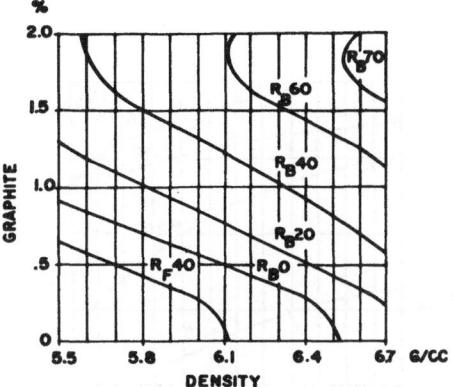

Fig. 10. Transverse rupture strength versus graphite addition and density. Iron: Ancor 80; graphite: Asbury FG; sintering: 30 min, dried purified exogas, 2050°F, atmosphere: 21% H_2, 9% CO, 70% N_2.

Fig. 11. Hardness, R_B, versus graphite addition and density. Iron: Ancor 80; graphite: Asbury FG; sintering: 30 min, dried purified exogas, 2050°F, atmosphere: 21% H_2, 9% CO, 70% N_2.

find that at least theoretically it would be entirely practical to sinter iron—carbon parts in plain cracked ammonia provided it is kept very dry. The interesting feature is that also in actual practice the method works well under certain conditions, namely that the furnace is run at equilibrium conditions; i.e., is completely dried out and that the load of iron—carbon parts is high and uninterrupted. Thus the parts themselves create the gas constituents that are necessary to give complete protection against decarburization. Cracked NH_3 also has the advantage of being strongly reducing and, therefore, takes over work from graphite.

2. Reaction Velocities (rich versus lean gas)

It has previously been mentioned that it is not enough that the gas is in equilibrium with the steel unless long sintering times or high sintering temperatures can be permitted. It is important also that the gas is as low as is practical in the inert constituent nitrogen.

Although the iron and the graphite are intimately mixed and pressed together, the diffusion of the carbon into the iron takes place primarily via the gas phase. Iron or iron oxide heated with graphite to normal sintering temperatures under vacuum are not far from inert to each other.

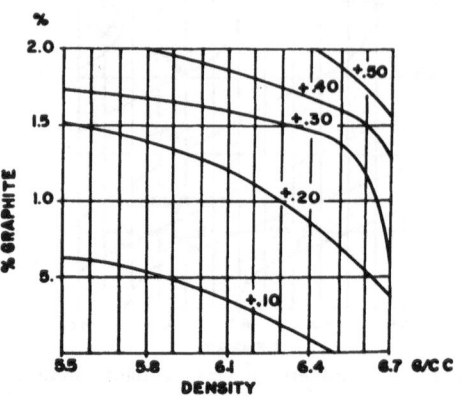

Fig. 12. Dimensional change, %, versus graphite addition and density. Iron: Ancor 80; graphite: Asbury FG; sintering: 30 min, dried purified exogas, 2050°F, atmosphere: 21% H_2, 9% CO, 70% N_2.

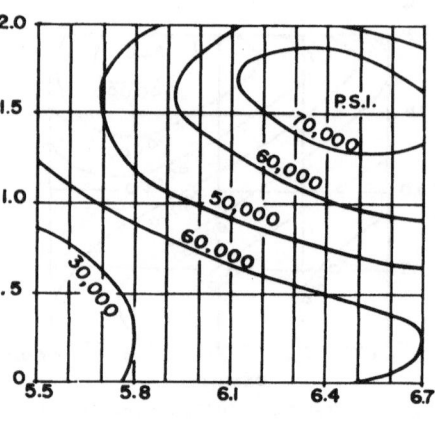

Fig. 13. Transverse rupture strength versus graphite addition and density. Iron: MH-100; graphite: Asbury FG; sintering: 30 min, dried purified exogas, 2050°F.

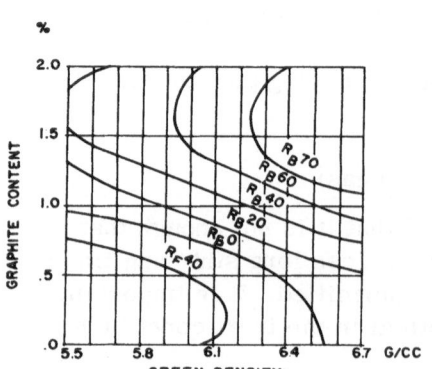

Fig. 14. Hardness, R_B, versus graphite addition and density. Iron: MH-100; graphite: Asbury FG; sintering: 30 min, dried purified exogas, 2050°F.

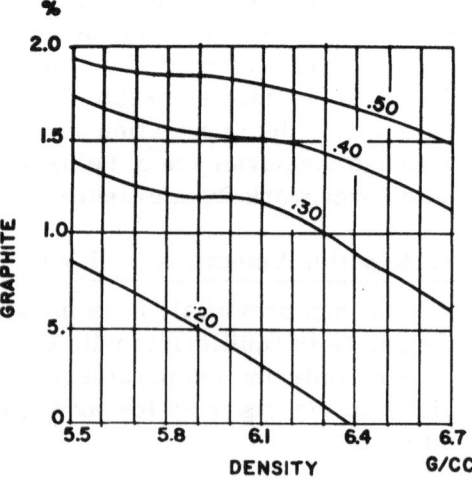

Fig. 15. Dimensional change, %, versus graphite addition and density: Iron, MH-100; graphite Asbury FG; sintering: 30 min, dried purified exogas, 2050°F.

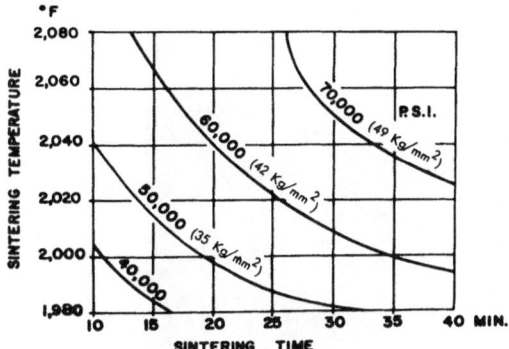

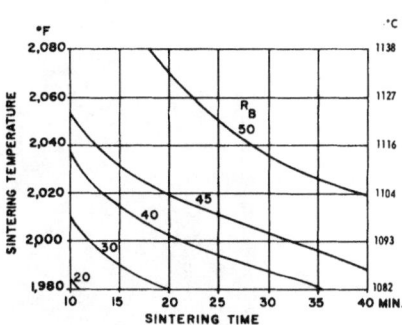

Fig. 16. Iron—carbon: Modulus of rupture, PSI, versus sintering temperature and sintering time. Iron: MH-100; graphite: Asbury FG 1.25%; green density: 6.10 g/cc; test bars: MPA standard; atmosphere: 21% H_2, 9% CO, 70% N_2.

Fig. 17. Iron—carbon: Hardness, R_B, versus sintering temperature and sintering time. Iron: MH-100; graphite: Asbury FG 1.25%; green density: 6.10 g/cc; test bars: MPA standard; atmosphere: 21% H_2, 9% CO, 70% N_2.

For example, purified exogas is a rather lean gas (high in nitrogen), but has the advantage of being easy to handle at high carbon potentials because of the rather low carburizing velocity. Rich gases (low in nitrogen), on the other hand, are more difficult to use at high carbon potentials because of the high carburizing velocity and because of their tendency to deposit soot. Figures 8 and 9 show the influence of inert nitrogen on results.

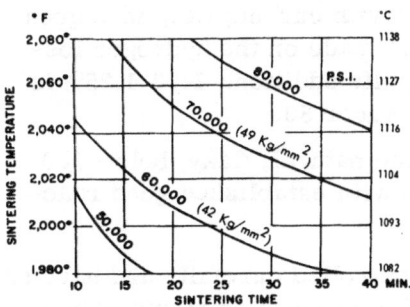

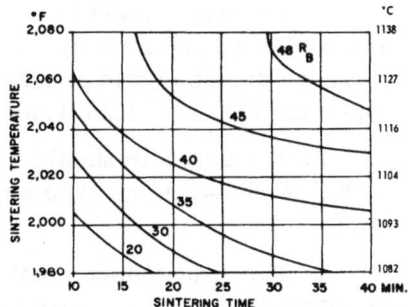

Fig. 18. Iron—carbon: Modulus of rupture, PSI, versus sintering temperature and sintering time. Iron: Ancor 80; graphite: Asbury FG 1.50%; green density: 6.10 g/cc; test bars: MPA standard; atmosphere: 21% H_2, 9% CO, 70% N_2.

Fig. 19. Iron—carbon: Hardness, R_B, versus sintering temperature and sintering time. Iron: Ancor 80; graphite: Asbury FG 1.50%; green density: 6.10 g/cc; test bars: MPA standard; atmosphere: 21% H_2, 9% CO, 70%, N_2.

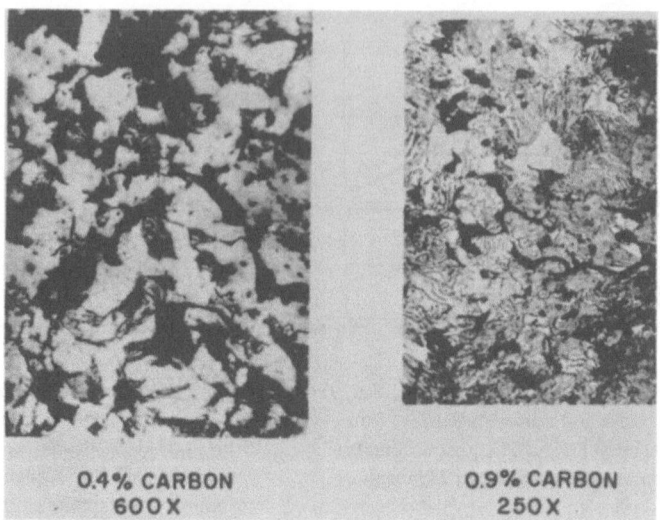

0.4% CARBON
600X

0.9% CARBON
250X

Fig. 20

Results show that gas high in nitrogen results in slower carbon pickup. It also shows that the one iron powder (MH-100) picks
up carbon faster than the other (Ancor 80) and, therefore, gets
harder in the early stages. In practice, MH-100 would be the preferable powder in most cases.

We can now sum up the desired conditions for manufacturing
iron−carbon parts as follows:

1. The iron should be reactive with carbon; i.e., have good
carbon pickup, and a check should be made on the hydrogen loss
content to decide on the proper graphite addition. 1.00-1.25% is
proper for MH-100, 1.25-1.50% for Ancor 80.

2. The graphite should be of the natural, flaky, below 100
mesh, low-ash-content type and one with established good reactivity.

3. The atmosphere must be controlled carefully and have as
low nitrogen content as possible. Dry cracked ammonia, dry endogas, or dry and purified exogas are all practical.

4. Furnace load should not vary too much from hour to hour.

To illustrate finally what can be expected if the above conditions are fulfilled, a series of graphs are included picturing iron−

carbon parts sintered for 30 min in a dried and purified exogas. See Figs. 10 to 15.

Influence of time and temperature is shown in Figs. 16 to 19.

Dimensions are very much determined by sintering condi-tions. As a rule the faster the reaction between the iron and the graphite, the smaller the dimensions will be. As can be seen, dimensions are rather stable over a range of densities. With con-trol at atmosphere, sintering time, and sintering temperature, di-mensions will consititute no problem.

Figure 20 shows two photomicrographs of sintered iron with 0.40% and 0.90% combined carbon, respectively.

For proper control of iron—carbon parts production, metallo-graphic test procedures are to be recommended.

Effect of Different Furnaces and Sintering Practices on Strength of Sintered Iron Powder*

Ernst Geijer

Höganäs AB

Höganäs, Sweden

The powder metallurgy industry has long been aware that large variations in strength of sintered iron powder parts are obtained when identical parts are sintered in different laboratory and production furnaces. This survey is an attempt to chart the variations obtained when identical parts with no addition to the iron powder other than a lubricant are sintered in different furnaces under presumably identical conditions.

Procedure

To eliminate as many as possible of the variables in the production of iron-powder parts, the following procedure was followed: From the same batch of a standard sponge iron powder (Ancor MH-100) mixed with 0.5% zinc stearate, tensile bars (according to the Metal Powder Industries Federation Standard 10-50T) were pressed in a production press (The Presmet Corp.) to two green densities, 6.08 and 6.46 g/cc.

*Reprinted from Materials Research and Standards, Vol. 2, pp. 124-126, 1962, by permission of the American Society for Testing and Materials.

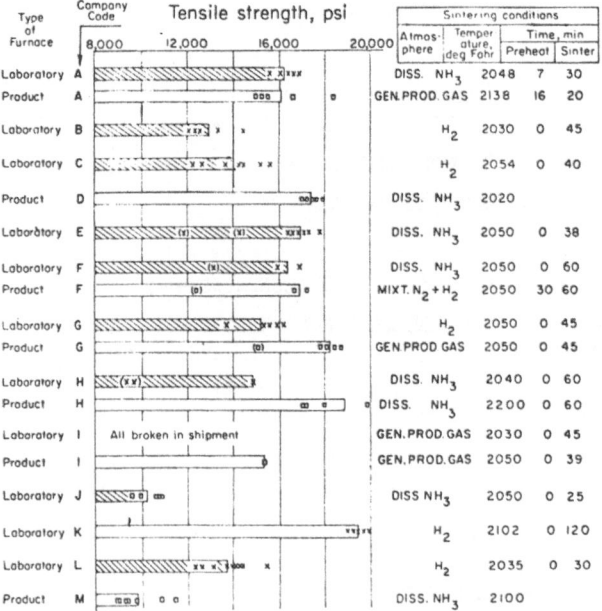

Fig. 1. Strength of bars having green density of 6.08 g/cc.

To obtain additional information on the importance of the press and the die used, parts were pressed at each participating laboratory from the same batch of mixed powder and in dies made according to Standard 10-50T. The green densities were kept as close as possible to those mentioned above. The participating companies were requested to sinter, when possible, in either hydrogen or dissociated ammonia.

After sintering in different laboratory and production furnaces, all bars were tested in the same laboratory (Höganäs Sponge Iron Corp.). Thirteen companies particpated in this survery.

Results

In Figs. 1 and 2, the tensile strengths of the test specimens pressed in the same press and sintered in the different production and laboratory furnaces are plotted. The sintering conditions (atmosphere, temperature of hot zone and preheat, and time in hot zone) are shown in Fig. 1.

The average tensile strength for each furnace is shown as

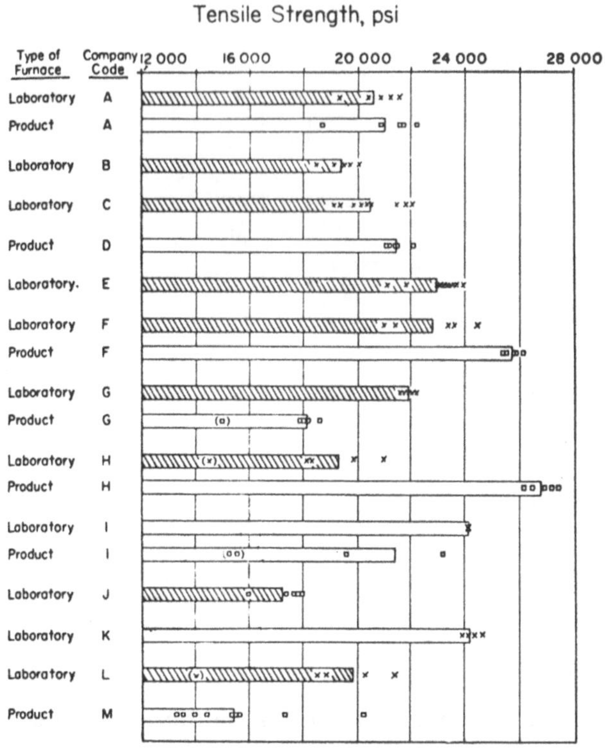

Fig. 2. Strength of bars having green density of 6.46 g/cc.

a bar diagram. In determining these averages, the unusually low figures were disregarded, on the assumption that these green bars were damaged during shipment to respective companies for sintering. They are shown within parentheses.

The companies are coded with letters A to M. In some cases tests were run at the same company in both a production and a laboratory furnace. The variations in tensile strength from a given furnace are reasonable, considering that we are judging individual figures, not averages. The variations in average tensile strength from furnace to furnace are, on the other hand, considerable. To a large extent this is caused by the varying sintering atmosphere, time, and temperature in the furnace. To investigate the variations obtained from furnace to furnace when the sintering is done under presumably the same sintering conditions, the tensile strengths for specimens from the nine laboratory furnaces from Figs. 1 and 2 representing sintering temperatures from 2030 to 2054°F and a

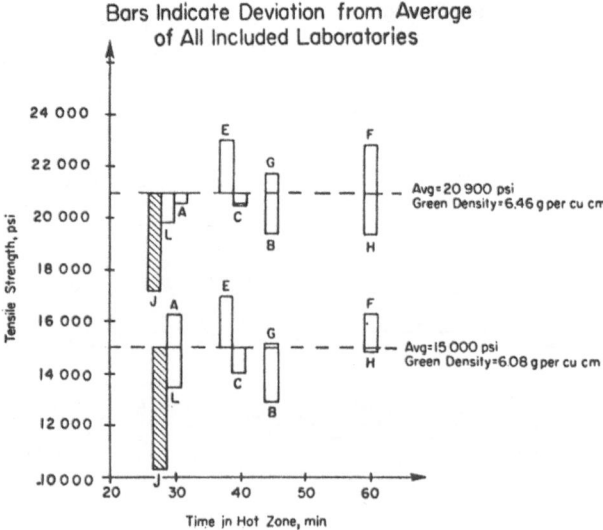

Fig. 3. Average tensile strengths for all laboratories.

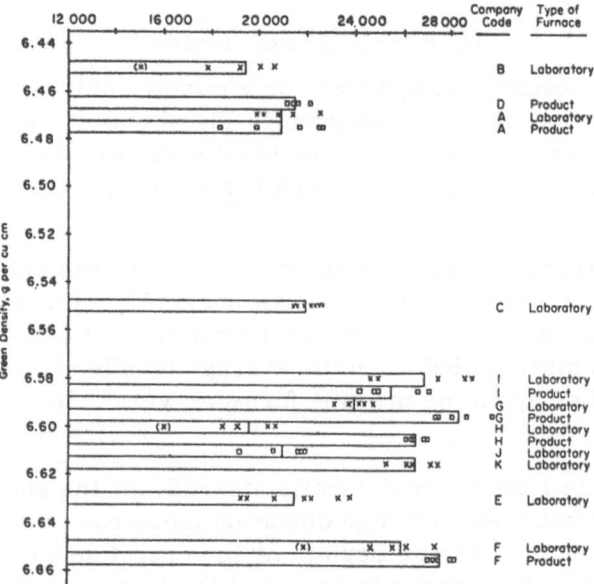

Fig. 4. Strength versus green density.

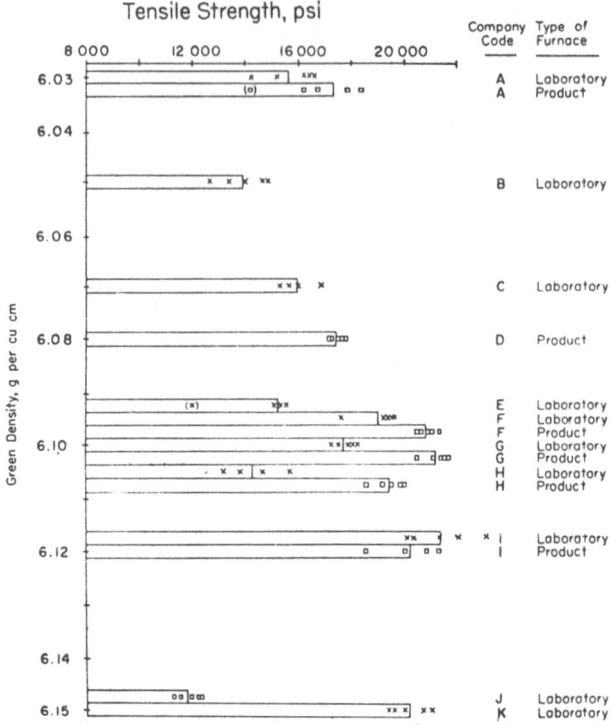

Fig. 5. Strength versus green density.

sintering atmosphere of hydrogen or a mixture of hydrogen and nitrogen were put in a bar diagram in Fig. 3 with the average as a zero point for each density. The tensile strength is plotted versus sintering time, which is the one remaining uncontrolled variable.

The strength figures obtained from one furnace (J) deviated considerably from the others. It was found that furnace J was new, and that this test was one of the first trial runs. Consequently, this furnace was disregarded when the average tensile strength and the deviations therefrom for the nine furnaces were calculated.

In Table 1 the average tensile strength for the eight furnaces, with the maximum and average deviation therefrom, are shown. The maximum and average deviations in percent seem reasonable considering the differences in design of the furnaces. A maximum deviation of 14% is quite satisfactory.

TABLE 1. Average Tensile Strength

	Green Density, g per cu cm	
	6.08	6.46
Average strength, all eight laboratories, psi	15 000	20 900
Maximum deviation, psi	2 100	2 100
Average deviation, psi..	1 200	1 200
Maximum deviation, per cent.............	14	10
Average deviation, per cent.............	8	6

Bars pressed on same press; sintering temperature 2030 to 2054°F; atmosphere, hydrogen, or hydrogen plus nitrogen

The sintering time is not of primary importance to tensile strength, within reasonable limits. There is little tendency toward higher tensile strength for longer sintering time.

A large part of the deviations shown in Table 1 and Fig. 3 may be due to differences between recorded and true temperatures in the eight laboratory furnaces.

The bars representing tensile strength for the nine furnaces included in Table 1 and Fig. 3 are crosshatched in Figs. 1 and 2. In Figs. 4 and 5, the tensile strength is plotted versus green density for bars that were pressed and sintered at the same company. The same powder mix was used in all cases.

A comparison of Figs. 1 and 2 and Figs. 4 and 5 shows that the variation in average tensile strength obtained from each furnace is about the same when the tensile bars are all pressed in one place as when they are pressed at the respective companies where they were sintered. This indicates that differences in sintering conditions affect strength more than differences in pressing procedure at the same green density.

Conclusions

Reasonable correlation can be obtained among laboratories when sintering straight iron powder parts with the existing laboratory furnaces and sintering practices. The survey should be expanded to cover parts made from iron powder with additives other than a lubricant, such as graphite or copper powder. Enough information should then be available to judge the importance of more closely specified laboratory sintering procedures or the introduction of a standard sintering furnace.

The Fatigue Strength
of Sintered Iron Compacts

J.M. Wheatley
Formerly Metallurgy Department, University of Cambridge

and

G.C. Smith
Metallurgy Department
University of Cambridge

Introduction

Although many investigations have been made of the mechanical properties of sintered ferrous metals, there is little published work on the fatigue behavior. As the porosity of a compact is increased, the ductility is reduced more rapidly than the tensile strength, and although the fatigue-strength behavior might be expected to parallel that of the UTS, the notching effect of the pores could cause an increasingly rapid fall in fatigue strength with increasing porosity. This might be particularly true for a metal such as iron, which under certain conditions can undergo brittle failure.

Goetzel and Seelig [1] showed that in the 6-32% porosity range the fatigue strength in bending of iron compacts decreased more or

less linearly, and that compacts of fine electrolytic powder gave
superior fatigue properties to those obtained from a coarse powder. Their results also demonstrated that S/N curves for porous
iron compacts could show a fatigue limit, with the knee occurring
in the range $1\text{-}5 \times 10^6$ cycles. Hempel and Wiemer [2] investigated
the fatigue behavior of iron compacts made from two different
powders, one produced by disintegration and the other by reduction.
Specimens were produced with porosities ranging from 11 to 23%
and the fatigue strengths showed a similar dependence on porosity
to those obtained by Goetzel and Seelig, but the strengths were
lower, owing possibly to larger particle sizes of powders. Hempel
and Wiemer [2] determined most of the fatigue strengths in bending,
but in addition tested compacts from one of the powders in alternating torsion. From the results of the two types of tests they obtained a mean value of 0.58 for the ratio of the fatigue strength in
torsion to that in bending and concluded that fatigue failure in the
compacts was being governed by the maximum shear-strain-en-
ergy criterion, as with wrought materials.

In both these investigations the final densities were achieved
by a single pressing and sintering operation, and the effects were
not considered of working operations or of changes in the sintering conditions. In the present case the fatigue behavior of iron
compacts has been studied in the 4-30% porosity range, applying
varying sintering and working treatments to achieve the required
final densities. The fatigue tests were carried out under axial-
loading push/pull conditions and also in alternating torsion.

Experimental

1. Powder

Electrolytic iron powder of –300 B. S. mesh size was chosen
for all the experiments; the analysis gave: carbon 0.01, manganese
0.03, silicon 0.01, sulfur 0.015, phosphorus 0.01%, with the remainder iron. Several 10-lb batches were used during the course of
the work and the pressing and sintering characteristics were reproducible.

2. Compaction

The compacts were prepared in a double-acting steel die.
Their shape was rectangular and two sizes, $1.25 \times 0.2 \times 0.2$ and

TABLE 1. Summary of Preparation Techniques for Torsion-Fatigue Specimens

Specimen Batch	Specimen Density, g/c.c.	Compacting Pressure, tons/in²	Sinter for 4 h in vacuum at:	Mechanical Deformation	Resinter for 4 h at:
			°C		°C
C1	5·50	10	1050	None	None
C2	5·82	15	1050	,,	,,
C3	6·16	20	1050	,,	,,
C4	6·54	30	1050	,,	,,
C5	6·73	35	1050	,,	,,
C6	5·57	10	850	,,	,,
C7	6·10	20	850	,,	,,
C8	6·54	30	850	,,	,,
C9	6·70	35	850	,,	,,
C10	6·79	35	1050	,,	,,
C11	6·86	35	850	,,	,,
C12	6·89	35	1250	,,	,,
C13	6·40	15	1375	,,	,,
C14	6·82	35	1050	,,	1050
C15	7·10	35	1050	CR	1050
C16	6·34	15	1050	CR	1050
C17	7·26	23	1050	CR	1050
C20	7·43	35	1050	HR+CR	None
C21	7·61	35	1050	HR	,,

CR = Cold rolled.   HR = Hot rolled.

1.75 × 0.28 × 0.28 in., were produced to provide specimens for the two different fatigue machines. The weights of powder for making compacts of different density were so adjusted that in all cases the compacts were approximately square in cross section.

3. Sintering

Sintering was usually carried out in vacuum (10^{-3} mm) in a 2.0 in. internal-diameter silica tube furnace. An inner furnace tube was loaded and evacuated and then the furnace, running at the required temperature, was placed in position. When a thermocouple indicated that the specimens had reached temperature, the furnace was left for the required time and then removed. A sintering atmosphere of nitrogen was used for some experiments, with a similar procedure to that outlined above.

4. Further Treatment of Compacts

The densities of the compacts increased by only a small amount on sintering and it proved difficult to arrive at high densities by a single pressing and sintering operation. Deformation treatments were, therefore, applied to obtain higher densities.

TABLE 2. Summary of Preparation Techniques for Push/Pull Fatigue Specimens

Specimen Batch	Specimen Density, g/c.c.	Compacting Pressure, tons/in²	Sinter* at:	Mechanical Defor- mation	Resinter* at:
			°C		°C
A1	5·90	15	820 (N₂)	None	None
A2	6·29	25	820 (N₂)	,,	,,
A3	6·72	35	820 (N₂)	,,	,,
A4	5·73	15	1020 (N₂)	,,	,,
A5	6·85	35	1020 (N₂)	,,	,,
A6	6·79	35	1050	,,	,,
A7	6·24	25	1050	,,	,,
A8	6·59	25	1050	,,	,,
A9	6·14	15	1050	,,	,,
A10	6·41	15	1050	CR	1050
A11	6·42	30	1050	None	None
A12	6·88	30	1050	CR	1050
A13	6·93	30	1050	HR+CR	None
A14	6·82	30	1050	HR	,,
A15	7·27	35	1050	HR	,,
A16	7·35	35	1050	HR+CR	1050
A17	6·67	30	1050	None	None
A19	7·08	30	1050	ReP	1050
A20	6·81	35	1050	None	None
A22	7·36	35	1050	ReP twice	1050 twice
A23	7·40	35	1050	ReP	1050 16 h
A24	7·01	45	1050	None	None
A25	7·44	45	1050	ReP	1050
A26	7·08	45	850	None	None
A27	7·15	45	850 20 h	,,	,,
A28	6·62	30	850	,,	,,
A29	7·31	45	None	ReP	1050

CR = cold rolled.   HR = hot rolled.   ReP = repressed.
* 4h in vacuum unless otherwise stated.

These consisted in: (a) reductions up to 15% by cold rolling, followed by resintering; (b) hot working to ~ 20% reduction; (c) repressing to ~ 10% reduction in height and resintering. Cold rolling was not particularly effective, owing to the difficulty of applying any appreciable cold deformation without causing cracking. Hot working was carried out by sealing compacts in a flat mild-steel sheath and then rolling at ~ 1000°C. The compact densities were higher than those obtained by cold rolling and resintering, but the method suffered from the disadvantage that all compacts in a batch did not elongate by the same amount, so that greater variations in density were present than existed between compacts treated in other ways. Repressing was usually carried out at the initial com-

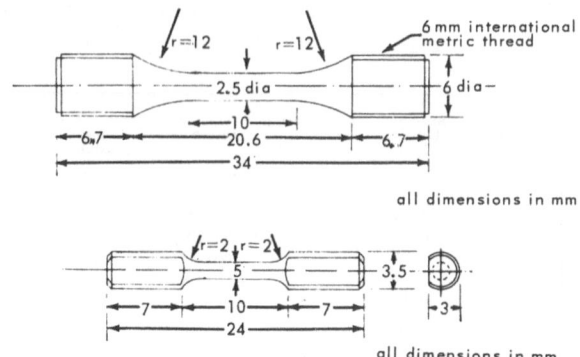

Fig. 1. Dimensions of fatigue-test specimens: (a) For Amsler machine; (b) for Chevenard micro-torsion machine.

paction pressure; a single repressing and sintering had a significant effect upon the density, but further repressing had relatively little effect.

As shown in Tables 1 and 2, specimens were produced using all three of the above techniques, but even so it was not possible to obtain compacts with densities in the 7.5 – 7.84 g/cc range, except for one isolated batch produced by hot rolling.

5. Fatigue Specimens

The dimensions of the fatigue specimens are shown in Fig. 1. The compacts were first carefully filed and turned to produce cylinders and from these the specimens were formed by plunge grinding on profiled carborundum wheels, and then finished to the correct diameter with successively finer grades of emery cloth. They were then degreased and annealed in vacuum for 30 min at 600°C to remove residual stresses.

Density measurements were made on finished specimens by a water–displacement method, a silicone–oil or polystyrene film being employed to prevent penetration of water into the pores. Sectioning revealed little variation in pore size and distribution along the length of specimens. Differences in density between specimens in a batch were usually small and the standard procedure was to determine the density of three specimens per batch.

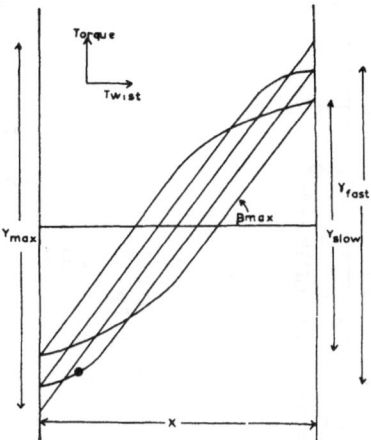

Fig. 2. The form of the hysteresis loop
obtained in the alternating-torsion tests,
and the way in which this is affected by
the rate of testing.

6. Fatigue Machines

The larger-sized specimens were tested at a frequency of
7000 c/m (cycles/min) with axial loading and zero mean stress in
an Amsler vibrophore fitted with a 0.4-ton dynamometer, and the
stress on the specimen was calculated directly from the applied
load and specimen diameter.

The smaller specimens were tested in a Chevenard micro-
torsion fatigue machine, which imposes a constant strain amplitude
on the specimen. To determine the torque on the specimen, the
torque/twist relationship during one fatigue cycle can be recorded
photographically at the low frequency of 1 c/m, compared with the
normal operating frequency of 1500 c/m. The form of this relationship is shown in Fig. 2, the distance X being a measure of the
total strain applied to the specimen, and the distance Y the corresponding torque. The linear portions of the loop show the behavior
of the specimen during elastic unloading from the maximum values
of applied strain at each end of the strain cycle, and from the slope
of these the modulus of rigidity of the specimen can be determined.
The torque/twist loop was found to be frequency-dependent in the
way indicated in Fig. 2, so that for the same applied strain, the

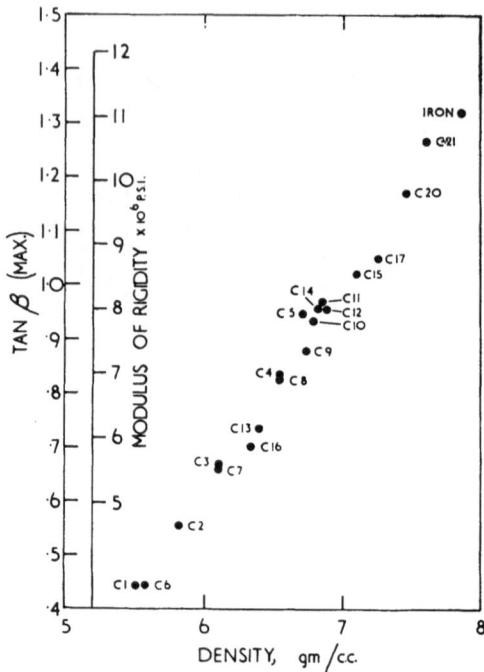

Fig. 3. Relationship between the rigidity modulus and the density.

torque carried by a specimen was greater at 1500 c/m than at
1 c/m, although the rigidity modulus remained substantially constant. However, owing to inertia effects the value of the torque at
high strains could not be determined accurately by the mirror
system normally employed to record the slow-speed torque/twist
relationship. Nevertheless, it was apparent from such measurements, and also from measurements of torque made by means of
a strain-gauge attached to the torque bar of the machine, that specimens behaved, at the high frequency, in a more or less elastic
manner up to the strain which gave failure in ~ 10[7] cycles. In view
of this, the conversion of the fatigue-limit strains to stress values
has been made by assuming elastic behavior and using the measured
values of rigidity modulus for the specimens. It is believed
that the error involved is small, and certainly less than that inherent in the use of the torque values determined from slow-speed
torque/twist loops, as would be the normal practice with this particular type of fatigue machine.

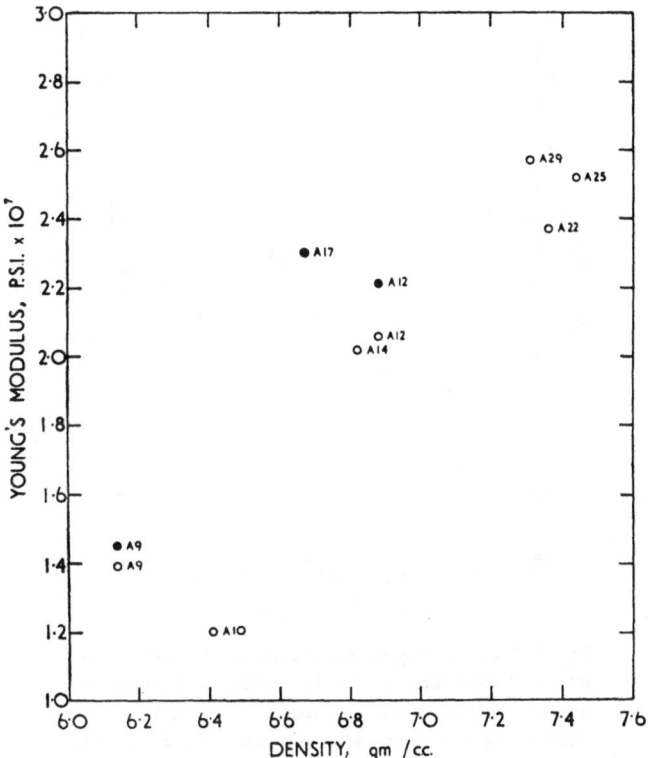

Fig. 4. Effect of density on Young's modulus of elasticity determined under static and dynamic conditions. ● Dynamic (7000 c/m) electrical gauge. ○ Static mechanical gauge.

Results

1. Elastic Modulus

Values of the rigidity modulus obtained from slow-speed torque/twist loops showed a linear dependence on density (Fig. 3). Extrapolation to the density of pure iron gives a modulus value that is in good agreement with quoted results for fully dense material.

Experiments were also made to determine whether the Young's modulus showed a similar dependence on density. Amsler fatigue specimens were loaded axially and the strain measured

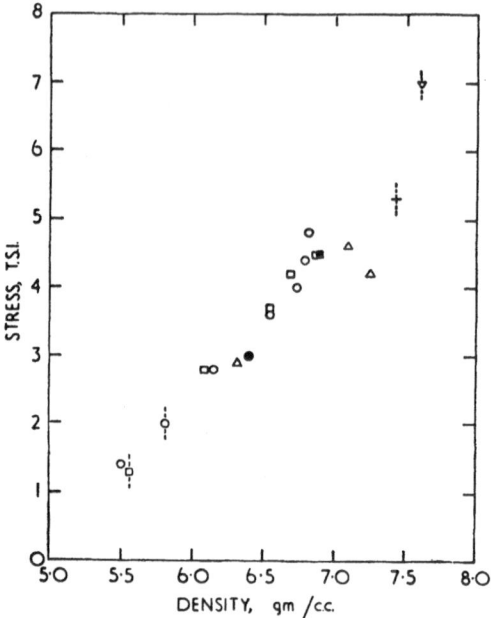

Fig. 5. Alternating-torsion fatigue. Surface shear
stresses for failure in 10^6-10^7 cycles as a function
of specimen density. □ Sintered in vacuum at
850°C; O sintered in vacuum at 1050°C; ■ sintered
in vacuum at 1250°C; ● sintered in vacuum
at 1375°C; Δ cold rolled; ∇ hot rolled; + hot and
cold rolled.

statically with a mechanical Johansson gauge, using determinations
on a silver steel specimen to standardize the gauge. In addition,
a few dynamic measurements were made at a frequency of 7000
c/m, with electrical resistance strain-gauges cemented to the
specimens to measure the strain.

The results are shown in Fig. 4 and are similar in character to those of Fig. 3. The Young's modulus falls with decreasing density and at a density of ~ 6.2 g/cc is only half the value
for pure iron; this density is about the same as that at which the
rigidity modulus is halved. The three points relating to dynamic
measurements show a slight trend to higher values than the static
results. This would be consistent with the idea that during slow

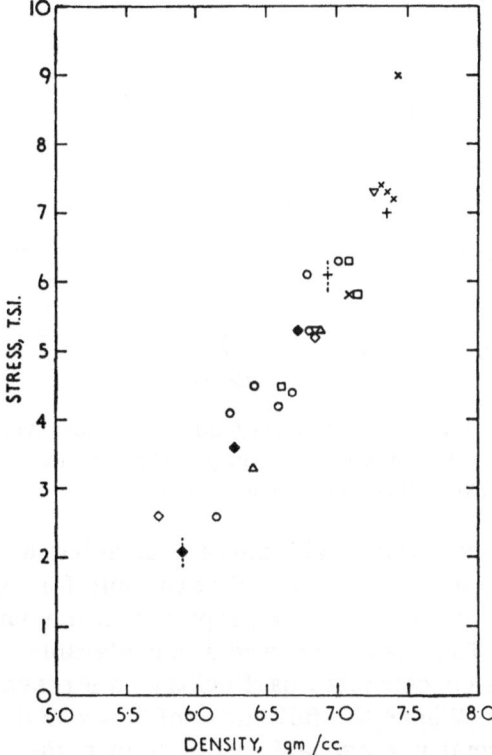

Fig. 6. Push/pull fatigue. Axial normal stresses
for failure in 10^6-10^7 cycles as a function of
specimen density. □ Sintered in vacuum at 850°C;
○ sintered in vacuum at 1050°C; ◆ sintered in nitrogen at 820°C; ◇ sintered in nitrogen at 1020°C;
△ cold rolled; ∇ hot rolled; + hot and cold rolled;
× repressed.

loading of porous specimens, in what is nominally the elastic range,
a small amount of time-dependent plastic deformation takes place
at the pores, resulting in slightly higher strains, and thus lower
modulus values, than with high-frequency dynamic stressing. However, owing to the small size of the specimens some difficulties
were experienced in attaching the resistance strain-gauges, so
that further experiments on larger specimens would be required to
establish the magnitude of this effect.

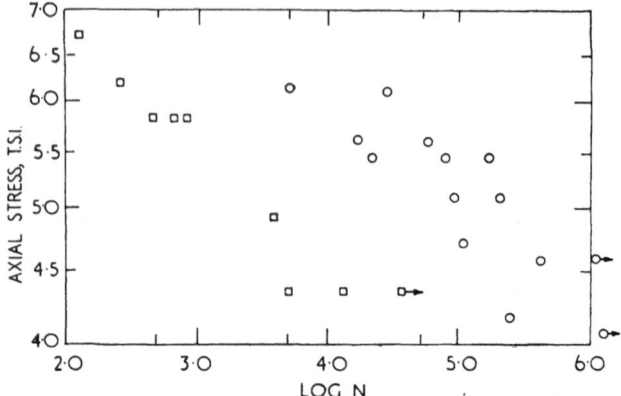

Fig. 7. Fatigue results in push/pull for specimens A17, showing the effect of testing frequency. ○ Specimen tested at
7000 c/m; □ specimens tested at 1 c/m.

An S/N curve was established for each batch of specimens.
About ten specimens were normally available for this, but in some
cases, because of difficulties in preparation, the number tested
was only five. The results showed a considerable degree of scatter
and testing was concentrated particulary on stresses giving lives
of ~ 10^6 cycles. Where the full curve of S/N was determined, the
shape was normally reasonably well defined (see Figs. 7 and 8)
with a sloping portion in the range of lives of up to 10^6 cycles and

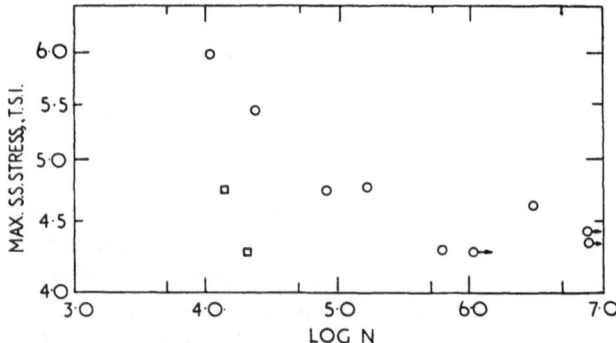

Fig. 8. Fatigue resulting in alternating torsion for specimens
C10, showing the effect of testing frequency. ○ Specimens
tested at 1500 c/m; □ specimens tested at 1 c/m.

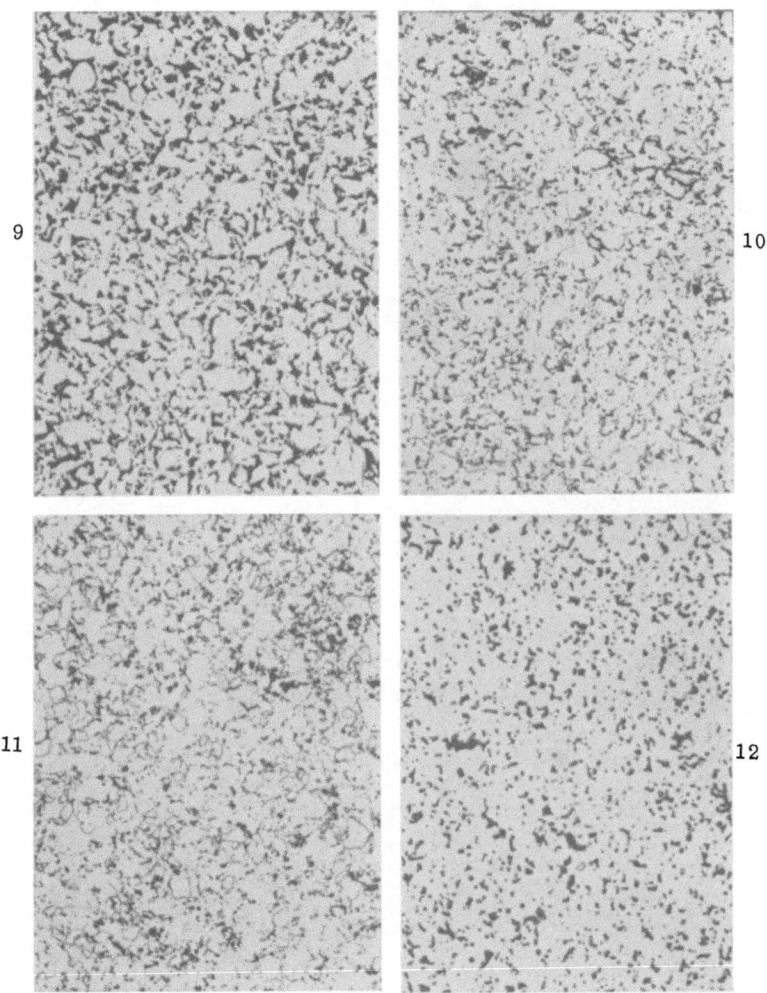

Figs. 9-12. Structures of specimens produced by pressing and sintering at various pressures and temperatures. $\times \sim 70$.

Fig. 9. C2; 15 tons/in^2 and 1050°C density = 5.82 g/cc.

Fig. 10. C10; 35 tons/in^2 and 1050°C density = 6.79 g/cc.

Fig. 11. C11; 35 tons/in^2 and 850°C density = 6.86 g/cc.

Fig. 12. C12; 35 tons/in^2 and 1250°C density = 6.89 g/cc.

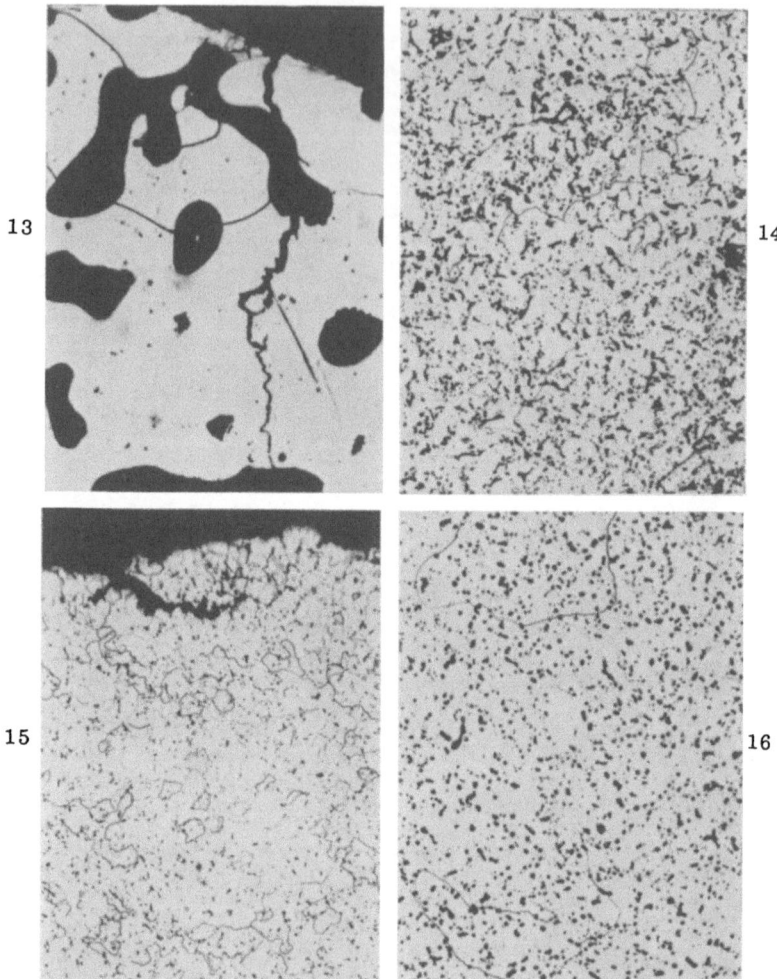

Figs. 13-16. Structures of specimens produced by pressing and sintering at various pressures and temperatures, followed for the specimens shown in Figs. 14-16, by the treatment indicated.

Fig. 13. C13; 15 tons/in^2 and 1375°C density = 6.40 g/cc. × ∼325.

Fig. 14. C15; 35 tons/in^2 and 1050°C followed by cold rolling and resintering. Density = 7.10 g/cc. × ∼70.

Fig. 15. C21; 35 tons/in^2 and 1050°C, followed by hot rolling. Density = 7.61 g/cc. × ∼70.

Fig. 16. A16; 35 tons/in^2 and 1050°C, followed by hot and cold rolling and resintering. Density = 7.35 g/cc. × ∼70.

a much flatter portion at longer lives. In many cases the longer-life portion appeared to be more or less horizontal, thus indicating a fatigue limit. The stress at which this flatter part of the curve appeared has been termed the fatigue limit and is the stress plotted in Figs. 5 and 6. The values have been plotted as single points and the accuracy is estimated to be $\sim \pm 0.25$ ton/in^2; dashed lines on some points indicate greater uncertainty. In both figures the fatigue-limit stress shows a dependence on density that is approximately linear. In Fig. 5 extrapolation to the density of pure iron leads to a value of ~ 6.8 tons/in^2, which fits into the range of quoted results for pure iron [3] subjected to alternating torsion, while in Fig. 6 the extrapolated value of ~ 9.5 tons/in^2 is in reasonable agreement with published values [3] for push/pull conditions.

3. Frequency Effect in Fatigue Testing

The difference in the torsion tests between the torque/twist relationship at high and low frequencies suggested that the fatigue behavior might also depend on the testing frequency in both torsion and direct stressing. A machine was, therefore, constructed which enabled Amsler specimens to be tested at 1 c/m, for comparison with the 7000 c/m results, and two specimens were also tested in the Chevenard machine at a frequency of 1 c/m for comparison with the results at 1500 c/m.

Figure 7 shows the behavior of different batches of A17 specimens, which were nominally identical. Considerable displacement of the S/N curve is produced by the decreased testing frequency, the average decrease in life being of the order of fifty times. A similar effect was found with the two slow-speed tests in alternating torsion, as shown in Fig. 8, relating to specimens C10.

4. Metallographic Examination

Specimens from the different batches of compacts were examined metallographically to determine the influence of the sintering variables upon the final microstructure.

Higher compacting pressures resulted in smaller pore sizes and a more coherent appearance in the structure, as can be seen by comparing Figs. 9 and 10 which refer to specimens pressed

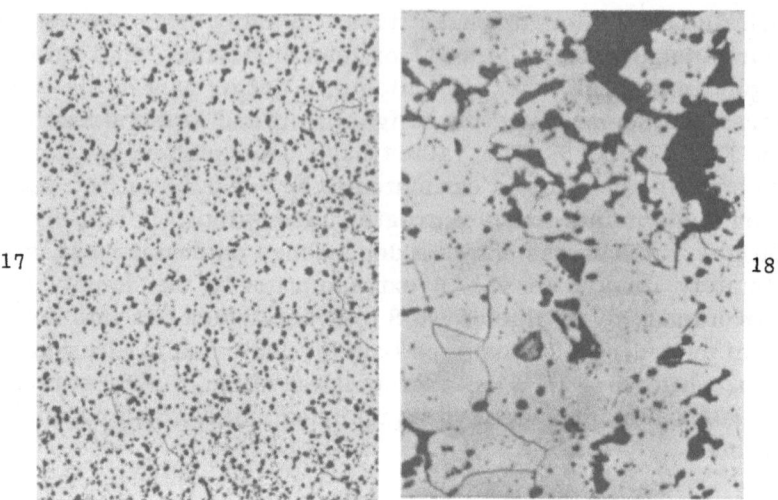

Fig. 17. Structure of batch A23 specimen produced by pressing at 35 tons/in^2 and sintering at 1050°C, followed by repressing and resintering. Density = 7.40 g/cc. x ~70.

Fig. 18. Fatigue crack in a specimen from batch A14. x ~ 335.

at 15 and 35 tons/in^2 but sintered at the same temperature of 1050°C. Sintering temperature did not significantly influence the density but had an effect on the grain size and pore shape, particularly in compacts produced at the higher pressures. Thus, the specimen in Fig. 11, sintered at 850°C, shows a smaller grain size than the specimen in Fig. 10. The sintering atmosphere, i.e., vacuum or nitrogen, did not appear to have any effect on the structures. Time was a much less important variable than temperature, so far as grain size was concerned, and extension of the sintering time at 850°C to 20 h produced no visible change in microstructure compared with that obtained in 4 h. With the two highest sintering temperatures considerable grain growth occurred together with a rounding of the pores (Fig. 12), and at 1375°C this rounding was marked (Fig. 13).

The effect on the structure of cold rolling followed by resintering was small; no directionality effects were observed and the overall appearance was very similar to the obtained by a single pressing and sintering, as can be seen by comparing Figs. 14 and 10. Hot rolling without resintering produced a slight refinement of the grain size and a more irregular appearance of the grains (Fig. 15). Resintering, following hot rolling, had the effect of increasing the grain size and spheroidizing the pores (Fig. 16). Similar structures to this also resulted from repressing and resintering (Fig. 17).

A few metallographic observarions were also made on tested fatigue specimens to investigated the relation of the fatigue cracks to the structure. The cracks appeared to grow from pore to pore, the crack propagation between the pores being intergranular, as in places in Fig. 18, or transgranular, as in Fig. 13.

5. Location of Fatigue Damage

In sections, no evidence was found of internal crack initiation from the surfaces of pores, although clearly there must be a probability of this happening in porous specimens.

To investigate this point further, experiments were made with the A17 specimens, tested at a stress that would normally result in fracture after 100,000 cycles. Instead of being run straight to fracture the tests were interrupted every 50,000 cycles and the specimen diameter reduced by 0.002 in, and the tests then continued at the original stress. It was thought that this treatment would increase the overall life of the specimens, if the fatigue damage was confined to the surface layers and not present at interior pore surfaces. Three specimens treated in this way had lives of 120,000, 195,000, and 225,000 cycles, so that the trend of the results was in the anticipated direction. However, in view of the inherent scatter in fatigue tests on porous specimens, more results are required to establish whether significantly increased lives can be obtained in this way, as with wrought metals, where it has been established that fatigue normally starts from the external free surface [4].

Discussion

The results in Figs. 5 and 6 show that the residual porosity is the most important variable governing the fatigue strength, and the rate of fall in fatigue strength with increasing porosity appears to be constant over the range of porosities investigated. Sintering temperature, time, and nitrogen or vacuum environment are of secondary importance, despite the changes which these variables may introduce in the grain size or pore shape. Earlier results on copper [5] showed a similar trend, but a somewhat greater influence of grain size on fatigue strength.

The method by which a specimen is prepared does not have any appreciable effect, as can be seen by comparing the fatigue strengths of once-pressed and sintered specimens with those of specimens densified by cold rolling, hot rolling, or by repressing and resintering. There is a tendency for the cold-rolling treatments to give slightly inferior properties, possibly because of cracking during the rolling; the two results relating to hot rolling could not be specified accurately owing to a larger than average amount of scatter.

There are no previous results with which to compare the push/pull fatigue strengths, but the present values agree quite well with the bending fatigue strengths determined by previous investigators, being slightly lower than those obtained by Goetzel and Seelig, who used electrolytic powders, and slightly higher than those of Hempel and Wiemer, who used disintegrated and reduced powders.

The presence of residual porosity does not have any sudden effect on the fatigue strength, but reduces it gradually in a similar way to the effect on tensile strength. This implies that the effect of the pores is simply to reduce the cross-sectional area of metal through which the crack has to grow, with the stress-concentration effects of the pores being negligible during growth in relation to the stress-concentration effects of the fatigue crack itself. It is possible that nucleation of cracks on a free surface may be made easier by the presence of the pores, but the effect cannot be large, otherwise the fatigue strength obtained by extrapolating to the density of solid iron would be lower than that of the fully dense material. With regard to the possibility of internal-crack nucleation, the total interior surface area of the pores is much greater than that of the free surface of the specimen and in addition there will be some stress-concentration effects associated with the pores, both factors tending to promote internal-crack nucleation. However, the length of crack that could be nucleated on the surface of a pore would be much less than that on the free surface and this would probably render the initiation process more difficult. In addition, cracks forming and growing on the surfaces of closed pores would do so under different environmental conditions and their growth rate would probably be lower than that at a free surface in the presence of oxygen, water vapor, and other gases in the atmosphere.

The ratio of the fatigue strengths in torsion and push/pull changes with density, the mean values at densities of 6.0, 6.5, 7.0, and 7.5 g/cc being 0.9, 0.79, 0.78, and 0.77, respectively. Hempel and Wiemer obtained an overall value of 0.58 for the ratio between the torsion and bending-fatigue-strength values, the individual values at densities of 6.2, 6.65, and 6.95 g/cc being 0.5, 0.53, and 0.7. Values of this ratio for wrought steels [6] usually lie in the range 0.52-0.69, and for cast irons [6] 0.79-1.01. The pores in the present specimens thus appear to be acting in a similar manner to the "holes" present in cast iron, due to the graphite flakes. Cox [7] has shown theoretically that a material having randomly distributed holes should show a ratio of between 0.75 and 1, depending upon the shape of the holes, so that the present results are consistent with this analysis.

The Young's modulus values agree with those plotted in Fig. 12 of McAdam's paper [8], in which the modulus change with porosity is represented by a smooth curve. However, the rate of decrease of modulus with porosity is much greater up to 30% than beyond. In this former range the behavior is approximately linear, and appears to be unaffected by changes in pore shape and size. The results in Fig. 3 show that a similar pattern of behavior is found with rigidity modulus, which means that Poisson's ratio is unaltered by increasing porosity.

Acknowledgments

The authors would like to thank Professor G. W. Austin, O.B.E., for his interest and encouragement in connection with this work, and also to acknowledge helpful discussions with Mr. I. Morris of the Ministry of Aviation. They also wish to thank Mr. S. D. Charter for his careful help with the experimental work.

References

1.   C. G. Goetzel and R. P. Seelig, Trans. Amer. Soc. Metals, 40:747 (1940).
2.   M. Hempel and H. Wiemer, Arch. Metallkunde, 3:11 (1949).

3.   H. J. Gough, Fatigue of Metals, Scott, Greenwood and Son, London (1924).

4.   G. F. Modlen and G. C. Smith, J. Iron Steel Inst., 194:459 (1960).

5.   O. J. Dunmore and G. C. Smith, Symposium on Powder Metallurgy, 1954 (Special Rep. No. 58), p. 209 (Iron Steel Inst.), London (1956).

6.   P. G. Forrest, Fatigue of Metals, p. 110, Pergamon Press, London (1963).

7.   H. L. Cox, Aeronaut. Research Council Monograph R. and M., No. 2704, p. 112 (1953).

8.   G. D. McAdam, J. Iron Steel Inst., 168:346 (1951).

The Temperature Dependence of the Mechanical Properties of Porous Iron

N.A. Filatova

Institute of Powder Metallurgy and Special Alloys
Academy of Sciences of the Ukrainian SSR

Cermet components (filters, bearings, etc.) can operate at high temperatures. It is, therefore, important to know how temperature affects the mechanical properties of materials with varying porosity.

It is a well-known fact that the mechanical properties of ordinary nonporous metals depend, to a large extent, on the temperature. In [1-3] data were given on the strength of carbon steels. It will be interesting to compare these data with the strength characteristics of porous cermet iron.

The following conclusions can be drawn from an analysis of [1-3]: (a) The general character of the curves showing the temperature dependence of strength for steels containing various amounts of carbon is the same, (b) the strength minimum on the curves lies at a temperature of 100°C; the maximum is at 300°C; with increase in the carbon content they are displaced toward lower temperatures, and (c) the strength of steel is the same at the recrystallization temperature and at 20°C.

Table 1. Strength and Plastic Properties of Porous Iron at High Temperatures

Test temperature, °C	Obtained characteristics	For commercial iron from data of [3]	For porous sintered material Porosity of specimen, % (± 1%)						
			3	5	11	19	30	38	45
20	σ_{B_1}, kg/mm²	43.2	34.0	29.1	21.5	14.9	7.9	4.8	3.2
	δ, %	25.0	30.5	–	9.3	7.0	–	3.3	–
100	σ_{B_2}, kg/mm²	42.36	31.7	27.0	20.9	13.9	7.6	4.2	2.9
	$\sigma_{B_2} : \sigma_{B_1}$	0.98	0.99	0.93	0.92	0.93	0.96	0.87	0.9
	δ, %	–	16.7	–	8.5	5.0	–	2.1	–
200	σ_{B_3}, kg/mm²	46.4	38.2	32.2	24.4	16.3	9.0	5.0	–
	$\sigma_{B_3} : \sigma_{B_1}$	1.08	1.13	1.10	1.14	1.09	1.14	1.05	*
	δ, %	2?.5	17.3	–	5.0	4.4	–	2.4	–
300	σ_{B_4}, kg/mm²	47.9	28.2	23.0	15.2	8.1	5.9	3.4	1.8
	$\sigma_{B_4} : \sigma_{B_1}$	1.12	0.82	0.79	0.71	0.55	0.75	0.71	0.5
	δ, %	–	55.0	–	16.7	9.7	–	3.1	–
400	σ_{B_5}, kg/mm²	39.8	21.6	17.2	10.7	6.6	4.2	3.0	1.0
	$\sigma_{B_5} : \sigma_{B_1}$	0.92	0.63	0.59	0.50	0.44	0.53	0.62	0.3
	δ, %	45.0	47.1	–	14.4	6.8	–	4.6	–

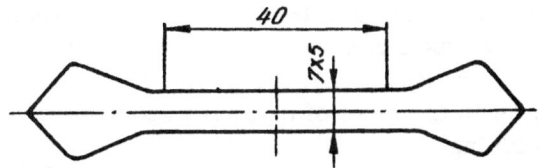

Fig. 1. Specimen for tensile test.

We performed the following investigations to study the mechanical properties of cermet iron at high temperatures.

The properties of the porous material during tension were determined with flat specimens (Fig. 1) prepared from reduced powder having the following content of impurities: iron in oxides, 3.2%; C, 0.05%; S, 0.006%; P, 0.02%; Mn, 0.030%; and Si, 0.05%.

During the test, the specimens were heated in a tubular furnace. The furnace, consisting of two halves, was suspended vertically so that the specimen, clamped by bolts, was in the middle part of the furnace space. The specimen was placed in the furnace heated to the temperature of the experiment. On reaching a given temperature, the specimen was heated for 10 to 15 min.

The results of the investigation are given in Table 1 and in Fig. 2. As can be seen from Fig. 2, the curves for the temperature dependence of the strength for specimens with porosity from 0 to 30%, and for the nonporous metal, have the same shape.

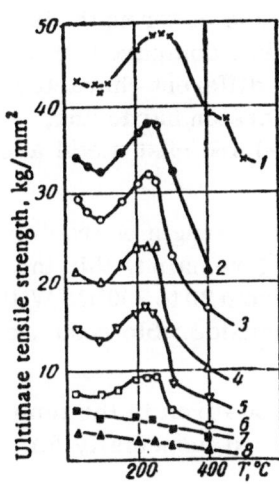

Fig. 2. Dependence of the ultimate tensile strength on test temperature: 1) from [2, 3]; 2) sintered iron with 3% porosity; 3) sintered iron with 5% porosity; 4) the same with 11% porosity; 5) the same with 19% porosity; 6) the same with 30% porosity; 7) the same with 38% porosity; 8) the same with 45% porosity.

With increase in the test temperature from room temperature to 100°C, there was a reduction in strength. With further increase in temperature (to 250°C), the strength increased; this may have been due, as in the case of ordinary steel, to processes of dispersion hardening.

Iron powder contains a number of elements which have variable solubility in alpha-iron. These are carbon, oxygen, nitrogen, phosphorus, etc. However, in this case, the strengthening will be affected by carbon, oxygen, and nitrogen since the amounts of the other impurities are much less than their limiting solubility at room temperature. The effect of increased temperature is not limited to these factors. Deformation during the test leads to work hardening, strengthening of the metal; recovery and recrystallization lead to softening of the material and an increase in its plasticity. Depending on which of these processes predominates, there is hardening or softening of the material.

There are no minimum or maximum sections on the strength-temperature curves for specimens with a porosity greater than 30%. With increase in the test temperature the strength of the specimens is uniformly reduced.

The plasticity, characterized by the relative elongation (see Table 1), is reduced in the range of temperatures of dispersion hardening. This reduction is expressed most strongly in specimens with low porosity (3÷11%). This behavior of porous iron is similar to that of nonporous soft steel produced by casting and rolling. The plasticity minimum, as in the case of strength, is somewhat displaced toward lower temperatures compared with nonporous steel. This is possibly due to the different character of impurities which separate out during dispersion hardening, since the methods for preparing the powdered and cast metal are different.

The experimental results show that the strength characteristics of iron with porosity between 0 and 30% remain within the limits of strength existing at temperatures from 20 to 300°C. With further increase in temperature there is a considerable reduction in strength.

The characteristics of plasticity of porous iron in the temperature range of dispersion hardening are reduced, especially for low values of porosity.

These features in the properties of porous iron should be taken into account when choosing a material for high-temperature operation.

Conclusions

It has been shown that the general character of change in strength and plasticity of porous iron between room temperature and 400°C is similar to the change in these characteristics for ordinary nonporous iron. The analogy in the change in properties is more clearly expressed in specimens with low values of porosity (3-30% for the ultimate strength, and 3-19% for the relative elongation). For higher values of porosity the properties change monotonically.

References

1.   T. I. Volkova, Vestn. Mashinostr., No. 4, 1951, p. 132.
2.   M. A. Zaikov, Zh. Tekh. Fiz., 19 (6):684 (1949).
3.   M. A. Zaikov, Zh. Tekh. Fiz., 18 (6):847 (1948).

Bibliography on Methods for the Production of Iron Powders

Henry H. Hausner

and

Helen C. Wenk

Introduction

There are many methods for the production of iron powders
for commercial purposes. Table I lists only the four main types
of commercially available iron powders. Actually there are many
more methods than those for the four mentioned in the table; modifications of these methods and others offer interesting aspects for
special purposes, as shown in Chapters 1 and 2.

The following bibliography with 142 references on the production of iron powders is far from complete, but a carefully selected one which permits one to obtain some ideas about the many possibilities for producing iron powder.

One has to bear in mind that the characteristics of the powder always depend on the manufacturing process. The term
"characteristics of a powder particle" is rather complex, and refers to many factors, which are listed in Table II.

TABLE I.   Basic Types of Iron Powders

1. Reduced powder
2. Electrolytic powder
3. Atomized powder
4. Carbonyl powder

TABLE II. Characteristics of a Powder Particle

1. Particle size (particle diameter)
2. Particle shape
3. Density (porosity)
4. Surface conditions
5. Microstructure (crystal grain structure)
6. Type and amount of lattice defects
7. Gas content within a particle
8. Adsorbed gas layer
9. Amount of surface oxide
10. Reactivity

TABLE III. Characteristics of a Mass of Powder

1. Particle characteristics
2. Average particle size
3. Particle-size distribution
4. Specific surface (surface area per 1 gram)
5. Apparent density
6. Tap density
7. Flow of the powder
8. Friction conditions between the particles
9. Compressibility (compactability)

Table III shows the characteristics of a mass of particles or powder.

How much the fabrication process affects the particle shape and the approximate surface area of a particle is shown in Table IV.

The effect of particle size and particle-size distribution on apparent density and flow of a powder is shown in Table V.

Of the following references, 60% are from the years 1950–1960; 30% are prior to 1950, and only 10% since 1960. Ninety-two references pertain to the production of iron powder by various re-

Table IV. Approximate Surface Area of Powder Particles
Fabricated by Various Methods

Process of Powder Production	Particle Shape	Approx. Surface Area
Carbonyl process	uniform spherical	D^2
Atomization	round irregular spheroids	$1.5\text{-}2\ D^2$
Reduction of oxides	Irregular spongy	$7\text{-}12\ D^2$
Electrolytic process	dendritic	$7\text{-}12\ D^2$
Mechanical comminution (crushing)	angular	$3\text{-}4\ D^2$
Mechanical comminution (ball milling)	flakes or leaves	varies over wide range

Table V. Apparent Densities and Flow Rate of
Electrolytic Iron Powder of Three Different
Particle-Size Distributions

Particle size, mesh	Percent particles		
	Powder A	Powder B	Powder C
+100	4	3	15
-100 +150	11	26	10
-150 +200	18	18	30
-200 +250	16	6	25
-250 +325	18	16	5
-325	33	31	15
App. density, g/cc	2.6-2.8	3.2-3.4	3.8-3.9
Flow rate, sec	29	24	20

duction processes, and twenty-two to the production by electrolysis
the others refer to other production methods.

1.   Anon., "Iron Powder from Low Grade Ore," J. Metals 16 (8):
       617(Aug. 1964).

       High purity Fe powder from low grade (30-35% Fe) ores in
       northwestern Alberta, Canada. Ore first roasted with coal
       in reducing atmosphere to convert ferric oxide and other

Fe-bearing materials to metal and ferrous oxide; roasted material dissolved in HCl and filtered to remove insoluble impurities. Hydrated ferrous chloride is crystallized out, dried, and reduced in H_2 atmosphere at elevated temperatures to produce nonpyrophoric Fe powder (99% purity) and hydrogen chloride.

2.   Anon., "Manufacture of Electrolytic Iron Powder," Deutsche Gold- und Silber-Scheideanstalt., F.D. Rep. No. 1671/48 (1948). Account in French (1945) of working of a small $FeCl_2$ bath.

3.   Anon., "Manufacture of Iron Powder," Am. Machinist 97(15): 164(1953); see also Iron Age 172(1):74(1953). Republic Steel Corp. production.

4.   Anon., "A New Electrolytic Iron Powder," Materials and Methods 32(3):136, 138(1950).

New process for electrolytic iron powder useful for core manufacture. Iron content approx. 99.8%, and apparent density between 1.6 and 2.8.

5.   Anon., "New Iron Powder," Materials and Methods 40(1): 152-3(1954); see also J. Electrochem. Soc. 105(7):413-17 (1958).

Plastic Metals Div. of National Radiator Co.'s reduced-oxide Fe powder, and electrolytic powder, resp.

6.   Anon., "Reduced Iron Powder," Metallic Materials 3(10):76-80(1963). (Japanese)

Description and discussion of direct reduction process, electrolysis, mechanical crushing, atomization, carbonyl, Höganäs, and other methods.

7.   Anon., "Report on First Meeting for Production of Iron Powders," Reichsminister f. Rüstung und Kriegsproduktion, U.S. Dept. Comm. PB 17391(1944).

Lectures and discussions on mechanically produced powders, sponge iron powders, centrifugally produced powders, atomized powders; the Berghaus chlorination process; the Walther process, the principles of the raw iron-cinder layer process.

8. Anon., "Pickle Liquor May Yield Cheaper Iron Powder," Iron Age 167(18):112(1951).

Pickle liquor treated with Al shavings. High-quality fine Fe powder and $Al_2(SO_4)_3$ result.

9. Anon., "Production of Iron Powder in Tunnel Kiln," Deut. Pulvermet. Ges., U.S. Dept. Comm. PB 22344 (1945).

Brief description of history of production of Fe powder, starting with Hametag process, followed with description of more efficient and economical methods used in Germany.

10. Anon., "Sponge Iron in Japan," U.S. Bur. Mines Report I.C. 7440 (1948).

Rotary kiln method, using mixture of 100 parts Fe ore, 56 parts anthracite, 10 parts limestone; reaction yields pellets of Fe mixed with slag. In the Kikuchi method a high frequency current passes through mixture of Fe ore, coal, and limestone.

11. N. N. Alekseev, "Preparation of Iron Powder," Zavodsk. Lab. 16:256(1950).

Fe_2O_3 is mixed with $1/2$ of its weight of transformer oil in a covered crucible in a muffle for 2 to $2^1/_2$ h to 750–800°C. After discarding the top layer, an excellent grade of Fe powder is obtained.

12. A. H. Allen, "Pure Iron Powder," Steel 104(15):43–54 (1939).

Five different types of reduction, including low temperature reduction of iron oxides.

13. F. K. Andryushchenko et al., "Kinetics of the Reduction of Iron Oxide in the Presence of Iron Powder," Vysshikh Ucheb. Zavedenii, Khim. i. Khim. Tekhnol. 2:219–24 (1959).

Reduction of Fe oxide with use of Fe powder in concentrations below stoichiometric requirements of equation $Fe + Fe_2O_3 = 3Fe_3O_4$, goes as far as Fe_3O_4 at temps. well over 570°C. Accumulation of lower Fe oxide occurs if more than stoichiometric amount of Fe powder is used.

14.   V. I. Arkharov et al., "Reduction of Iron Oxides by Graphite," Zh. Fiz. Khim. 29:272-9(1955).

FeO, Fe_2O_3, and Fe_3O_4 reduced by graphite in a SiO_2 tube at 1100-1150°C at a pressure of approx. 0.001 mm Hg.

15.   K. Asai, "Bench Scale Manufacture of Iron Pentacarbonyl and Tests on Some of the Properties of Iron Pentacarbonyl and Carbonyl Pure Iron," Bull. Chem. Research Inst. Non-aqueous Solutions, Tohoku Univ. 4(1):17-68(1954).

Manufacture of $Fe(CO)_5$ with CO and pilot plant runs with water gas.

16.   H. Asamura and K. Kobayashi, "Powder Metallurgy, I. Production of Iron Powder from Mill Scale by the Reduction with Commercial Carbon Monoxide-Hydrogen-Nitrogen System Gas," Osaka Furitsu Kogyo-Shoreikan Hokoku 7(2): 1-4(1955).

Reduction of -200 mesh Fe mill scale by heating at 900°C in CO-H-N. Rate of reduction found to be between rates of H and CO. Excess gas flow does not accelerate reduction velocity. Principal reduction reaction progresses between H and FeO at < 800°C; reduction by CO is promoted at > 900°C.

17.   J. Astier, "Low Temperature Reduction of Iron Ores and Enrichment of Sponge Iron," Inst. Hierro Acero 9:109-17 (1956).

Experiments on French ores from Lorraine, Normandy, and the Pyrenees, to obtain enriched Fe sponge. This appears to be a little easier than enrichment of the ore, at least when temp. of reduction does not exceed 1200-1300°C.

18.   E. P. Barrett, "Iron Sponge Production," U.S. Bur. Mines. Bull. No. 519 (1954).

Difference methods of producing sponge Fe in the U.S. and abroad Metallic Fe and O contents. Economic considerations for use of sponge Fe, compared to those for scrap Fe in various steel production methods,

19.   E. P. Barrett and C. E. Wood, "Production of Sponge Iron," U.S. Bur. Mines R. I. 4305 (1948).

       Gaseous reduction of Fe oxide glomerules in shaft furnace. Reduction could not be completed in hot H_2 unless the column was heated externally.

20.   E. P. Barrett et al.," Investigation of Bubble Hearth Process for Production of Sponge Iron," U.S. Bur. Mines R. I. 4092 (1947).

       Reduction furnace and experimental work described.

21.   A. Boulle et al., "The Electrolytic Reduction of Iron Oxide," Congr. Intern. Chim. Pure Appl. 16 (Paris, 1957), Mem. Sect. Chim. Minerale, 91-6 (1958).

       Only magnetic oxide Fe_3O_4 can be reduced by atomic H during electrolysis. Method of preparation and types of impurities are improtant. Rate of reduction, but not yield, depends on particle size.

22.   V. Ya. Bulanov et al., "Process for Obtaining Iron Powder from Alloy Scale by Reducing with Converted Natural Gas," Poroshkovaya Met. 5(10):1-4(1965).

       Most satisfactory results with scale from rolled, pre-alloyed steel, reduced for 60 min at 1100 and 1000°C in converted natural gas without additions.

23.   E. H. Carman, "Method of Preparing Iron Powder for Permanent Magnets," Metallurgia 52(312):165-8(1955).

       Fe oxides reduced in H at temps. between 250-350°C. Apparatus described. Relation between reduction time and temp., and between particle size and reduction temp. given. Smallest Fe particle size obtainable is size of oxide particles employed.

24.   P. E. Cavanaugh, "Sponge Iron Production in Tunnel Furnace,"
      Inst. Hierro Acero 4:93-7(1951); see also J. Can. Ceram.
      Soc. 19:62-7(1950); see also Brit. Clayworker 60:238-9,
      274-8, 310-14, 345-6 (1951); ibid. 61:30-33(1952).

      Furnaces of the Milton type used to made bricks were adapted
      to produce sponge Fe.

25.   P. E. Cavanaugh, "Commercial Production of Sponge Iron,"
      Iron Age 163(22):67-71, 82 (1949).

      Test runs conducted in Sweden, employing Wiberg-Soderfors
      process have indicated excellent performamce of Canadian
      Steep Rock ore and high production rates.

26.   M. Clasing and F. Sauerwald, "Electrowinning of Heavy Met-
      als from Fused Electrolytes. IV. Electrolysis of the Fused
      Ferric Chloride–Sodium Chloride System, " Z. Electrochem.
      54:358-61(1950).

      Electrodeposition of Fe in powder form from fused $FeCl_3$-
      NaCl melt at temp. above 158°C on rotating cathode in an H_2
      tube with Fe anodes.

27.   T. P. Colclough, "I. G. Farbenindustrie, Ludwigshafen-
      Oppau," B.I.O.S. Rept. XXIV-12 (1946); see also Metal
      Progr. 50(2):332, 342 (1946).

      Carbonyl iron manufacture at I. C. Farbenindustrie carbonyl
      nickel plant was later incorporated in B. I. O. S. Final Report
      263.

28.   C. S. Cronan, "Direct Reduction of Low Grade Iron Ore,"
      Chem. Eng. 69(5):62-4(Mar. 1962).

      Reduction of low-grade magnetite ore by Arkota Steel Corp.
      at Coolidge, Ariz. Uses magnetic beneficiation and reduc-
      tion by reformed natural gas to convert ore of 7% Fe content
      into "ultraclean" sponge iron which is then pulverized or
      arc melted.

29.   Lü Da-Min, "Production of Iron Powder by Atomization of
      Cast Iron," Gangtie No. 20, 964-71 (1959); Ref. Zh. Met. No.
      10, 23201 (1960).

Air or water atomization of molten cast Fe produces Fe
powder containing 97.5-98.5% Fe and 0.060-0.15% C. Air
atomization gives better results, but requires more complex equipment.

30.   G. I. Damskaya et al., "Direct Production of Iron Powder
       from Pyrite Cinders in a Fluidized Bed," Tsvetn. Metal.
       No. 4, 28-33 (1960).

       Laboratory preparation of Fe powder from cinders resulting
       from roasting pyrite concentrates in fluidized bed furnace.
       Fluidization of dust itself for reduction purposes required
       granulation.

31.   G. Dan and O. Ivanciu, "Electrolytic Production on a Mercury
       Cathode of Iron Powder for Magnetic Applications," Studii
       Cercetari Met. 4(1):111-20(1960).

       Cathode = Hg layer in an electrolyte, anode = stainless steel
       in form of annulus or perforated plate; current density = 3.4-
       10 A/dm^2, cell boundary = 2.25-4 V, bath temp. = 20.70°C, time
       of electrolysis 2 h.

32.   I. A. Deryugin et al., "Electron Microscope Investigations
       of the Cathode Deposits of Dispersed Iron," Nauk Zap.,
       Kiivs'k Derzh. Univ. 14(8); Zbornik Fiz. Fak. No. 7, 223-30
       (1955).

       Ice-cooled two-layer electrolytic bath deposits highly dis-
       persed Fe powders on cathode. Anode immersed in lower
       layer, FeCl$_3$ solution, and is separated from cathode by a
       porous diaphragm. Rotating cathode is placed above anode.
       Deposited Fe particles are continuously washed off, and sur-
       face of cathode is continuously covered with a surface-ac-
       tivation adsorption layer, causing a strong polarization of
       cathode. High rotation speed of cathode and low temp. of
       bath promote formation of finer particles.

33.   T. D. DeSouza-Santos, "Production of Sponge Iron in Iron
       Receptacles Containing 26% Chromium," ABM (Bol. Assoc.
       Brasil. Metais) (Sao Paulo) 13(47):67-128(1957).

Effect of variables on reduction of Fe ore and subsequent carbonizing of sponge Fe. Dimensional variations of sponge Fe as function of temp., and inflence of the mold on state of surface of sponge-Fe cylinders.

34. E. Diepschlag, "The Reduction of Iron Ore under High Pressure," Arch. Eisenhüttenw. 10:179-81(1936).

Effect of high pressure on rate of reduction.

35. A. Domsa, "Kinetics of the Direct Reduction of Ferrous Oxides with Methane Gas," in "Modern Developments in Powder Metallurgy," Plenum Press (1966).

By introducing methane directly in reaction space, both reduction of oxides into metallic Fe and continual regeneration of reducing agents can be achieved.

36. A. Domsa and A. Palfalvi, "Obtaining Frem-Type Iron Powder," Acad. Rep. Populare Romine, Studii Cercetari Met. 4: 59-72(1959).

Frem-type powder (Fe reduced by CH_4) is produced from rolling mill scale by drying, sifting, and grinding.

37. B. T. DuPont, "Electrolytic Iron Powders," Presented at First Annual Meeting Metal Powder Industries Federation, May 1944; abstracted in Iron Age 153(22):55(1944).

38. B. T. DuPont and R. Fulton, "Five Ways to Make Iron Powder," Iron Age 169(17):135-9(1952).

Description of techniques of production and equipment for manufacture of Fe powder by electrolysis, reduction, and other methods.

39. E. D. Eastman, "Reduction of Iron Oxides by Fuel Gases," U.S. Bur. Mines R. I. 2485 (1923).

Factors of reduction form two classes: those which affect equilibrium between reacting substances and products; and those which affect rate of reaction.

40.   J. O. Edström, "The Mechanism of Reduction of Iron Oxides,"
       J. Iron Steel Inst. (London) 175:289-304(1953).

       Pure, natural, single crystals of hematite and magnetite
       reduced by CO between 800-1000°C. Reduction by H_2 between
       450-1000°C also studied.

41.   J. O. Edström, "Reduction Structures of Iron Powder,"
       Jernkontorets Ann. 140(2):116-29(1956).

       Reduction of Fe_2O_3 by CO and H complete at 1000°C. Reduc-
       tion of Fe_3O_4 not complete at about 1000°C, but can be reduced
       by CO and by H containing CO. At 500-600°C, both Fe_2O_3 and
       Fe_3O_4 concentrates are reduced by pure H. Reduction prod-
       ucts of Fe_2O_3 have greater porosity than those of Fe_3O_4.
       Number and size of pores in resulting iron powder can be
       adjusted by selection of suitable reduction temp. and gas
       composition.

42.   E. Edwin, "Gaseous Reduction of Iron Oxides," Tek. Tidskr.
       Bergvetensk 56:41-51(1926); Tek. Ukeblad 73:225-35(1926).

       The "Norsk-Staal" process.

43.   F. Eisenkolb, "Progress in Powder Metallurgy of Iron
       Materials," Forsch. Fortschr. 28(6):164-71(1954).

       New iron powder production methods, new investigations of
       mechanism of sintering, and of potential improvements of
       sintered steel.

44.   F. Eisenkolb and G. Ehrlich, "Investigation of Chemically
       Prepared Iron and Iron Nickel Powders," Monatsber. Deut.
       Akad. Wiss. Berlin 1:12-20(Jan. 1959).

       Preparation of metal and oxide powders by decomposition
       of Fe and FeNi salts (oxalate, formate, carbonate, nitrate,
       and sulfate) in air and nitrogen at 350-800°C, and subsequent
       reduction in hydrogen at 300-400°C. X-ray investigation of
       lattice structure, particle size, compound formation, and
       solubility.

45.   F. Eisenkolb and E. Strobel, "Production of Metal Powders
       for Coatings of Welding Electrodes," Tech.-wissenschaft-

liche Abhandlungen des Zentralinst. Schweisstech. der DDR, No. 13 (1959).

Oxygen content of Fe powder used for welding electrodes must be kept as low as possible, though production of completely O-free Fe powder is not economically feasible. Particle size of powder is important, and size analysis of Höganäs grade W-40 powder and a finer grade is given. In Germany, Fe powder for welding and other purposes is produced mainly by atomization (RZ process). Rotating-disc method is used in Czechoslovakia. In East Germany, Hametag process is used.

46. A. S. Eisurovich, "Permanent Magnets with a Fine Powder Base," Izv. Akad. Nauk SSSR, Otd. Tekhn. Nauk, Met. i Toplivo No. 4, 130-4 (1959).

Fe powder produced by reduction of iron formate in H at approx. 300°C. Coercive force of the Fe powder obtained changes with reduction temp. and time and size of particles. Magnetic properties are also influenced by addition of CaO or Co.

47. S. Eketorp, "Höganäs Sponge Iron Process," Jernkontorets Ann. 129(12):703-10(1945); Iron Steel Inst. (London) Transl. No. 275 (July 1946); Met. Powd. Rept. 1(1):4(1946).

Use of very high grade magnetic concentrate which is packed in ceramic containers with alternate layers of coal, heated to 1200°C in furnace of type used in production of firebrick.

48. S. Eketorp, "Success and Failure of Sponge Iron Methods," Can. Metals, Preprint (Feb. 1950).

Comparison between Wiberg-Soderfors and the Höganäs processes.

49. C. V. Firth, "Iron Powder," Information Circ. No. 3, Univ. Minnesota Mines Expt. Sta., Minneapolis (May 1943).

Production at experimental station based on concentration and reduction of Mesabi Range carbonate slate, a highly siliceous ore.

50.   E. Fornander, "Direct Production of Iron," Chem. Met. Eng.
       30(22):864-5, 907-8(1924).

       Reduction processes of Chenot, Gurlt, Husgafvel, Wiborgh,
       Gröndal, Sieurin, Bourcoud, Berglöf, Wiberg, Basset, de-
       scribed.

51.   A. D. Franklin and R. B. Campbell, "Low Temperature Re-
       duction of Iron Oxides," J. Phys. Chem. 59(1):65-7(1956).

       Fe oxides reduced with CaH_2 at 125-450°C. It was found that
       each oxide crystallite produces one single crystal Fe par-
       ticle, whose diameter is determined by the crystallite rather
       than by the particle diameter of the oxide. With Fe particles
       of < 300 Å, sintering begins at temps. as low as 200°C. In
       reduction process, nucleation of the new phase is much slow-
       er than growth. Only one nucleus is formed per oxide crystal.

52.   I. N. Frantsevich and I. D. Radomselski, "The Technology of
       Iron Powder Production with the Aid of Converted Natural
       Gas," Powder Met. Yaroslav, USSR 1956, pp. 129-52.

53.   H. Freeman, "Direct Iron in Canada," Trans. Can. Inst.
       Mining Met. 58th Ann. Meeting (1956), 326-9; Can. Mining
       Met. Bull. 49(532):566-9(1956).

       Process begun in 1947 at Cap de la Madeleine for manufac-
       turing Fe powder, concerned with direct reduction of Fe
       oxide at temps. below melting point of Fe. High grade pow-
       der is prepared from concentrates, rating 70% Fe.

54.   A. E. Fridenburg et al., "A Fine Carbonyl Iron Powder for
       High Frequency Magnetic Powder Materials," Soviet Powd.
       Met. and Metal Ceramics No. 1, 23-9 (Jan.-Feb. 1963),
       Consultants Bureau, New York, N. Y.

       Method for continuous preparation of very fine fractions of
       carbonyl iron powder, which combines the process of Fe
       pentacarbonyl decomposition and separation of the resultant
       powders by the outlet gases. Powder production with simul-
       taneous extraction of very fine fractions is investigated. It
       is established that fine carbonyl iron powder has improved
       electromagnetic characteristics in a wide frequency range.

55. A. Gallo, "Elimination of Chlorides in Manufacturing Electrolytic Iron Powder," Met. Ital. 54(3):105-7(Mar. 1962).

Difficulties in eliminating chlorides present in electrolytic iron scale; after attempting wet treatment of scale, final method adopted consisted in distillation of chlorides in reducing atmosphere.

56. G. E. Gardam, "Production of Iron Powder by Electrodeposition," in "Symposium on Powder Metallurgy, Iron Steel Inst. Spec. Report No. 38 1947, p. 1-7.

Electrolysis of 10% ferrous ammonium sulfate solution at 35°C, pH of 2.5, with half current being fed through iron anodes and half through lead anodes in porous pots, at 200 A/ft^2. Cathode scraped off every two h, collected sludge washed with sulfuric acid, and then with citric acid. Surrounding liquid made ammoniacal and powder washed, drained, and rapidly dried under reduced pressure or in H. Powder with O content of about 1% is obtained. Annealing of powder in H at from 600-700°C for 1 h improved properties of compact.

57. G. E. Gardam, "Production of Iron Powder by Electroposition," Selected Govt. Research Repts. (Gt. Brit.), Powder Met. 9:75-84(1951).

Electrolysis of a 10% ammonium ferrosulfate solution.

58. V. H. Gottschalk, "Making Iron Powder in the Tunnel Kiln," U. S. Bur. Mines I. C. 7473 (Aug. 1948).

Report based on four German reports translated from FIAT microfilm, and represents production at Metallgesellschaft, Inc.

59. M. Guedras, "Production of Iron Powder by Reduction," Metallurgie (Paris) 86(12):913-4(1954).

Fe_2O_3 prepared by precipitation from salt solution and reduced, resulted in purity similar to that of electrolytic Fe, but its sintering properties were not investigated.

60.   Z. Hara, "Reduction Rate of Iron Ore in a Fluidized Bed,"
       Seisan-Kenkyu 10(10):24(1958).

       Rate of reduction of hematite ore and iron mill scale pow-
       der by H.

61.   Z. Hara, "Formula for the Reducibility of Iron Ore," Seisan-
       Kenkyu 8(5):22(1956).

       Factors included such as degree of oxidation, particle size,
       and porosity of iron ore. (Formula deduced from experi-
       mental data).

62.   Z. Hara and S. Fukutake, "Electrolytic Deposition of Iron
       Powder," Japan Inst. of Metals, Spec. Rep. No. 12, 101
       (1955).

       Local increase in pH value around the cathode causes precip-
       itation of $Fe(OH)_2$. High current density and high electrolyte
       concentration are beneficial to homogeneous Fe-particle
       size.

63.   J. Hui, "Reactivity of Iron Powders Prepared by Reduction
       of Ferric Oxide," Compt. Rend. 252(9):1325-7(Feb. 27, 1961).

       Refers to the influence of nature of anion of ferric salt used
       for preparation of oxide. Anion eliminated by washing pre-
       cipitate, which is then dried at 80°C for 48 h, ground, and
       sieved. Only fraction < 170 μ used in reduction experiments.

64.   E. Iwanciw, "Mechanism and Kinetics of Reduction of Iron
       Oxides," Arch. Hutnictwa 4(2):122-59(1959); H. Brutcher
       Transl. No. 4820.

       Pure Fe_2O_3 in form of sintered briquettes and of magnetite
       in form of natural crystal aggregates reduced by $CO + CO_2$
       in presence of C.

65.   J. Jackson, "Iron by Direct Reduction," Industrial Chemist
       37(441):537-41(Nov. 1961).

Processes to produce pure iron direct from ore without going through intermediate stage of impure product "pig iron"; H-iron process: iron ore fines reduced in fluidized bed using H obtained from natural gas. Iron obtained as powder which is pyrophoric. Nu-Iron process: fluidization also used. Powder is briquetted. Other processes described are Esso-Little, HyL, R-N, Strategic-Udy, and Cyclosteel.

66.　F. Jaeger and H. Winterhoff, "Production of Sponge Iron by Direct Reduction: HyL Process," Stahl Eisen 82(5):290-3 (Mar. 1962).

Process of Hofalata y Lamina S. A., Monterrey, Mexico. Partly magnetic hematite ore reduced by H and CO obtained by catalytic cracking of mixture of natural gas and steam. Flow diagram of reduction process given.

67.　G. Jangg et al., "Recovery of Iron Powder Through Low Temperature Heating of Iron Amalgam," Z. Metallk. 50(8): 460-5(1959).

Continuous process to produce larger amounts of Fe residue of uniform quality. Low temps. gave porous, easily pulverized, Hg-free Fe powders. When temp. was increased to 600°C, products were less porous and rather hard.

68.　O. Jensen and E. B. Louxs, "Procedure for the Fabrication of Sponge Iron by the Reduction of Concentrated Agglomerated Ores with Hydrogen or Water Gas," Compt. Rend. Congr. Intern. Chim. Ind., 27th Congr., Brussels Vol. 2 (1954); Ind. Chim. Belge 20, Spec. No. 386-9 (1955).

Sponge Fe production using H or water gas as reducing agent. Reduction of ore is endothermic. Heat supplied to process by electrical heating; sponge serves as electric resistance. Degree of reduction is about 95%.

69.　T. L. Johnson, "Producing Sponge Iron in Rotary Kiln," Steel 117(20):128-9, 140, 142(1945).

Pilot plant operated by Bureau of Mines employs coal to reduce iron in ore to metallic form.

70.   B. M. S. Kalling and F. Johansson, "Reduction of Iron Ore in Rotary Furnace without Melting," Jernkontorets Ann. 138: 253-70(1954); J. Iron Steel Inst. (London) 177:76-85(1954).

Reduction at pilot plant at Domnarfvet Steel Works, Sweden. Coke breeze used as reduction agent. 95% reduction easily obtained.

71.   B. M. S. Kalling and I. Rennerfelt, "Decarburization of Granulated Pig Iron: The 'R. K. Process'," Jernkontorets Ann. 123:115-54(1939); Iron and Coal Trades Rev. 89:359 (1939).

Granulated Fe is continuously fed in a rotary furnace and decarburized without melting in a gaseous mixture of CO and CO_2 in such proportions that no surface oxidation takes place.

72.   B. M. S. Kalling and J. Stalhed, "Manufacture of Sponge Iron," Steel 125(12):72-5, 102, 106(1949).

Wiberg-Soderfors process described. Requires less reducing agent than other reduction processes.

73.   M. R. Kalyanram, "The Production of Iron Powder by Gaseous Reduction," J. Met. Club, Royal College of Sci. and Tech. (Glasgow) No. 11, 39-44 (1958).

High purity powdered hematite reduced by CO. Reduction temp., time, gas-flow rate, and particle size studied.

74.   R. M. Khandwala, "Production of Iron Powder from Indian Iron Ore and its Properties," Bombay Technologist 3:82-3 (1953).

Batch reduction carried out with H in electric furnace. Continuous reduction carried out in rotary kiln.

75.   R. M. Khandwala and G. S. Tendolkar, "Studies in Reduction of Powdered Hematite. I. Batch Reduction. II. Continuous Reduction in a Rotary Kiln," J. Sci. Ind. Research (India) 13B(8):561-71(1954).

Effect of operating variables on rate of reducing high purity hematite in H. Rate proportional to rate of supply of H.

76. O. Knacke, "The Reduction of the Iron Oxides, in Particular of Wüstite," Arch. Eisenhüttenw. 30:581-4(1959).

Sequence of stages of reduction: hematite, magnetite, wüstite, iron. Expression found for heterogeneous reaction velocity.

77. V. F. Knyazev et al., "Production of Iron Powder at the Sulin Metal Works," Powder Metallurgy, Yaroslav (USSR) (1956) pp. 153-8.

Process resembling Höganäs process.

78. E. J. Kohlmeyer and H. Spandau, "Ultrafine Iron Powder," Arch. Eisenhüttenw. 18:1-6(1944/45).

Vaporization of Fe by air or O blast on C-rich molten metal, or by adaptation to Fe of method of vaporizing in electric arc.

79. V. V. Kondakov and Z. F. Chukhanov, "The Reduction of Iron Oxides with Gases. The Kinetic Equation of FeO Reduction with Hydrogen," Dokl. Akad. Nauk SSSR 106:697-700(1956).

Stream of reducing gas passed through carefully sized Fe-ore particles; gas flow rate and temp. kept constant throughout test. Solids then cooled in stream of inert gas, and metallic Fe content tested. Equations derived for average relative steam concentration in vapors and for amount of FeO in relation to traction time.

80. R. Koprizhiva, "Iron Powder from Scale," Novosti Neft Tekhn. 5(13/14):25(1940).

Scale ground to 200 mesh and reduced in H at 650-800°C for 1-2 h. Product containing 60-75% Fe is then oxidized by heating to 600-700°C in air; oxidized powder is subjected to a second reduction with H at 800°C for 2-3 h. Final product contains 92-99% metallic Fe.

81.   J. Kubelik, "The Production of Iron Powder," Hutnicke Listy
      13(12):1129-31(1958).

      Blast-cupola method (for powder for sintered machine parts)
      is cheapest one that provides Fe powder with suitable prop-
      erties.

82.   N. T. Kudryavtsev and N. I. Mikhailov, "Electrolytic Pro-
      duction of Finely Divided Iron Powder," J. Appl. Chem.
      USSR 33:1342-6(June 1960), Consultants Bureau, New York,
      N. Y.

      Effect of current density, $FeSO_4$ concentration, and electro-
      lyte pH on current efficiency and dispersity of powder pro-
      duced by electrolysis of a FeO_4 solution saturated with K_2SO_4.

83.   M. S. Kurchatov, "Reduction of Iron Oxide and Iron Ores
      with Different Forms of Solid Carbon when Heating Takes
      Place Stepwise," Compt. Rend. Acad. Bulgare Sci. 9(3):41-4
      (1956).

      Reduction under conditions in which reacting substance were
      well mixed and heated in stages.

84.   M. S. Kurchatov, "Reduction of Iron Oxide and Iron Ores by
      Different Forms of Solid Carbon When the Reacting Sub-
      stances are not in Contact with One Another," ibid., pp. 49-52.

      Reduction of Fe oxide and Fe ores studied with carbon black,
      wood charcoal, coke, and graphite.

85.   M. S. Kurchatov, "The Possibility of Producing Sponge Iron
      in Chamber Furnaces of the Coke Oven Type," Izv. Khim.
      Inst. Bulgar. Akad. Nauk 5:321-58(1957).

      Solid C used for reduction. Ore-coal mixture finely ground
      and compressed into bricks. Rate of heating is fast because
      of high thermal conductivity of mixture. Laboratory experi-
      ments show that industrial application of process is practical.

86.   L. L. Kuzmin and V. L. Kiseleva," Preparation and Proper-
      ties of Electrolytic Iron Powder," J. Appl. Chem. USSR 22:
      311-18(1949).

Experiments with sulfate bath in which effect of pH, temp., and cathode current density were determined with object of producing pure iron powder.

87. J. F. Kuzmick and K. W. Bruland, "Iron Powder Made by H-Iron Process," Metal Progr. 73(3):92-6(1958).

Process developed in laboratory of Hydrocarbon Research Corp. for reduction of magnetite ore at high temp., yielding high grade concentrates of over 72% Fe.

88. C. J. Leadbeater et al., "Some Properties of Engineering Iron Powders," Selected Govt. Research Repts. (Gt. Brit.), Powder Met. 9:33-74(1951).

Test results on 28 different types of iron powders produced by different methods.

89. P. Leasge-Bourdon and A. Michel, "Preparation of Finely Divided Iron and Iron Alloys by Electrolysis with a Mercury Cathode. Application to the Study of Hexagonal Iron Carbide," Compt. Rend. 249(17):1675-7(1959).

A pasty mass of Fe in Hg is formed at the cathode, which can be quickly washed, dried, and kept under H. Hg is eliminated by heating in H, and gives finely divided Fe. When finely divided Fe is heated at 200-250 h at 170°C, with mixture of CO and H, a hexagonal Fe carbide is obtained. Finely divided and homogeneous ferroalloys may be obtained in the same way from an amalgam mixture.

90. I. G. Lesokhin, "Preparation of Powdered Iron from Pyrite Ash in a Fluidized Bed," Zh. Prikl. Khim. 32:60-5(1959).

Degrees of reduction for pyrite ash, with or without S content, given for various gas rates.

91. M. B. Lev et al., "Production of Spherical Iron Powder by Atomization of Molten Metal," Poroshkovaya Met. No. 2, 89-98 (Mar.-Apr. 1964).

Description of equipment and method for industrial-scale production of spherical iron powder; data on effect of burner.

design, carbon content of iron, preliminary annealing, air pressure, and equipment geometry on powder properties; experimental equipment for atomization by water and preliminary operating results.

92.   A. I. Levin and S. A. Pushkareva, "Adsorption Phenomena and Cathode Processes for Iron Electrodeposition in Powdered and Compact Forms," Zh. Fiz. Khim. 31:1983-90(1957).

Study of cathode reactions during Fe deposition investigated at constant potential of polarization curves and determination of current efficiency.

93.   H. Ley, "Reduction Rates of Iron Oxide and Mill Scale," F. D. Rep. No. 670/48 (1948).

Retarding effect of water vapor investigated in 1940.

94.   I. Ljungberg, "Electrolytic Production of Straight and Alloyed Metal Powders. I. Iron Powder," Powder Met. No. 1/2 (1958) pp. 24-9.

Production at Husqvarna Vapenfabriks, Sweden of electrolytic iron powder by direct precipitation of the powder on the cathode, and the deposition of a brittle product which is subsequently ground to powder. Also, fused salt electrolytic production from alkaline solution and from an amalgam.

95.   I. Ljungberg, "Production and Properties of Electrolytic Iron Powder," Stahl Eisen 74(5):279-85(1954).

Methods used at Husqvarna Vapenfabriks A-B described.

96.   C. C. Lu et al., "Rate of Reduction of Iron Oxide with Charcoal," Chin Shu Hsueh Pao 3(2):111-19(1958).

Mill scale from rimmed steel reduced by charcoal results in high grade Fe powder suitable for use in powder metallurgy.

97.   F. E. Luborsky, "The Formation of Elongated Iron and Iron-Cobalt Particles by Electrodeposition into Mercury," in "Ultrafine Particles," John Wiley, New York (1963) pp. 236-61; see also J. Electrochem. Soc. 108(12):1138-46 (Dec. 1961).

Initial formation of the elongated Fe and Fe-Co particles in an unusual ordered structure during electrodeposition into Hg. Control of process results in formation of a gel of Fe particles in Hg with a pronounced texture.

98. R. Mitsche and E. M. Onitsch-Modl, "Experiments with Granulation of Cast Iron and Pig Iron," Berg-Hüttenmänn. Monatsh. Montan. Hochschule Leoben 100(3):121-6(1955); see also Powder Met. Bull. 7:134-7(1956).

Experiments on predisintegration of molten cast-Fe stream made using breakers. Stream passed through a heat resisting screen. Stream then divided by passage through a multiple trough.

99. A. G. Moskvicheva and G. I. Chufarov, "Iron Oxide Reduction Kinetics with Gaseous Reducing Agents at Low Temperatures," Dokl. Akad. Nauk SSSR 105:510-13(1955).

Fe_2O_3 and Fe_3O_4 powders reduced with CO and H, and gas recycled in closed cycle. A porcelain boat containing a 0.5 g sample was heated within a tube of high melting glass. Gaseous reduction products separated by freezing with liquid N. Temp. range of reduction reaction was 200-300°C. Mechanism of oxide reduction is proposed.

100. G. Naeser, "New Process for Iron Powder Manufacture," Met. Powd. Rep. 3(1):12(1948); Intern. Powd. Met. Conf. Graz, Ref. No. 52.

RZ powder, made by atomizing molten cast iron consists of porous hollow spheres or bowl-shaped particles. The inside walls are of porous iron due to decarburizing, upon which a layer of sponge iron is sintered.

101. G. Naeser et al., "A New "RZ" Process for Production of Cheap High Quality Iron and Steel Powder," Mannesmannwerke Forschungsinst. F.D. Rep. No. 883/47(1947).

Correct ratio of C to O must be maintained so that both are removed as CO_2 when the powder is subsequently annealed. Cost is one-quarter that of Hametag, and one-third of DPG powder.

102. A. Nishioka, "Manufacture of Reduced Iron Powder for
      Microwave Attenuator," Repts. Elec. Commun. Lab., Nippon
      Telegraph-Telephone Public Corp. 4(8):16-19(1956).

      Fe powder prepared by reducing ferric oxalate, results in
      powder of proper particle size, shape, and purity for micro-
      wave attenuator use.

103. T. Okamura et al., "Magnetic Properties of Sintered Iron,"
      Sci. Rept. Res. Inst., Tohoku Univ. Ser. A 3(6):748-54(1951).

      Production of very pure Fe powder through direct oxidation
      of Fe carbonyl and reduction of resulting Fe oxide in a
      stream of H.

104. R. N. Okuno and L. H. Mott, "New Sources for Electrolytic
      Iron Powder," Iron Age 174(8):132-3(1954).

      By-product of processing of electrolytic sheet Fe. High
      purity fragments broken from cathode sheet during handling
      are converted into powder with apparent density of 2.34-
      2.38 g/cc.

105. E. Pelzel, "Studies of the Particle Shapes and Structures of
      Atomized Powders," Metall 9(3/4):81-8(1955).

      Influence of alloying additions on particle shape. Powder
      produced by air atomization of melts. Reaction between
      liquid oxide and C, in case of Fe leads to formation of hol-
      low balls; small additions of Al or Si had no influence.

106. E. Pelzel, "Contribution to the Question of the Production of
      Iron Powder Through Reduction of Crystalline Oxide," Ver.
      Oesterr. Stahlwerke Jashrb. 125-8 (1950-51).

      Economic considerations for production of Fe powder by
      reduction of mill scale or Swedish Fe ore in vertical chamber
      furnaces or continuous heating furnaces.

107. M. Petrdlik, "Production of Sponge Iron Powder from $FeSO_4 \cdot$
      $7 H_2O$" Ber. u. die II Int. Pulvermet. Tagung in Eisenach,
      Akad. Verlag, Berlin (1962) pp. 453-9.

Methods of production of sponge iron in Czechoslovakia.
Economy of process.

108. G. R. Polgreen, "The Production and Application of Magnetic
Powders," G. E. C. Journal 19:152-69(1952).

Technical production: various grades of Fe powders by
electrolytic processes, mechanical disintegration, carbonyl
process, and chemical reduction.

109. I. D. Radomysel'skii et al., "Investigation of the Process of
Grinding Reduced Sponge Iron and Development of the Technology of Producing Iron Powders of Different Bulk Density,"
Soviet Powd. Met. and Metal Ceram. 5(11):351-2(Sept.-Oct.
1962), Consultants Bureau, New York, N. Y.

Recommendations for grinding to increase fine-fraction yield
of powder: (a) grinding of reduced briquettes in a hammer
mill and sifting of powder into coarse and fine fractions; (b)
grinding of coarse fraction in ball mill for over 16 h and
sifting of powder by means of a vibrating screen; (c) mixing
of fine fractions obtained in first and second screenings.

110. S. Ranganathan and B. R. Nijhawan, "Iron Powder by Electrodeposition," J. Sci. Ind. Res. (India) 14 A:333-4(1955).

In trial runs with different baths and natures of deposits, it
was noted that the higher the Fe content of the bath, the
greater was its throwing power. Tendency of deposits to
form cracks and blisters was less. The harder and more
compact the deposit, the easier it was to get a fine powder.

111. B. Razumoki, "Production of Iron Powder from Rolling Scale,"
Prace Inst. Min. Hutnictwa 6(4):188-99 (1954).

Optimum conditions determined for production of Fe powders
by reduction of rolling scale with H.

112. V. A. Roiter et al., "Mechanism of Reduction of Iron Oxides
with Hydrogen, Carbon Monoxide and Mixtures of these
Gases," Zh. Fiz. Khim. 25(8):960-70(1951); H. Brutcher
Transl. No. 3313.

Investigation of processes of reduction of chemically pure
Fe_2O_3 with H and CO and mixtures of the two gases. Experiments made at temps. as low as possible to eliminate influence of macrofactors.

113. D. W. Ross, "Production of Sponge Iron in a Shale Brick
      Plant," U. S. Bur. Mines R. I. 3822 (1945).

      Reduction in tunnel kiln, and reduction in periodic down draft
      kiln.

114. S. T. Rostovtsev and A. P. Em, "Kinetics of Low Tempera-
      ture Reduction of Iron Ores, Reduction of Chemically Pure
      Ferric Oxide with Hydrogens," Dokl. Akad. Nauk SSSR 93(1):
      131-4(1953); H. Brutcher Transl. No. 3217.

      Reduction kinetics at temps. of 275-400°C.

115. S. T. Rostovtsev and A. P. Em, "Kinetics of Low Tempera-
      ture Reduction of Iron Ores, Reduction of Synthetic and
      Natural Iron Ores with Hydrogen," Dokl. Akad. Nauk SSSR
      93(2):329-32(1953); H. Brutcher Transl. No. 3218.

      Continuation of preceding article.

116. K. S. Sanvordankar and G. S. Tendolkar, "Electrodeposition
      of Iron Powder," J. Indian Chem. Soc., Ind. News Ed. 17:13-
      19(1954).

      Operating variables in electrodeposition from $FeCl_2$ bath
      studied. Bath concentration of 30-5 g/l gave satisfactory Fe
      powder. More dilute bath gave spongy powder; higher con-
      centrations gave mixture of coherent and loose powder. Aver-
      age particle size and apparent density increased with Fe
      concentration, temp. interval of powder removal and current
      density up to 70 A/ft^2, beyond which particle size and density
      decreased with further increase in current density. Flow
      rate and specific surface decreased with increase in particle
      size and apparent density.

117. W. Schreiter, "Recovery of Powdered Iron from Iron-Rich
      Intermediates in Nonferrous Metallurgy," Die Bergakad.,

Freiberger Forschungsh. B. Hüttenw. 3:45-50 (1953).

Residue from Sn kilns, containing 2.5-4.5% Sn, 44-69% Fe, 14-27% S, is obtained from feed burned in air; this residue is reduced by solid reductants. Bulk specific volume of product increases with reduction temp.

118. H. Siepmann, "A New Process for the Reduction of Iron Powder," Stahl Eisen 73(6):360-64(1953); see also Metal Progr. 65(3):192, 194(1954).

Cites disadvantages of present reduction processes for Fe powders. Possibilities of improvement through eddy processes and their effectiveness described together with apparatus construction and method for process.

119. A. F. Silayev and V. I. Prossvirin, "Granulometric Composition and Shape of Powder Particles Produced by Solidification," Vestn. Mashinostr. 35(6):61-4(1955).

Ferrous alloys melted and allowed to flow downward through a circular hole while an air stream was applied to the metal stream at right angles. Particle-size distribution varied with nature of alloy when air stream pressure was constant. Generally particle size decreased with increased air pressure up to 4 atm; further pressure increases did not produce smaller particles. Most alloys formed globules, though FeMn produced chiplike particles. Particle shape depends principally on surface tension of metal, solidification time, and viscosity just before solidification.

120. A. F. Silayev and V. S. Rakovskii, "Some Problems of Obtaining Iron and Ferroalloy Powders by Atomization," Soviet Powd. Met. and Metal Ceram. 4:291-4(July-Aug. 1962) Consultants Bureau, New York, N. Y.

Investigation of atomization process for producing powders of any metal or alloy. It can be used for materials with Melting point up to 1700°C. Some problems involved in decarbonization of cast Fe powders are discussed in light of modern ideas for obtaining powder by pulverization. Description given of technology of obtaining ferrosilicon from metals.

121. A. F. Silayev et al., "Production of Iron, Steel, and Iron-
      Alloy Powders by Pulverization," Tr. Tsentr. NITTMASH
      No. 56, 124-47 (1953).

      Report on several investigations of powder production by
      various mechanical methods at USSR Scientific Res. Inst.
      for Machine Construction.

122. O. G. Specht, Jr. and C. A. Zapffe, "The Low-Temperature
      Gaseous Reduction of Magnetic Ore to Sponge Iron," Trans.
      AIME 167:237-80(1946).

      Examination of previous work; experimentation on reduction
      of magnetite with H; development of thermodynamic relation-
      ships in reduction.

123. A. M. Squires and C. A. Johnson, "The H—Iron Process,"
      J. Metals 9(4):586-90(1957).

      Direct reduction of Fe ore fines by H in fluidized beds at low
      reduction temp. and elev. pressure. Resulting powder has
      same impurities as were present in ore.

124. J. L. Stähled, "The Production of Sponge Iron by the Wiberg-
      Söderfors Process," Stahl Eisen 72:459-66 (1952); see also
      Document. Met. Sondernr. 1:60-74(1956).

      Charcoal originally used for formation of reducing gas.
      Later producer gas utilized. Lump ore or sintered ore can
      be used as starting material. Sponge Fe suited especially
      for production of electrical steel.

125. P. O. Stelling and I. G. M. Pereswetoff-Morath, "Method of
      Direct Reduction of Iron Ore Concentrate by Carbon Monox-
      ide Without Melting," Jernkontorets Ann. 141:237-60(1957).

      Fe oxides reduced by CO at high temp. on fluidized bed have
      tendency to stick together when certain amount of metallic
      Fe is produced. If reduced at about 600°C, Fe_3C is formed
      instead of metallic Fe, and no sticking occurs.

126. E. W. Stewart et al., "Influence of Additives in the Produc-
      tion of High Coercivity Ultrafine Iron Powder," Trans. AIME
      203:152-7(1955).

Effect of several additives on reduction characteristics in reducing $Fe(HCO_2)_2$ by H described. Various additives inhibit sintering of reduced Fe particles by several mechanims. Low density compacts produced from the resulting ultrafine Fe powders exhibited improved magnetic properties.

127. V. G. Syrkin, "Production of Ferropowder by Thermal Decomposition of Atomized Iron Pentacarbonyl," Soviet Powd. Met. and Metal Ceram. 3(21):232-9 (May-June 1964) Consultants Bureau, New York, N. Y.

Production of Fe powder for magnetic cores by atomization of liquid $Fe(CO)_5$ by a centrifuge device in a chamber and dissociation of Fe carbonyl at about 300°C in presence of ammonium. Recommendations given for optimum centrifuge feed rates and optimum dissociation temps. for production of carbonyl Fe powder with maximum magnetic permeability.

128. E. P. Tatievskaya et al., "The Rates of Reduction of Iron Oxides," Zh. Fiz. Khim. 24(4):385-93 (1950).

Although calculated equilibrium pressure of O_2 over Fe_2O_3 is much greater than over Fe_3O_4, it was found that all three oxides of Fe are reduced by H at similar rates. It is concluded that reactions are independent of dissociation and that reduction of FeO is not the slowest stage in reduction of higher oxides.

129. M. Tenenbaum and T. L. Joseph, "Reducing Iron Ores under Pressure by Hydrogen," Trans. AIME 135:59-72 (1939).

Cubes of Fe ore reduced by H under comparable conditions at several pressures. Pressures greater than 10 in. of Hg increased rate of reduction.

130. M. Tigerschiöld, "Sponge Iron," Blad Bergshanteringens Vönner 20, 240-67; 21:219-29 (1932); Stahl Eisen 52:1244-6 (1932).

Reduction methods used in Sweden (Kalling process) for production of sponge iron from FeO.

131. H. Timmerbeil, "Manufacture of Iron Powder by DPG Disc

Process," Met. Powd. Rept. 3(1):11(1948); Int. Powd. Met. Conf. Grax, Ref. No. 9.

SC process in which Bessemer steel is atomized, as in DPG disc process, but powder is more highly splattered, and acquired a porous structure.

132. P. M. Tyler and W. M. Pollitzer, "Krupp-Renn Process for Low Grade Iron Ores in Germany," FIAT Rept. No. 799 (1946).

Reduction process for sponge iron. Krupp-Renn process; washing and gravity concentration; and Lurgi process.

133. H. Uchida and T. Minegishi, "Iron and Nickel Carbonyls. II. Syntheiss of Iron Carbonyl. III. Thermal Decomposition of Iron Carbonyl," Tokyo Koggo Shikensho Hokoku 45:1-8, 9-19 (1950).

Reduction of Kamaishi magnetite with H on large scale basis, for formation of $Fe(Co)_5$, which is decomposed to form Fe powder when vaporized at 250-400°C, and to cottonlike fibers at 400-500°C.

134. K. C. Wassberg, "Manufacturing Iron Powder by Hydrogen Reduction of Iron Oxide," Tidsskr. Kjemi, Bergvesen Met. 14:147-52(1954).

Mill scale in a semipilot plant pulverized to -85 mesh in ball mill and reduced in a Mo furnace with a cracked ammonia atmosphere. At 900°C reduction was 90%; at 700°C rate of reduction was slow.

135. L. Weber, "The Reduction of Iron Ore with Petroleum Products," Erdöl Kohle 11:241-4(1958).

Possibility of producing Fe sponge from ore and residues from petroleum industry discussed in light of possible shortage of high grade coke.

136. J. Wiberg, "New Methods for Production of Sponge Iron," Jernkontorets Ann. 142:289-355 (1958).

Processes reviewed, and suggestions for modernizing Wiberg procedure to permit use of natural gas.

137. M. Wiberg, "A New Method for the Production of Iron Sponge," Trans. Am. Electrochem. Soc. 51:279-304 (1927).

Gaseous reduction can be carried out continuously with very low consumption of fuel and electrical energy.

138. H. Wiemer, "Production of Iron Powder in North America and England," Stahl Eisen 63:30-1(1943).

Review of the different production methods.

139. C. E. Williams et al., "Process for Production of Sponge Iron," U. S. Bur. Mines R. I. 2578 (1924).

Process and furnace described; externally and internally heated furnaces compared.

140. B. Wranglen, "Production of Electrolytic Iron Powder," Jernkontorets Ann. 132:501-16(Dec. 1948).

Anode consists of scrap iron fed into wooden baskets which are faced with filter cloth and lined with graphite or Duriron plates. Process then operates similarly to electrolytic detinning of tinplate scrap.

141. J. Wsieklica, "Possibilities of the Application of Hydrogen for the Reduction of Powdered Domestic Ores," ABM (Bol. Assoc. Brasil. Metals) (Sao Paulo) 11:389-409(1955).

Literature review on reduction of hematite; economic and technical advantages over reduction with coal are demonstrated. Laboratory experiments made with local hematite in powder form showed that stationary furnace used did not permit equal reduction of ore at all locations. Ore grain size influenced progress of reduction.

142. P. M. Zarelevitch, "Preparation of Iron Powder," Trans. Inst. Pure Chem. Reagents (USSR) 15:51-7(1937).

Fe_2O_3 and $Fe(OH)_3$ reduced with H in electric furnace with best results by heating to 700°C for 40 min with 4.5 parts H in excess of theory. Purification of H by removal of O by passing through alkaline solution of pyrogallol and drying with H_2SO_4.